The False Promise of Superiority

The False Promise of Superiority

The United States and Nuclear Deterrence after the Cold War

JAMES H. LEBOVIC

OXFORD
UNIVERSITY PRESS

Oxford University Press is a department of the University of Oxford. It furthers the University's objective of excellence in research, scholarship, and education by publishing worldwide. Oxford is a registered trade mark of Oxford University Press in the UK and certain other countries.

Published in the United States of America by Oxford University Press
198 Madison Avenue, New York, NY 10016, United States of America.

Library of Congress Control Number: 2022921533

ISBN 978-0-19-768087-2 (pbk.)
ISBN 978-0-19-768086-5 (hbk.)

DOI: 10.1093/oso/9780197680865.001.0001

1 3 5 7 9 8 6 4 2

Paperback printed by Marquis, Canada
Hardback printed by Bridgeport National Bindery, Inc., United States of America

To Holly

CONTENTS

PART III CASE STUDIES

PREFACE

I finished this book at a painfully difficult time in world history. After invading Ukraine, Russian forces are leveling that country, backed by a Russian president who aspires to redraw state boundaries across Europe. With his threats to escalate beyond conventional force use, a nuclear conflict between the United States and Russia remains a realistic possibility. These developments test much of what we know, or think we know, about nuclear deterrence and the robustness of the international order that grew from the wreckage of the last world war. As history unfolds, I hope that readers can dismiss these claims as the alarmist fears of an author who was too close to events to view them in appropriate perspective.

ACKNOWLEDGMENTS

I started thinking seriously about deterrence and nuclear weapons as a student of international politics over a half-century ago. My interest was provoked, then, by the pretense with which the best and brightest minds of the period confidently sold bold solutions to pressing security threats. My interest was reignited in the post–Cold War years, first with the events of 2001, when the concept of deterrence fell into disrepute and then again, a decade later, when strategic nuclear arms control seemed a dying enterprise. I choose now to revisit the topic to address a new generation of scholarship. In my view, it fosters dangerous illusions about the benefits of US nuclear superiority and accompanying bargaining tactics.

Over the years, I benefited from advice and comments from numerous people, too many to mention here. Some were mentors or good colleagues; many others were unwitting provocateurs. I cannot thank them all. So, I will thank those who contributed most directly to this volume. Its earliest incarnations appeared in an article on the Iran nuclear deal published in *Strategic Studies Quarterly*. I wish to thank Jacques Hymans, Matthew Gratias, and Mike Guillot for their comments on that piece which I revised (as Chapter 8) for inclusion in this book. I also thank Andrew Payne for his comments on my Cuban Missile Crisis paper prepared for the 2021 (virtual) International Studies Association Conference. A revised version of that paper appears as Chapter 7 in this volume. The book benefitted from very useful recommendations, for revision, by Keith Shimko, for which I am grateful. Portions of the book also benefited enormously from careful readings by my (present and former) George Washington University colleagues—Charles Glaser, David Shambaugh, and Caitlin Talmadge. James Morris deserves recognition, too, for inspiring the book's cover art. I reserve a strong expression of gratitude for Greg Koblentz, a Washington-area colleague. His thorough and incisive comments and recommendations on topics, big and small, throughout the manuscript immeasurably improved the quality of the work.

I would be remiss if I failed to thank Dave McBride, my editor at Oxford University Press, for his support for this manuscript in its eventful journey to publication. Of course, I also want to thank my wife, Holly, to whom this book is dedicated. We will likely never discuss the book. In the world of Gill-Lebovic, by contrast, I deserve only second authorship.

1

The United States and Nuclear Deterrence after the Cold War

Contemporary scholars engage in fanciful thinking about US "nuclear superiority." Perhaps this is understandable. Even Cold War–era policymakers flirted with strategies and technologies to mitigate the catastrophic consequences of nuclear weapons use. Both the strategies and the technologies seem more viable now that the United States apparently enjoys quantitative and qualitative nuclear advantages over contenders. Yet these advantages are largely illusionary. Try as it might, the United States cannot escape oppressive facts and possibilities that govern war and peace in the nuclear age.

During the Cold War, international politics theorists devoted much thought to how, where, when, and why to employ nuclear weapons. They were inspired, in part, by nuclear strategists who battled in the "great debates" of the period. Where some strategists saw threats to deterrence, others did not. Where some envisioned solutions, others feared grave threats to the stability of US–Soviet relations. They were not alone in their skepticism. Proposals that often shined in the abstract tarnished significantly under social-scientific scrutiny.

The ideas that enthralled Cold War–era strategists resonate in the works of contemporary scholars who, in their focus on US nuclear superiority, reveal the same biases and blind spots. Consequently, they highlight capability asymmetries or improvements that only promise to override undesirable deterrence constraints; and they accept too readily that coercive bargaining tactics can compensate, rather than worsen, conditions when capability alone appears insufficient. In this, they fail to attend to an underlying reality: assumptions about adversary *intent*—conjectures influenced by political, social, psychological, and organizational factors—will determine how the parties act in a conflict, and whether they seek to avoid one. Too often, however, these assumptions are buried in policy justifications and analysis. There, they hide from needed criticism.

The False Promise of Superiority. James H. Lebovic, Oxford University Press. © Oxford University Press 2023.
DOI: 10.1093/oso/9780197680865.003.0001

Superiority, by Implication, in Current Policy Debates

We might dismiss nuclear superiority as an oxymoron—the fanciful obsession of a select few who, with the distance that think tanks and universities afford, inhabit the abstract world of assumption. But we must take the concept seriously. Beyond its seductive appeal, it continues to impact scholarly writings, policy planning, and government rhetoric regarding nuclear weapons.

Of course, recent signs of the concept's appeal—the nonsensical bluster of Donald Trump's presidency—are admittedly hard to take seriously. Trump's unschooled fascination with the power of the atom, subject exclusively to *his* command, reduced the credibility of all claims related to the use of nuclear weapons. Trump opined about nuking hurricanes,[1] and he bragged about obliterating Afghanistan (presumably with nuclear weapons),[2] winning nuclear arms races,[3] and the relative size of his nuclear button.[4] That button only seemed bigger when Trump proclaimed the US development of a "super duper missile" that dwarfed its rivals in performance. In his words, "You take the fastest missile we have right now—you've heard Russia has five times, and China is working on five or six times. We have one 17 times. And it's just gotten the go-ahead."[5] Lost on him was that the math did not make the United States at least three times more powerful than its competitors. Indeed, his rhetorical record—as when he advocated a tenfold increase in the US nuclear stockpile, countering decades of movement toward a smaller (more efficient) arsenal—provides no clear sense of whether or not the United States enjoys nuclear superiority, or even what that requires.[6]

The musings of a living strawman thwart serious discussion and debate, but even Trump's words mean *something*. They appealed to Trump, and thus presumably appealed also to the like-minded percentage of the US population that knew (and cared) little about international affairs and believed that US global challenges stemmed largely from weak US leadership. Even if not taken literally, Trump's words reflected a cavalier embrace of nuclear weapons and a false sense of immunity from the disastrous effects of a nuclear conflict.

More problematically, such words—in some dressed-up form—continue to influence planners and strategists who see danger—and opportunity—arising from new technologies and bigger arsenals. Arms control was among the casualties, then, when the George W. Bush administration formally withdrew, in 2002, from the (1972) Anti-Ballistic Missile Treaty, giving Russia and China reason to increase their nuclear force capabilities; and, decades later, when the Trump administration withdrew from the (1987) Intermediate Nuclear Forces Agreement—a crowning achievement of Cold War–era arms control—ostensibly to counter a Russian and Chinese push to acquire intermediate-range missiles. Efforts to build on the force reductions of the (Obama administration's) New START (Strategic Arms Reduction

Treaty) agreement languished as administrations sought new nuclear weapons (cruise missiles and low-yield warheads), and new weapon applications. Whereas the George W. Bush administration's Nuclear Posture Review (NPR) broadened the list of potential US nuclear targets to include biological and chemical weapon targets (an option scaled back in the NPR of the Obama administration to exclude parties to the Nuclear Nonproliferation Treaty "in compliance with their nuclear non-proliferation obligations"), the Trump administration's 2018 NPR sanctioned a possible nuclear response to foreign cyberattacks on the US infrastructure.[7] As presidential candidate, Joe Biden promised a retraction: He embraced "deterrence" and "retaliation for a nuclear attack" as the "sole purpose" of nuclear weapons (Biden 2020). As president, however, he demurred. The 2022 NPR adopted a less restrictive position—accepting these missions as the "fundamental purpose" of nuclear weapons—which harked back to Obama-era standards in allowing for the first use of nuclear weapons.[8]

The United States is obviously not the sole culprit here. In 2018, Russian President Vladimir Putin announced five new technologies designed to penetrate US defenses before adding another the following year.[9] Indeed, Russia, despite its Cold War–era "no-first-use" policy, professed plans to employ the country's nuclear forces in various non-nuclear contingencies.[10] Revisions in Russian military doctrine permit the use of tactical nuclear weapons to augment the depleted Russian conventional force.[11] Thus, Putin's 2022 claim to have alerted Russian nuclear forces, and Russia's thinly veiled threats of a global nuclear war[12]—in response to NATO's support for Ukraine after its invasion by Russia—raised the risk of a nuclear conflagration, as in no time since the 1962 Cuban Missile Crisis.

For its part, China, which long maintained a relatively small nuclear force for deterrence, is expanding and diversifying its nuclear arsenal, fueling calls for a US response. The US Department of Defense—reminiscent of its glossy *Soviet Military Power* report of the Cold War years—now shines its light on *Military and Security Developments Involving the People's Republic of China*, implying, at least, that China's larger and more diverse nuclear capabilities are potential game-changers in the US–China nuclear relationship. Predictably, perhaps, Vice Chairman of the Joint Chiefs of Staff General John Hyten dubbed the Chinese hypersonic missile a first-strike weapon that could give China the capability to launch a surprise attack on the United States.[13]

Yet looming adversary deployments are not necessarily cause for US nuclear offsets. Rather than fret about what adversaries *could* do with their weapons, we should ask what they *would* do given US retaliatory assets, which limit gains relative to costs from an attack. The failure to look beyond weapons—or their threatened use—reflects a pervasive blindness. It requires a careful look at the assumptions behind past US nuclear strategies, as bequeathed now to the present.

Nuclear Superiority, in Theory

For one high-profile scholar—Matthew Kroenig (2018)—US nuclear superiority is real: the relative advantages that the United States enjoys in sheer numbers of warheads could pay off in a strategy of brinkmanship: the United States can take other countries to the edge of war knowing that they will back down. For other scholars—Keir Lieber and Daryl Press (2006a, 2006b) most notably—the United States had, or will soon obtain, nuclear "primacy." It could then disarm its most powerful nuclear adversaries—Russia and China—in a first strike.

What should we make of these concepts? At first glance, "superiority" seems the more politically charged of the two terms. In Kroenig's analysis, it appears to flaunt relative US nuclear power. Yet, in principle, superiority could apply only to a subset of units. Then, superiority relative to some party (or state) could mean *inferiority* compared to others. Kroenig (2018: 16) himself softens the term by using it synonymously with "advantage." The term suggests, then—albeit not by his reading—that US adversaries possess or could acquire offsetting or neutralizing advantages. By contrast, "primacy" assumes that the United States stands—as the one and only—at the top of the heap, though Lieber and Press equivocate, more than Kroenig does, on the implications. In their view, US primacy might give the United States a coercive edge, but it could also fuel instability. Indeed, both Russia and China have sought to build up their nuclear forces to offset an apparent US advantage. Thus, these scholars view the coercive and political implications of the US nuclear position with greater uncertainty.[14]

Still, the two concepts are similarly limited. They wrongly attribute influence more to relative nuclear capabilities than to the contesting parties' intentions. These immediate and long-term goals are shaped by the parties' beliefs about the likely course of war and evaluation of the costs and benefits of acting (or not acting)—based, in part, on assumptions about each other's objectives. We undercut the impact of terms like "superiority" and "primacy," then, when we concede that the United States would use nuclear weapons only under truly exceptional conditions or that the weakest of nuclear powers still presents a *potential* nuclear threat to the United States in a conflict.

The conflating of capability with influence remains problematic when strategists employ seemingly less loaded terms in making the case for a "robust" or "flexible" nuclear force. As Keith Payne (2020: 39) puts it, "Having a *spectrum of deterrence options* [italics added] and focusing on threat credibility seem only prudent in the contemporary threat environment given the diversity of opponents and their nuclear threats, the potential variability of their decision-making, and the range of possible deterrence goals." With these options, the United States can presumably tailor US deterrence to the unique, extreme, or opaque goals of a potential challenger—or match the US response to the level at which it occurs (Mahnken and Evans 2019).

With the language of "preparedness," we still travel the same road, thinking that influence comes from the barrel of a gun. The answer to nuclear challenges remains more weapons, of greater varieties, in more places to plug deterrence "holes"—or to fill "gaps"—where they might unexpectedly arise. As we shall see, however, these straightforward "solutions" should invite great skepticism. Claims of superiority—by any name—rest on questionable math, dubious assumptions about how a nuclear conflict might unfold, and insufficient regard for the aversions of the parties and conditions that increase or limit restraint.

The Cold War Roots of Contemporary Thinking

Cold War–era analysts fought hard over the requisites of deterrence, the significance (and insignificance) of nuclear advantages, and the implications of each for policy.[15] Their disagreements yielded sharply contrasting interpretations of the same evidence and episodes. In the lore of the era, the Cuban Missile Crisis emerged as a defining moment—indeed, a parable of sorts. The nuclear powers had danced on the precipice of incineration and lived to tell the tale. For some analysts, the message was "stick to your guns"—the big and the little ones. For others, it was "act decisively but seek a way out." With time, the core arguments gained nuance—and attention, through innovative wrinkles.

The competing arguments reduced nonetheless to simple premises. Hawks saw a world in which states achieved peace through strength. They thus advised US policymakers to anticipate, answer, and preferably best Soviet nuclear acquisitions and ensure that the Soviets understood that the United States was prepared, if necessary, to use its nuclear weapons. Doves maintained, instead, that deterrence was secure because the United States and Soviet Union shared an aversion to nuclear war.[16] The United States and Soviet Union were like two scorpions in a bottle (Wohlstetter 1959): each feared striking the other, because either, though mortally wounded, could inflict a fatal sting. Doves thus spread the word that a US–Soviet nuclear exchange would *assuredly* result in mutual catastrophe—what, in the Cold War period, was widely known as "Mutual Assured Destruction" (or MAD)—while seeking to maintain a (unilateral) US retaliatory posture—Assured Destruction—that would intentionally produce that catastrophe. They recognized, however, that policymakers could lose control of events in times of crisis. They advised US policymakers to maintain open channels of communication, leave room to maneuver, dutifully consider options, and avoid knee-jerk reactions to events.

Hawks did not all subscribe to the same view. Some sought plans and weapons that would permit the United States to *fight* wars, as traditionally understood. Other hawks accepted that MAD *might* occur but fought its dire implications. To boost deterrence, they hoped to convince the Soviets that the United States had retaliatory options short of triggering catastrophe. If war occurred, they expected to ply these options coercively to convince the Soviets to back down. Yet doves also differed in their individual views. Some thought the mere existence of nuclear weapons proscribed nuclear war: even a small number of nuclear weapons, given their catastrophic effects, could deter a nuclear attack. Others rejected the principles of such "existential deterrence" and pushed for nuclear capabilities in huge quantities (adherents of the so-called assured destruction position)—aimed directly at Soviet cities—to reinforce the message that nothing good could come from a nuclear exchange. We could certainly question their "dovish" credentials. Perhaps they were "doves" only when compared to "hawks." Their dovishness stemmed, however, from an overriding belief that both the United States and the Soviet Union recognized they would suffer prohibitively in a nuclear exchange.

Things obviously get messy conceptually, then, when placing viewpoints into distinct boxes. They get messier still when recognizing that many analysts accepted the essential logic of assured destruction but looked to insure against catastrophe by relying on the nuclear capabilities and tactics that enthralled some hawks. We can be forgiven, then, if we place the assertions of many Cold War–era theorists into a box marked "nebulous middle." Those in the group would share basic assumptions: that the US–Soviet nuclear deterrence relationship was stable, that all-out nuclear war would be catastrophic, and that the United States needed options—in capabilities and tactics—to reinforce deterrence and avoid the worst should deterrence fail. We would likely disagree, however, over whether certain theorists were more dove than hawk or more hawk than dove.

We can profit, then, when we move beyond simplistic labels like "hawk" and "dove" to appreciate the full logic (and illogic) of past and current nuclear debates but also in recognizing preoccupations that have always afflicted thinking about deterrence. These longstanding biases—centered in capabilities and tactics—undergird contemporary beliefs that the United States can overcome traditional challenges to either impose its will or coerce adversaries to achieve favorable outcomes.

Deference to Capabilities

Despite rivalries among Cold War–era analysts, and the unforgiving rhetoric of arcane policy debates, the protagonists shared a basic assumption. For most nuclear analysts, the essence of deterrence—and its confounding challenges—lay in the phenomenal destructiveness of nuclear weapons.

Whereas doves tended to see security in the massive amount of capability—on land, in the air, and at sea—that the United States could deliver, hawks imposed a more exacting standard. They sought security in numbers of US warheads available per Soviet targets, the relative damage that US and Soviet arsenals could inflict, the size of the US arsenal vis-à-vis its traditional levels, and so forth. They warned that various "gaps" of advantage, and openings and loopholes in arms-control agreements, favored the Soviets should they launch an unprovoked strike on US nuclear forces. Going even further, some hawks argued that the Soviets had engineered these unfavorable asymmetries to coerce the West and profit, at some point, from an actual attack. The Soviets would presumably strike because they read victory into ratios of advantage, whatever costs a nuclear war would bring.

"Denial strategies" thus took shape in the 1960s and 1970s and gained force in reaction to the mainstreaming of AD principles which centered instead on "punishing" the Soviets should they launch a nuclear strike. Denial proponents emphasized, not the imposing of costs, but rather acquiring the capabilities necessary to deny an adversary benefits from an attack. They posed a simple question with a seemingly obvious answer: "Why would an adversary strike if it had nothing to gain?" In their thinking, that question acquired urgency from a related one: "What could the *United States* gain from carrying through on its deterrence threat?" They expressed fears that stability then—at the level of an all-out nuclear exchange—might induce instability at some lesser level.

The "stability–instability paradox," as it was called,[17] reflected concern in particular that, with robust Soviet nuclear capabilities, the United States could not deter a Soviet conventional attack on Western Europe.[18] Some analysts—henceforth "soft warfighters"—saw, in the irksome scenario, reason to develop limited options with the hope of controlling and stemming the march to all-out nuclear war. Other analysts committed fully to a denial strategy. These "hard warfighters" feared that the Soviets had positioned themselves militarily to seize the initiative when the time was right. They *would* then violate arms-control agreements, by outfitting their missiles with large number of warheads. They could coerce the United States from a position of strength and might even launch a disarming nuclear offensive against the US homeland. From this perspective, Soviet risks were virtually irrelevant: The United States had no rational reason to impose costs in retaliation. (On these strategies, see Chapter 2.)

A legacy of the Cold War remains today in thinking about the use, and threatened use, of nuclear weapons. All such thinking begs the essential, Cold War–era question, "What would a state gain—and what would it lose—from an actual attack?" The question is sidestepped, however, as analysts focus on materiel issues—the number, nature, and positioning of weapons systems—more than the contesting parties' intentions and how they might change due to political, social, psychological, and organizational influences over the course of a nuclear confrontation.[19] These influences will determine a party's level of uncertainty, propensity for risk-taking, valuation of wartime costs, and all the rest that determine what leaders seek, and how and when they pursue it.

Deference to Tactics

Analysts did not rely on nuclear capability–based reasoning alone. In drawing from works in strategy and bargaining, they sought to reinforce, augment, and close gaps in deterrence theorizing. Whereas a "first wave" of theorists—Bernard Brodie preeminent among them—grappled with the revolutionary impact of nuclear weapons on global politics and warfare, a "second wave" of theorists of the late fifties and early sixties—Thomas Schelling, Herman Kahn, Glenn Snyder, Albert Wohlstetter, and others—provided the concepts and theoretical underpinnings for much of our thinking about nuclear deterrence.[20]

As a group, these analysts highlighted "the paradoxical nature of deterrence in which each side hopes to gain security, not by being able to protect itself, but by threatening to inflict unacceptable damage on the other" (Jervis 1979: 292). They thereby acknowledged the seemingly incongruous logic of deterrence applied to parties with a "second-strike" capability, which permits them to retaliate with devastating force whatever the size and nature of an adversary first strike. By this logic, a party—if attacked—essentially threatens to blow *itself* up by launching a suicidal retaliatory strike against the attacker. Thus, these theorists supposed that nuclear stability might require tactical support. The United States somehow had to *convince* the Soviets that the United States would deliver on its threat if attacked. They saw a need, then, to make commitments, manipulate risks, establish resolve, and build reputations for acting to overcome deficiencies that might undermine deterrence.

It was left to a third wave of theorists to acknowledge, and test, the influence of various political, sociological, psychological, and organizational factors on deterrence relationships. The mechanisms of credibility enhancement—commitment, risk manipulation, resolve, and reputation—remained central to the research enterprise. Third-wave theorists stressed the dangers of over- and under-reliance, however, on these concepts. (On these three waves, see Jervis 1979.)

Whereas capability-based analyses too often ignored adversary intent or read intent—broad purposes, goals, or motives—into capability balances and imbalances, those who worked on the tactical front too often focused on the intentions *behind the tactic* rather than the targeted party's intentions, though the latter would determine how the tactic was viewed and handled. Analysts thereby ignored or downplayed warnings from third-wave theorists who cautioned that the best of strategies might flounder under the weight of domestic politics, time pressures, fear and stress, intelligence lapses, organizational failings, and the adversary's changing valuation of gains, losses, costs, and risks. An adversary might respond in ways, then, that exacerbate conflict, reduce the room for both parties to maneuver, or concede control of events to anonymous forces or subordinates unequipped to understand the situation or handle the responsibility. Threats, risky action, and bold commitments might not ameliorate conflict; instead, they might reinforce the target's propensity to resist, or even fuel, an action–reaction process that could "spiral" out of control. In that sense, analysis could benefit when viewing intentions as variable and conditional—that is, as the "actions that a state plans to take under certain circumstances" (Rosato 2014/15: 52).[21]

Implications of Alleged Superiority for a "New" Nuclear Age

A fourth wave of theorists now wrestle with well-worn deterrence concepts,[22] even as some policy analysts deny the relevance of past thinking in championing the benefits of superiority. Implying that the "old rules don't apply," these analysts focus instead on advantageous ("asymmetrical") relationships between the United States and its contemporary nuclear adversaries or, conversely, the dangers posed by even small nuclear arsenals.[23] They suppose that the United States can—or must—rely more on sheer military might to address contemporary challenges.

Contemporary analysts are not wrong to question the contribution of past thinking. Deterrence theory has well-documented limitations (Jervis 1979). It places the focus on coercive mechanisms with little regard for whether incentives permit cooperation, not just war avoidance. It generally assumes that disputing parties rationally acquire, evaluate, and process information pertaining to the availability of options, and their accompanying benefits and costs. No less significantly, it neglects the strategic "context." That is, it emphasizes the materiel or tactical underpinnings of deterrence over the impact of a party's intentions and how these intentions might change counterproductively over the course of a conflict—perhaps in response to another's tactical manipulations.

Why this emphasis? We can find the answers in social scientific works across disciplines. Psychologists who focus on decisional heuristics conclude that people rely on salient and familiar referents rather than acquire holistic understandings of policy problems (Kahneman 2011; Kahneman and Tversky 1979; Levy 1997). Their arguments dovetail with those of organizational theorists who observe that individuals employ immediate goals as surrogates for more abstract ones, perhaps from a (rational) desire to establish clear and observable criteria for assessing organizational performance (March and Simon 1993; Steinbruner 1974).

These ideas converge nicely with works in national-security decision-making. Aaron Rapport (2015: 2) argues that policymakers assess short-term goals by assessing how "feasibly they can be executed" but assess temporally distant goals—which, by nature, are abstract and devoid of context—by evaluating their desirability. In his words, "as feasibility rises to the fore, [decision makers] become less prone to examine whether immediate costs are justified by overarching objectives." If so, we should expect analysis, and deliberations, to center on whether options satisfy defined short-terms goals—for example, the destruction of identified targets to meet an acceptable level of success—apart from broader concerns, including the risks of an all-out war (which can increase with assaults on those targets). In a similar vein, I claim elsewhere (Lebovic 2019) that policymakers myopically "ground their decisions in the immediate world of short-term objectives, salient tasks and tactics, policy constraints, and fixed time schedules" with the effect that "they exaggerate the benefits of preferred policies, ignore their accompanying costs and requirements, and underappreciate the benefits of available alternatives." I show that these tendencies drove US policymaking in the early through the late stages of the Vietnam, Iraq, and Afghanistan wars. In Iraq and Afghanistan, for instance, US policymakers initially focused their attention on regime change—a short-term goal well within reach of US military capability—paying little heed to the potential challenges that would await the vanquishing of the leaders of these two countries.

Yet these ideas do not require that we start from scratch. Rather than questioning our reliance on deterrence or doubting its overall stability, we have much to gain from acknowledging its robustness and questioning the value of potential "fixes" that capitalize militarily or politically (coercively) on ostensible US nuclear advantages. The important question for research, then, is not how we should reinforce nuclear deterrence. Instead, it is how and why our efforts at reinforcement might go *dangerously* wrong.

The Book

To make its argument, then, this book draws from Cold War and post–Cold War writings and examples to establish what we know, and do not know, about how deterrence works, and might fail. Each chapter concludes by listing potential "perils and pitfalls" that distort thinking about the benefits of nuclear capability or various compensatory coercive tactics. The book proceeds as follows.

The first two chapters that follow expose the liabilities of capability-centered thinking. They show that strategists put far too much emphasis on the materiel aspects of US relationships with other nuclear-armed states and show too little regard for assumptions about the parties' intentions that determine why war occurs and how it will unfold. Chapter 2 thus examines nuclear advantages in the Cold War–era theories of (dovish) Assured Destruction and (hawkish) warfighting proponents. It shows that nuclear capabilities—and the military and political implications of accompanying asymmetries—dominated thinking about the stability of deterrence in the Cold War years. It reveals then that contemporary thinking about superiority—far from a break with the past—has origins in the unfounded concerns and ungrounded ambitions of that period. Chapter 3 examines the unwarranted influence of nuclear capabilities in current thinking about US nuclear superiority. It critically assesses the reasoning behind the case for US nuclear superiority, probes the actual costs and risks the United States and its adversaries would accept in order to achieve their goals, and concludes that superiority is neither a viable nor a necessary policy objective.

The next three chapters assume that the United States—despite its alleged capability advantages—will not simply impose its will through nuclear force. It assumes, then, that the overall costs to the United States of employing nuclear weapons generally (i.e., almost always) outweigh the benefits. The chapters suppose, then, that the United States might rely on coercive tactics—backed by relative US nuclear strength—to make credible threats. With these tactics, US leaders can communicate their willingness to act, with nuclear force, if necessary. More specifically, US leaders can signal (a) that their goals are sufficient to justify nuclear weapons use and (b) that US nuclear advantages reduce the costs relative to the benefits of employing nuclear force. The chapters show that efforts to increase the credibility of US threats to employ nuclear force—by making commitments (Chapter 4), manipulating risk (Chapter 5), or demonstrating resolve and establishing a reputation for acting (Chapter 6)—rest on deficient logic; indeed, they court disaster. These chapters conclude that much of the policy (prescriptive) literature is profoundly one-sided. It assumes a compliant target that shares the tactician's understanding of the "game" and plays, then, by the tactician's rules.

Two cases studies follow, in successive chapters. Together, they provide broader perspective on the liabilities of a preoccupation with capabilities and tactics, as revealed in the prior chapters.

Chapter 7 expands on the discussion, in Chapter 5, of the loss of control in a crisis. It assesses US decision-making during the Cuban Missile Crisis to highlight the dangers that ensue when capabilities and tactics consume debate. The chapter establishes that US officials too quickly looked for solutions in *available* options—and blindly accepted risk. More specifically, it shows, through an analysis of the transcripts of key meetings of President Kennedy and his advisors, that officials restricted their thinking to salient options and failed, then, to scrutinize assumptions about the purpose and value of Soviet missiles in Cuba or the likely Soviet responses to US actions. The chapter suggests that such option-based thinking could easily prevail in a military confrontation between nuclear-armed states—when time is precious, and information is scarce.

Chapter 8 builds on the assessment of Cold War–era strategic failings, in Chapter 2. The chapter shows that hawks and doves diverged in their preferred capability-based metrics—much like their Cold War–era counterparts—when assessing constraints on Iran's nuclear program. It shows US officials focusing their attention on how Iran could circumvent, overcome, or evade program constraints while attending far less to Iran's incentives and disincentives, for pursuing a bomb. A perverse casualty of the preoccupation was that hawks gave little attention to the alleged benefits of US nuclear superiority as a constraint on Iran's options or ambitions.

Chapter 9 offers conclusions and recommendations. It discusses the propositions drawn from the conclusions in each chapter and reflects on the bias—the overemphasis on nuclear capabilities and tactics and underemphasis on intentions—in US thinking about deterrence. It then challenges the assumption of US nuclear superiority in arguing for a "sober" nuclear strategy. It maintains that deterrence works best when policymakers avoid precipitous moves to create or resolve a crisis and, instead, allow the danger of nuclear escalation to speak for itself.

PART I

ASSESSING NUCLEAR CAPABILITY

The History and Implications of Alleged Nuclear Advantages

US Cold War–era strategists lamented asymmetries that allegedly gave the Soviet Union a military or coercive edge. They sought, then, to close or reverse these "gaps." As US strategists now look to military and political advantages that stem from US "nuclear superiority," their logic reveals the flaws that beset Cold War–era strategies.

As before, strategists focus far too much on issues of capability and far too little on why nuclear war might occur and how it might unfold. That is, they give insufficient attention to the conflicting parties' intentions, and their susceptibility to destabilizing influences over the course of a conflict.

2

The Cold-War Nuclear Force Balance

The Challenge and Promise of Asymmetry

Despite the fervor of Cold War–era nuclear debates, the key strategies of the period—assured destruction, soft warfighting, and hard warfighting—had fragile conceptual and theoretical underpinnings. Consequently, all three strategies exaggerated the consequences of failing to meet established capability standards. The most ardent warfighting proponents took their concerns to the extreme. They insisted that "victory" in nuclear war was possible or, at least, the Soviets *believed* that victory was possible. The United States was obliged, then, to pursue offsetting measures to deny the Soviets potential advantages. The concept of "superiority" lived in these warfighting claims and, more generally, in a preoccupation with asymmetries—force balances and imbalances—that supposedly favored one side or the other.

The issues of the Cold War period remain relevant today. Ideas drawn from these three strategies continue to stoke controversy over the role of defense in strategy, the utility of arms control, and the appropriate response to the Russian nuclear threat. But the dubious logic backing the claims and focus of Cold War–era strategy provides strong reasons to question the current utility of designating "superiors" and "inferiors" in relationships between nuclear-armed countries.

Prelude

The Eisenhower administration oversaw a massive expansion of the US nuclear arsenal. The arsenal grew to serve the expanding target list which swelled, in turn, anticipating increases in the US stockpile and delivery capability (Rosenberg 1983: 23). The Soviet nuclear threat obviously grew, too, with the Soviet acquisition of thermonuclear weapons and capability to deliver ballistic missiles against US targets. But the US targeting plan, bequeathed by the Eisenhower administration to that of John F. Kennedy, emerged fundamentally as a "capabilities plan." It reflected the capabilities that were within reach of the US industrial complex, not the necessity

The False Promise of Superiority. James H. Lebovic, Oxford University Press. © Oxford University Press 2023.
DOI: 10.1093/oso/9780197680865.003.0002

of serving any objective other than delivering a massive, decisive blow against Soviet military and urban targets: "The plan made no distinction among different target systems, but called for simultaneous attacks on nuclear delivery forces, government control centers, and the urban-industrial base" (Rosenberg 1983: 7). Cities—big, open, and vulnerable—were tempting targets given the weapon inaccuracies of the day, challenges in locating and destroying military targets, and US dependence on relatively slow and vulnerable strategic bombers. From a military standpoint, pummeling the enemy by hitting an "optimal mix" of governmental, military, and industrial targets (Lambeth and Lewis 1983: 130) made sense: after all, the purpose of *any* war was to "win" (Trachtenberg 1988/89: 37).

The Eisenhower administration addressed the growing Soviet-nuclear threat by embracing a preemptive response to a Soviet attack. By attacking Soviet war-making capabilities, with signs that the Soviets were preparing to attack, the United States could at least blunt a Soviet nuclear offensive to *limit* the damage inflicted on the US homeland. In US thinking, less damage was certainly better than more damage to the US target set. Indeed, "by the time Eisenhower left office, the Strategic Air Command had been preparing and training for nearly a decade not only for massive retaliation, but for massive preemption" (Rosenberg 1983: 66).

US policymakers accepted that US–Soviet deterrence existed at some level. After all, the Soviets had not tested US forces even in Europe. Still, US officials remained gravely concerned about a loss in stability ensuing from ever-increasing Soviet nuclear might. Indeed, by the 1960s, the United States could no longer prevent the Soviets, preemptively or otherwise, from inflicting significant damage on the United States.

The Kennedy administration struggled for a workable response. The US Department of Defense, led by Robert McNamara, flirted early with doctrines that might give the United States a usable nuclear edge—or options, at least—that would contain the impact of Soviet nuclear capabilities (Freedman 1983: 234–239). In 1961, President Kennedy seriously considered employing US nuclear weapons to knock out Soviet nuclear capabilities, as the United States and the Soviet Union vied over the future of Berlin.[1] US officials eventually accepted that they could no longer disarm the Soviet Union in a preemptive strike nor limit damage to the United States satisfactorily in a nuclear exchange. Thus, by the mid-1960s, the United States had opted for Assured Destruction (AD) as declaratory doctrine. It amounted to a statement that nuclear weapons were *exceptional*: unlike other weapons, they were meant solely for deterrence purposes. Deterrence was secure, then, if the United States retained the capability to destroy the Soviet Union—specifically, its cities—assuredly under any, and all, conditions, including a surprise "bolt-from-the-blue" Soviet attack.

With this new reality, the policy focus shifted decidedly toward elucidating the requisites of *deterrence*. Mainstream analysts now insisted that the United States required the capability—in the form of an AD capability—to retaliate decisively for any Soviet attack. Hard warfighters argued, by contrast, that the fundamentals of warfare still applied. With weapons in the right quantities and qualities, a combatant could limit the Soviet capability to fight back or inflict retaliatory damage on the United States. Soft warfighters split the difference between these strategies. By manipulating risk and imposing costs through nuclear combat, they hoped the Soviets would back down.

The various strategies left essential questions unanswered. Under what specific conditions might deterrence fail? More specifically, how might these conditions influence the adversary by shaping its intentions, that is, its goals and cost acceptance? Absent answers, deterrence theory suffered in both logic and practical utility.

Assured Destruction

The logic of assured destruction had significant allure. US officials assumed the Soviet Union would not attack the United States if the United States could strike back fatally in return. They sought, then, to assure catastrophic damage to Soviet cities, in response to any Soviet nuclear attack. What AD doctrine meant in practice had less intuitive appeal.

The ostensible US policy goal was to make any nuclear war too costly for the Soviets to attempt. For Secretary of Defense McNamara, and his team of in-house experts, this was not a problem of Soviet valuation: that is, figuring out what the Soviets might want and the price they would pay to achieve it. Instead, it was a math problem involving calculations of marginal utility. Assuring retaliation with more than 400 (equivalent) megatons simply offered a poor return on the US investment given the large number of Soviet citizens living in small cities and rural areas. McNamara multiplied that magic number by three, investing the "assured destruction" potential in each leg of the "triad," hedging against the possible disablement of one or two legs—ICBMs, submarine-launched missiles, and bombs delivered by bomber (Kaplan 1982: 53).

His math played also to bureaucratic realities. McNamara hoped to employ the standard to deflect pressure from the military service chiefs who could use Soviet acquisitions, and the imperatives of war fighting, to push for newer and larger service arsenals. It was no coincidence, then, that the AD standard required no increase in the destructiveness of the US nuclear force. With a standard in hand, McNamara could resist service entreaties by insisting that the United States already had "enough" capability to deter the Soviet Union (Ball 1980: 168–177). When congressional critics expressed their deep concerns about a Soviet arsenal that favored highly destructive weapons (bombs with unfathomable explosive power), McNamara would retort that US advantages in numbers, accuracy, and delivery capability—in quantities that could survive any Soviet attack—easily compensated for these alleged disadvantages. In sum, by establishing an absolute standard—a weapon ceiling of sorts—he need not yield to the services' relativist (warfighting) logic.[2]

The Pentagon thus publicly set AD requirements around the destruction that US forces could efficiently inflict. Department of Defense annual reports of the late 1960s and the 1970s translated these requirements into damage estimates: should the United States need to retaliate, at least a third of the Soviet population would die and two-thirds of Soviet industry would lay in ruins (Schilling 1981: 59–60). US officials could always defend these high numbers by claiming they reflected "worst-case" thinking (of realist theory) given non-transparent Soviet goals. By making nuclear war a horrendous prospect, the United States could make nuclear combat an unthinkable proposition for the Soviet Union. What more dovish critics derided as "overkill"—an excessively destructive US force—protected, then, against the possible US underestimation of Soviet pain tolerance. It protected further against possible Soviet temptations. Making nuclear war truly unthinkable could help make peace an imperative.

Still, the United States could arguably have "destroyed" the Soviet Union—assuredly—when lowering the alleged damage "requirements." Under all conceivable circumstances, the United States had far more capability than needed to accomplish identifiable objectives, including the destruction of invading Warsaw Pact troops in Europe, the Soviet military infrastructure, and eventually (with technological advances) the residual (unlaunched) Soviet ICBM force. Indeed, the calculations made no allowance for the *actual* cost tolerance of Soviet leaders: US officials focused on *US* capabilities and efficiency considerations, not Soviet capabilities nor values. They effectively allowed technical—and political—considerations to define unacceptable Soviet costs.

Despite its pretense, AD rhetoric hid a fragile commitment. AD advocates understood that national suicide, for US leaders, was not the rational course. A challenge, then, was convincing the Soviets that the United States would do just that by making the US response to a Soviet attack virtually automatic. For example, a military aide with a briefcase containing the US nuclear codes shadowed the president, presumably binding him, through routine, to his nuclear responsibilities and communicating to the Soviets that he would respond instantaneously—maybe overwhelmingly—to an attack. By engraining that message in organizational procedures, doctrine, and public rhetoric, US officials hoped the Soviets would show necessary restraint. That the act of retaliating seemingly defied rational logic was irrelevant: presumably, when the time came, US leaders would respond, as they had *prepared* to respond, with little time to think, deliberate, and philosophize about the rationality of the fateful decision.

For many AD advocates, that was still not enough. They nominally adhered to AD principles which focused solely on attacking Soviet targets of "value," that is, cities. Yet many sought to compete—at least, qualitatively—with Soviet armaments; and they sought "options," short of catastrophe, should nuclear war occur. Consequently, they coveted enhanced US counterforce capabilities—accurate, responsive weapons for destroying Soviet hard targets. Even the McNamara-led Defense

Department planned to engage primarily in counterforce strikes against Soviet military targets—that is, to match US against Soviet military capabilities in the event of war (Friedberg 1980: 53). Thus, the doctrinal commitment to Assured Destruction did not trickle down to the operational level of the US Single Integrated Operations Plan (SIOP), US targeting packages and priorities in the event of a US preemptive or retaliatory strike.

The "flirtation" with counterforce was hard to reconcile with AD principles. Viewed accordingly, nuclear options beget nuclear temptations—and illusions that feed temptations. With options, parties might exaggerate the accompanying payoffs and understate the danger stemming from *any* use of nuclear force. Staunch AD proponents had little faith that the parties would exhibit the restraint and conflict-management skills necessary to keep a nuclear war "limited." From their standpoint, nuclear options that promised to contain nuclear wars, short of the catastrophic, only made nuclear war more likely, and no less deadly. The US commitment to assured destruction was equivocal *at best*. Soviet nuclear forces remained US targeting priorities. At least in the short term, Soviet urban-industrial targets might have remained out of the US retaliatory mix.[3]

By trying to do it all, AD advocates ended up burying, not addressing, the all-important issue of Soviet intent. Their standards played to levels of societal destruction the United States *could* inflict over the levels of destruction it *should* inflict given the valuation of the Soviet leadership. Then, many AD advocates hedged on their commitment by looking to counterforce options to reinforce deterrence while providing an alternative to a suicidal war should nuclear deterrence fail. Indeed, US officials in the Kennedy and Johnson administrations planned to use nuclear weapons, if necessary, to thwart a Soviet conventional offensive in Europe. Contravening AD principles, they assumed implicitly that the United States could "win" at the tactical nuclear level, limiting the fighting to Europe, without triggering the nightmare AD scenario.

In sum, AD advocates placed *means before ends* when determining the necessary size and character of the US nuclear force. They looked to existing US capabilities to answer the fundamental question, "What price would the Soviets willingly pay to achieve their goals?" For some, their capability focus had a perverse consequence: although they denied the possibility of winning a nuclear war, they still saw benefits in acquiring counterforce capabilities. Their thinking reduced, then, to a mishmash of strategic principles.

Soft Warfighting

Soft warfighters shared the fears of AD proponents that a nuclear war would end in catastrophe but feared that impending catastrophe alone would not deter one. They rejected the stilted three-stage (AD) vision of war—a first strike, a suicidal second strike, and then oblivion for both parties—in search of a strategy amounting to less than a full embrace of hard-warfighting logic.

Soft warfighters conceded that nuclear superiority was unachievable in a military sense. They envisioned nuclear conflict, instead, as a competition in "risk taking" that resembled a game of "chicken." The United States could vary its weapons and targets—hitting ever closer to key Soviet assets or upping the firepower—to send appropriate messages. The ideal message would signal both US resolve and restraint—leaving the Soviet Union with more to lose—and thus reason to reconsider and reverse course.

Their arguments were backed by compelling metaphors and vivid scenarios that would only lose luster had soft warfighters probed more deeply. Would the conflict's trajectory depend on the conditions that provoked the conflict? Under what circumstances would the exchange of nuclear fire promote rather than preclude restraint? The resulting logic nonetheless inspired successive US administrations to revise the standards for scaling and configuring the US nuclear force. The soft-warfighting innovations of the Nixon and Ford administrations did not amount to a frontal assault on past thinking. However, the Carter administration took bigger steps, in the direction of soft (and hard) warfighting, hoping to realize the full potential of US capability improvements across the technological spectrum.

The Nixon-Ford Administrations: Moving toward Soft Warfighting

The Nixon and Ford administrations expressed their dissatisfaction with AD by offering revised standards for assessing the deterrence capabilities of the US force. The aspirations, and failings, of the new approach surface when evaluating these standards and the administration's plans for "Limited Nuclear Options."

A New Standard

The Nixon administration ostensibly sought to meet the interrelated force standards of "sufficiency" and "essential equivalence." "Sufficiency" meant the United States would possess the retaliatory capability necessary to deter a Soviet attack. With "essential equivalence," the United States had to match *aggregate*—but not necessarily specific—Soviet nuclear capabilities. Through these principles, President Nixon and his National Security Advisor, Henry Kissinger, sought to deprive the Soviets of a coercive edge from capability counts—especially in land-based missiles and missile throw-weight—where the Soviets held an "advantage." Whereas the new criteria spoke, then, to the administration's political focus (Gavin 2012: 111), they also increased the definitional and measurement challenge.

Sufficiency, for one, arguably differed from AD in name only.[4] Yet it also appeared to undercut the notion that the United States required the capability to inflict unacceptable damage. The administration claimed, in part, that sufficiency required "preventing the Soviet Union from being able to inflict considerably more damage on America's population and industry" than US forces could inflict on the Soviet populous and industries in return (Smith 1981: 24). It thus appeared to stress matching destructiveness, not inflicting damage of some magnitude. Either way, the term provided limited actual guidance for scaling and configuring the US force: how could US officials know whether they met the matching criteria without knowing exactly how a nuclear war would unfold?

Although "essential equivalence" seemed to offer more precise guidance to force planning, it nonetheless biased thinking toward matching aggregate Soviet nuclear capabilities. Then, it directed attention to prewar numbers, not the military or coercive effectiveness of an arsenal over the course of a nuclear exchange. Equivalence in a prewar balance could amount to post-exchange superiority or inferiority, by some measures (and vice versa). Much would depend on force vulnerabilities, offensive strategies, and the military and political weight the parties might place on surviving weapons systems.

Analysts could reasonably ask whether sufficiency and essential equivalence were interchangeable concepts, or perhaps even contradictory ones. Analysts could interpret sufficiency and essential equivalence to mean the same thing: if the United States requires an essentially equivalent force to deter the Soviets, then sufficiency amounts to equivalence. Or else, they could conclude that the guidelines conflicted. If the Soviets only respect superiority, then an essentially equivalent force is wholly insufficient.

Sizing and Configuring US Forces for "Limited Nuclear Options"

Although the revised strategic principles fell short of a workable standard, the Nixon and Ford administrations made their strategic mark by pushing for *flexibility* in nuclear-weapons employment. For that purpose, they sought to introduce "limited nuclear options" into US war plans (Davis 1975).

Chastened by US operational doctrine that tilted decidedly toward an "all-or-nothing" response to a Soviet attack, the administration pursued options to effectuate the "tit-for-tat" tactics of "escalation control." Henry Kissinger, who "provided the impetus for the revived interest in U.S. nuclear strategy and limited nuclear options" (Terriff 1995: 69), feared that US threats to unleash Armageddon might lack the credibility needed for deterrence, and could leave the United States without real options should deterrence fail. Up to that point, the United States could deliver only massive counterforce attacks, or some combination of massive counterforce and countervalue attacks (Friedberg 1980: 55).

The administration was not the first, however, to pursue selectivity in targeting. The Kennedy administration successfully introduced distinctions between nuclear, other military, and urban-industrial targets into US war plans (Sagan 1987: 38), having inherited from its predecessor a plan to deliver a "single massive blow" to eviscerate the Soviet Union in war (Rosenberg 1983: 63). The 1960 SIOP called for striking the Soviet Union, China, and their allies with thousands of nuclear weapons despite—indeed, because of—the likely human toll: the deaths of hundreds of millions of people (McKinzie et al. 2001: 6). Killing people in unfathomably horrific numbers was not the acceptable cost of war but, rather, the point.[5] That was the planned outcome had the 1961 Berlin Crisis gone nuclear—ironic, to say the least, given the humanitarian concerns (about the freedom of exit of the oppressed peoples of East Germany) that sparked the crisis. Millions of oppressed Germans, Poles, and others would be among the casualties.[6] The plan amounted to the indiscriminate destruction of the Soviet bloc in the event of a nuclear conflict. Still, the "selectivity" that the Kennedy administration got for its efforts was undermined by the truly massive nature of any US nuclear strike. Even attacks restricted to nuclear and conventional force targets, outside of cities, would deliver thousands of US warheads against Soviet targets.[7]

Following the guidance of National Security Decision Memorandum (NSDM) 242, limited nuclear options were the cornerstone of the Schlesinger Doctrine, named for Secretary of Defense James Schlesinger (who served in the 1973–1975 period). The underlying aspiration, in war, was to coerce a Soviet retreat while not breaching the critical threshold that would spark an all-out nuclear conflagration (Gavin 2012: 112). By choosing the right targets, and configuring and scaling the US arsenal appropriately, US officials hoped that the cooler heads—unable to prevent a nuclear engagement—would now somehow prevail. Their hope: a nuclear war that they could not prevent, they could at least control, and possibly end short of catastrophe.

Soft warfighters, in and outside of government, looked, then, for a sweet spot between the logical extremes of hard warfighting and AD doctrine. Their focus was on acquiring weapons in various sizes, shapes, and capabilities for *political* effect, in *potential* contingencies. Some soft warfighters suggested, for instance, that the Soviets would view a warhead delivered by a tactical weapon, or a theater-based weapon, differently (due to its firing location) than one launched from a distance by a US-based ICBM.[8] With such thinking, leaders could select from among a meaningful range of alternatives, mix and match or sequence them for maximum effect, or temporize by choosing options that would buy some time until better information was available.

Soft warfighters assumed, then, that leaders were sufficiently rational to evaluate and manipulate risks in a nuclear conflict but also to understand when these risks were prohibitive, and "intra-war" deterrence (which had prevented all-out, nuclear conflict) would fail. But would they—or could they? Soft warfighters hoped that the United States could communicate its message of resolve and restraint to press the Soviets to reduce their levels of violence *under the worst of conditions*: after missiles started flying, destruction mounted, and each side thought the worst of its opponent's (and the best of its own) intent. What worked in principle, then, could easily fail in practice. What constitutes a useful weapon for signaling purposes is a matter of perspective. Once Soviet targets were hit, for instance, why would the Soviets care about delivery-system specifics at all?[9] Even if they cared, why would they read messages as the US officials intended? Might selecting targets of strictly military consequence cloud the countervalue message? Conversely, might choosing targets of value suggest that the United States would observe no restraint?

Targeting distinctions were equally problematic. As we shall see, distinctions between value and force targets were specious at best. They reduced, in fact, to word parsing when the Nixon and Ford administrations countenanced striking urban-industrial targets but not population centers "per se." NSDM 242 prioritized the hitting of "economic recovery assets" (Sagan 1989: 44–48), albeit as "withhold targets" (held in reserve for possible attack), in calling for the "destruction of the political, economic, and military resources critical to the enemy's postwar power, influence, and ability to recover at an early time as a major power." The striking of these recovery assets, however, could justify attacks on virtually any target—from government leaders and police and soldiers who enforce order to fertilizer plants (Lebovic 2013: 71–72).[10] But why stop there? All products can contribute, in some way, to recovery or military performance. After all, soldiers eat food, wear clothes, and require medical care. For that matter, soldiers and their governments depend on *societies* to survive. Thus, the underlying logic easily allows "value" targets to become "force" targets, and no solid basis exists for distinguishing between the two.[11] Indeed, "targeting 'economic recovery' targets was a good way of giving military value to what might otherwise have been construed as civilian targets" (Nolan 1989: 112).[12]

Unsurprisingly, the Joint Chiefs of Staff pushed back hard: they claimed that limited options would weaken (not strengthen) deterrence and expose US forces to a Soviet counterattack. From the military's perspective, the bargaining purposes that enthralled civilian officials were too vague and undeveloped to guide nuclear planning,[13] and all such gamesmanship risked surrendering the initiative should the Soviets choose to forgo limits. After all, the purpose of a nuclear war, like any war, was to win—or, at least, to achieve the best possible military outcome. For the military, then, a nuclear war required careful choreographing, necessitating significant advance preparation to prioritize, sequence, and coordinate the contributing parts.[14] Even limited attacks required "establishing rates of fire, timing of attacks, and weapons to be used, among other factors, in such a way that the execution of limited nuclear options did not disrupt the capability to implement the plan in its entirety" (Terriff 1995: 206)—but still risked compromising the "plan."

NSDM 242 thus delivered only modest changes—*at best*—to the targeting script. What the Nixon and Ford administrations got for their efforts hardly befitted the precision, scale, and selectivity found in writings on nuclear bargaining. US targeting plans offered US leaders relatively little latitude when using nuclear weapons: "The lion's share of the strategic forces continued to be rigidly committed to preplanned SIOP missions and to fulfilling the corresponding requirements of damage expectancy" (Blair 1993: 42). Indeed, evidence indicates that military personnel (the Joint Strategic Target Planning Staff) creatively defined targets to neutralize the directive's prohibitions on striking Soviet cities and national command-and-control targets (Sagan 2021: 137–138). When the SIOP was finally delivered, Kissinger thought the whole effort, to enforce limitations, a waste: SIOP-5 "mainly consisted of large nuclear options that were useless for the kind of signaling that Kissinger and [Secretary of Defense James] Schlesinger had sought."[15] True, the resulting reevaluation yielded a range of attack options—"major," "selective," "regional," and "limited" (Burr 2005: 73)—and, along with them, variations in the possible magnitude, targets, and sequence of attacks (Richelson 1986: 159; Sagan 1989: 44–48; Terriff 1995: 207–208). The SIOP incorporated only the major and selective attack options, however, and the latter—involving hundreds, up to a thousand, US warheads—were designed nonetheless to contribute to the full execution of the SIOP (Terriff 1995: 208–209). Urban-industrial targets continued to constitute most of the potential aim points.

Soft warfighters arguably offered the worst of both worlds (strategies), then, in attempting to overcome perceived deficits in strategic thinking. Their focus was on weapons and targets, not the likelihood of various contingencies or human (and organizational) capacities to make refined judgments, or communicate subtle messages, in the blunt world of a nuclear exchange. At best, their tactics might only have postponed the "kill-and-be-killed" scenario of AD doctrine (with a slow-motion variant). At worst, they would have triggered catastrophe if US officials believed they had limited options when a nuclear war was "unlimited" by design due to fixed war plans or the Soviets seizing the initiative by attacking in full. The latter was more than a remote possibility: the Soviets intended to go *big, early* in a nuclear exchange.

The Soviet incentive to go on the offensive, reducing their vulnerability to a devastating US attack, only increased as US first-strike capabilities grew: "Evidence shows that Soviet perceptions of the nuclear balance and changes in their nuclear policy correlated in time with the development of American counterforce capabilities" (Green and Long 2017a: 608). Soviet fears of a US first strike were not misplaced (see Green and Long 2017a: 610–612; Lieber and Press 2006a, 2006b). To cripple a largely land-based Soviet nuclear force and substantially reduce its capability to retaliate, US forces could capitalize on a multitude of accurate high-yield warheads; attack blind spots in Soviet early warning systems; employ high-altitude, nuclear bursts to disrupt Soviet command, control, and communications; and so forth.

True, US counterforce capabilities could not reduce Soviet retaliatory capabilities sufficiently to make a US strike acceptable to US leaders, except under dire circumstances. Critical in a crisis, however, was what Soviet leaders *thought* US leaders would accept—and what Soviet leaders would do, then, aware of their own military vulnerability.

The Carter Administration: An Equivocal Commitment to Soft Warfighting

Compared to its immediate predecessors, the Carter administration was even more conflicted about where it stood in the nuclear debate. It seemed lost in an unhappy middle ground between soft- and hard-warfighting principles. The administration struggled accordingly in developing a useful force standard, coherent strategy, and logical response to the Soviet introduction of new missiles into Europe.

Securing a Standard

For the Carter administration, the notion of "sufficiency" failed both as a deterrence principle and as a guideline for nuclear-war preparation. It left the United States unable to capitalize on growing US capabilities in command and control, reduced US platform vulnerability, and increased US weapon accuracy. By contrast, the administration embraced the concept of "essential equivalence." It suited Secretary of Defense Harold Brown's belief "that nuclear forces have a political impact influenced by static measures" pertaining, for instance, to warhead numbers and throw-weight (US Senate 1979a: 137). Administration officials joined their predecessors, then, in worrying about the US–Soviet "coercive balance." A coercive imbalance could shield the Soviets should they threaten the United States, or attack.

His arguments were linked problematically to relativist concerns with a "psychological balance"—a preferred US arsenal that "is not *seen* as inferior in performance" to Soviet nuclear forces [italics added] (US Senate 1979b, 1: 12). But performance to do what? Despite its superficial appeal, the relational logic of a coercive balance was far from compelling when probed and when pushed. Its deficiencies are glaring, in fact, when we view the logic in more personal terms.

Imagine a standoff between two feuding rivals—a modern-day Hatfield and McCoy—each able to kill members of the other's family. We can envision these rivals murdering the family of the other out of spite, or revenge. But could we conceive of a Hatfield—say, the father of three children—seeing himself advantaged, somehow, because the McCoy—the father of five children—had a larger family to kill? Or that the McCoy would see himself disadvantaged, somehow, by the situational arithmetic? Admittedly, the *math* would advantage the McCoy, if the two killed their hostages in succession—trading one (tit-for-tat) for another. The arithmetic suggests that McCoy would hold a "coercive advantage,"[16] for the Hatfield would be the first to run out of hostages to kill. For the McCoy to realize his coercive advantage, however, requires that he play the game out, that is, accept the death of *most* of his children. Would he regard that as a "winning" proposition?

Even the Hatfield–McCoy scenario simplifies the underlying problem, for the scenario focuses on numbers of hostages and ignores the fighting that preceded and accompanied the killing. The showdown is presumably messy. The Hatfields and McCoys are trading bullets, but their guns are misfiring; they cannot tell a hit from a miss; and they are consuming the available ammunition. If so, speculation about the fate of the hostages depends on how the conflict will unfold.

Issues pertaining to hostage valuation, and assumptions about how a conflict might unfold, challenged efforts to ascertain the coercive balance in the US–Soviet context. Soft (and hard) warfighters assumed that, with a first strike, the Soviets would attempt to knock out US nuclear retaliatory capability. For soft warfighters, that would help cement the Soviet coercive edge: the attack would leave some fraction of the Soviet force available for coercive employment in a subsequent (threatened) attack wave. Without exception, warfighters believed that that scenario would give the Soviets an overwhelming edge in the coercive balance. Yet even this dire scenario had a large fraction of the less vulnerable (even if less capable) legs of the US triad (aircraft and submarines) available for coercive strikes. Why should US decision makers feel at a coercive disadvantage? They could credibly threaten to hit Soviet population centers, fearing the Soviets would otherwise control the world or enslave the surviving US population. Would US leaders accept surrender over suicide if they viewed surrender as tantamount to suicide?

More to the point, where do these complex considerations leave official efforts to define the coercive balance? Like Nixon and Kissinger before them, Carter administration officials took the easy route: they assessed the balance crudely from the relative size of prewar arsenals.

The Countervailing Strategy

Secretary Brown did not believe, however, that deterrence rested on perception alone. Through the administration's "countervailing" strategy (Nolan 1989: 138),[17] he sought to acquire the means to fight a "protracted" nuclear war in which US forces could survive *and perform effectively* in a nuclear war environment.[18] The strategy shift provoked a critical question: in war, would US nuclear forces strive to win—in some meaningful military sense—or would they engage in a conflict that would escalate, or end, depending on the *will* of the parties? A clear answer was not forthcoming. That was apparent, for instance, when Brown testified before Congress, "a credible deterrent is achieved when our enemies believe that if they start a course of actions that could lead to war, they will either pay an unacceptable price *or* be frustrated in their attempt to achieve their objective" (US Senate 1979a: 137, emphasis added).

He was not uniquely ambivalent in that regard. Walter Slocombe (1981: 21), a high-ranking Pentagon official and articulate spokesperson for the new doctrine,[19] could not escape contradiction when outlining the countervailing strategy: "the

United States must have *countervailing* strategic options such that at a variety of levels of exchange, aggression would either be defeated or would result in unacceptable costs that exceed gains." Presumably, US leaders would opt to punish the Soviets *or* deny them gains in a nuclear exchange—hitting military targets *or* value targets—while tailoring the severity and geographical scope of the response. With US leaders seemingly unable to decide whether to punish the Soviets or to deny them possible gains, critics could justifiably ask whether time and information in a nuclear conflict would permit the orchestration of an "appropriate" strategy. After all, flexibility in response was no friend of decision makers absent the opportunity, resources, and knowledge to choose wisely among options.

To be sure, Slocombe (1981: 21, 24) ruled out "winning" a nuclear war. He nonetheless sought, through the strategy, to convince the Soviets that they could not win a nuclear war. That would presumably require the (considerable) means necessary to deny gains to a Soviet nuclear force committed to winning. But the countervailing strategy offered poor guidance for choosing between denial and punishment in targeting. Whereas the administration appeared receptive to the notion of nuclear-war *fighting*, and valued weapons for their military virtues (accuracy and responsiveness), it sought not to seek the military edge that could decide a war in the United States' favor. Urban-industrial sites still constituted half of the SIOP targets, which included "economy recovery" targets linked to the production of coal, steel, aluminum, cement, and electric power (Sagan 1989: 47). Indeed, the administration, by defining force *as value targets*, suggested that US goals in a nuclear conflict were fundamentally political. Slocombe (1981: 22–23) pushed for "plans and capabilities" that would permit the United States to "exact a high cost from the things the Soviets valued most—political and military leadership and control, military forces both nuclear and conventional, and the industrial economic capacity to sustain military operations."

In the final analysis, the nature and scale of imposed destruction, as promised by soft warfighters, might little have influenced war outcomes. Soft warfighters built their logical edifice by sidelining *the* essential questions. Were nuclear weapons intended to serve military *or* political purposes? If the latter, what level of punishment was the Soviet adversary fundamentally *unwilling* to accept, or willing to accept given its (hardening) perceptions of US intentions over the course of a nuclear conflict?

Escalation Control: The Soviet Challenge in Europe

The challenges of evaluating US nuclear capabilities for deterrence were palpable when soft warfighters turned to the European theater. Their efforts to boost the US alliance commitment ill served soft-warfighting principles.

In the late 1970s, the United States and its NATO allies sought to answer the Soviet deployment of (three-warhead) intermediate-range (SS-20) missiles in

Europe. These missiles could quickly destroy NATO nuclear and conventional forces across Western Europe.[20] For NATO, the missiles created an "assurance" problem. Absent an "equivalent" capability in theater, NATO's European members had to rely, for deterrence, on a US strategic nuclear force that might prove "unavailable," or that the Soviets might believe was unavailable, if needed. The Soviets would thereby calculate, or miscalculate, their way into a nuclear conflict.

For NATO members, a necessary response was the US deployment of Pershing II land-based, ballistic missiles in Germany and ground-launched cruise missiles throughout Europe. By the soft-warfighting logic behind NATO thinking, these missiles allowed the United States to match the Soviets at their level of provocation while implicitly threatening to bring the full weight of the US strategic nuclear force to bear to increase Soviet costs in the conflict. Soft warfighters remained unclear nonetheless about the purposes of the new weaponry.

Did the United States intend the missiles to show that the United States would engage fully in a nuclear conflict? That is, did it intend these weapons as a visible "trigger" to an all-out strategic nuclear exchange? As political instruments, these weapons—placed in vulnerable positions on the front lines of conflict—could signal to the Soviets that the United States had effectively tied its own hands. NATO would employ its nuclear weapons early, then, rather than lose them to a Soviet conventional offensive. It would do so with full knowledge that the use of nuclear weapons in theater could trigger a strategic nuclear exchange. If that was the logic, however, why were *these* missiles necessary for that purpose? For that matter, would the availability of additional nuclear options, in theater, reduce the credibility of the US threat to back NATO with US strategic nuclear weapons? After all, if the United States had withdrawn all US nuclear weapons from Europe, it would have had *no option* but to answer Soviet aggression with US strategic nuclear strikes.[21] US strategic forces were more than equipped for that purpose.

Or did the United States intend these missiles, instead, to limit any nuclear war to Europe?[22] After all, soft warfighters formulated their overall strategy with the hope of controlling the spread and intensity of a nuclear conflict. Yet the capabilities and placement of the weapons could easily have provoked an unlimited conflict. The Pershing II missiles in Germany—by virtue of their accuracy, prompt-delivery, and proximity—served, for all intents and purposes, as strategic nuclear weapons. These missiles could hit Soviet strategic targets—most perilously, Soviet command and control sites—quickly from their European positions.[23] NATO's European members "believed that it was essential to hold Soviet, rather than Warsaw Pact, targets at risk" with the goal of engaging the superpowers fully in the conflict (Anderson and Nelson 2019: 94). The assurance these countries sought was that they alone would not bear the costs of war.

The situation was thus ripe for an explosion. With Soviet plans to take the (nuclear) offensive if a NATO nuclear strike appeared imminent, and NATO's plans to take any nuclear conflict initially to the Soviet Union (Schulte 2012: 55–56), conditions could create a "first-strike" dilemma. Although neither side desired a nuclear war, both sides feared the implications of striking "second." Inasmuch as the Pershing II and cruise missiles could evade Soviet early warning in a surprise attack (Green and Long 2017a: 630), the NATO effort to "match" the Soviet deployment could have inadvertently sped nuclear-weapons employment at the theater level that could quickly expand to the strategic level.

In short, the United States made a *political* decision to placate US NATO allies. Accordingly, the United States sent 108 Pershing II missiles to West Germany, *on the front lines of the conflict*—to be paired with the 108 launchers for the Pershing 1A missiles that the new missiles would replace.[24] (West Germany—alone—was willing to accept the accompanying costs in political isolation and domestic resistance that the missiles would invite.)[25] The implications were significant. Soft warfighters had explicitly rejected AD thinking for its hopelessness, automaticity, and finality. Ironically, they had now *extended* its reach to encompass any, and all, confrontations—conventional or otherwise.[26] The search for advantages—and the desire to neutralize them—brought them full circle. They sought to surmount AD dilemmas but could not override them.

Schlesinger, himself, tried to straddle that debate. He conceded nonetheless that the US weapons in Europe, apart from deterrence, were intended to *contain* a nuclear conflict (Terriff 1995: 197). If that was his thinking, he was likely horribly wrong.

Hard Warfighting: Denial Strategies

It was left to the hard warfighters of the Reagan administration to put AD fully to rest. It, like the Carter administration, sought to capitalize on increased missile accuracy, upgraded warning, monitoring, and tracking systems, and command and control systems. It, too, prepared, then, for a protracted nuclear conflict—as measured in hours, maybe days, not months and years—fought, at the very least, for *immediate* military objectives. The two administrations differed, however, in one important respect: under Presidential Directive 59 (PD 59), the Carter administration explicitly denied that a nuclear war was "winnable," though that concession did not make for a more coherent strategy.

Background

In the view of hawkish critics, alleged Soviet advantages—particularly in heavy land-based missiles—made a mockery of the term "equivalence." They had a point. In embracing equivalence, the Carter administration failed to determine how much military requirements mattered when sizing and configuring the US nuclear force (Lebovic 2013: 93–94). Still, the criticism ran deeper. Hawks maintained that prior administrations had left the United States dangerously unprepared to deter a Soviet attack or reduce the Soviet nuclear threat in the event of an attack: with their heavier and more numerous land-based missiles, the Soviets could deliver more warheads promptly on target once the Soviets acquired MIRVs (multiple independently targetable reentry vehicles). With many warheads launched accurately against US targets from single missiles, hawks feared that the Soviets would eradicate the US land-based force in a first strike.

These critics were deeply troubled that the Soviets had deployed land-based missiles at a much higher rate than expected through the 1960s and into the 1970s (Wohlstetter 1974a, 1974b) and had not mimicked US deployments by distributing nuclear weapons more evenly among air-, land-, and sea-based legs.[27] That the Soviets seemed unconcerned that their large, land-based missile force was vulnerable to a US attack suggested, to these critics, that the Soviets planned to strike first (Pipes 1977, 1980).[28] Some hawks concluded, in fact, that the Soviets sought a "war-winning" capability (Gray 1979a; Gray and Payne 1980). With the decimation of the US land-based intercontinental ballistic missile (ICBM) force, the United States would lack the counterforce capability to destroy, promptly, remaining Soviet land-based missiles (that could threaten US cities) in their hardened silos.[29]

For these critics, arms control was now a dangerous farce: unlike US negotiators, the Soviets saw arms control "as a means to gain strategic advantage and to restrict U.S. programs that might deny them that advantage" (Van Cleave 1989: 33). Stated in even starker terms, "the United States has been negotiating with a country which not merely fails to share the orthodox American concept of strategic stability, but has been working very energetically to undermine whatever stability now exists" (Gray 1979b: 202). The Soviets recognized what US policymakers did not: "that war at any level can be won or lost, and that the distinction between winning and losing would not be trivial" (Gray and Payne 1980: 14).

The critics had vented their displeasure with the terms of the 1972 Strategic Arms Limitation Talks (SALT) I Treaty, which essentially froze numbers of US and Soviet weapon launchers in place, supposedly giving the Soviets a decided advantage. They continued to press their case. An officially sanctioned, Team B exercise permitted prominent hawks to critique the working assumptions of Soviet analysts in the US Central Intelligence Agency (then directed by George H. W. Bush).[30] Key participants subsequently pooled their efforts for domestic political influence. Paul Nitze, Albert Wohlstetter, Richard Perle, and others united to form the Committee on the Present Danger. Hawks descended in mass, then, to rally opposition to the 1979 SALT II Treaty.

The hawkish advocacy of major figures in strategic arms control—Paul Nitze, Edward Rowny, and Senator Henry Jackson—featured heavily in the debate.[31] With the treaty at center stage, and malign Soviet motives the common denominator of opposition, critics focused on imbalances and loopholes that the Soviets had presumably crafted to their advantage and intended to exploit.[32] Although the SALT II Treaty capped the number of MIRVs atop ICBMs, US treaty critics warned that the Soviets could double—maybe triple—the capabilities of their land-based force by exceeding the SALT II limitations that were programmed to expire in the early 1980s. A "window of vulnerability" would open then when the Soviets could target their MIRVs against the considerably smaller number of US land-based missiles.

The Reagan administration did not mute its displeasure. In the 1980s, it decried the Soviet encryption of telemetry information—electronic transmissions that could reveal the number of warheads the Soviets tested on missiles—even questioning whether the Soviets were providing false information on open transmission channels.[33] It read all Soviet violations of the Anti-Ballistic Missile (ABM) treaty as evidence that the Soviets were building a nationwide defense to break out from the ABM treaty. It stuck to its assessment though Soviet actions were far too limited and the challenge of building an effective missile defense far too great to position the Soviets to thwart a US nuclear offensive,[34] and though the United States, too, had committed treaty violations.[35] The failure to find evidence in obvious places only reinforced hawkish misgivings. It suggested to critics that the Soviets knew what they were doing: they were hiding illicit activities by feeding the misconceptions of the US intelligence community (Katz 1979).

Hard warfighters, among the critics, were well represented in the first term of the Reagan administration. They were unimpressed with soft warfighters' fanciful logic. The danger, for them, was that the Soviets, with their heavy, land-based missiles, possessed superior *warfighting* capabilities. The high-profile advocates—Colin Gray, Keith Payne, and William Van Cleave, to name a few—populated universities and think tanks and influenced the thinking of strategists and practitioners (both within and outside of the US government) about when and how a nuclear war would (indeed, must) be fought.

Strategic Deficiencies

Hard warfighters prized US weapons for their capability (in numbers, accuracy, and quick delivery) to destroy hard targets—specifically, the Soviet ICBM force—and the ancillary capabilities for command, control, communications, and intelligence (C^3I) that would permit the expeditious delivery of these weapons against those targets. Whereas some hard warfighters argued that a US–Soviet strategic nuclear conflict was winnable, most hard warfighters maintained, at least, that the Soviets believed victory in nuclear war was possible. In their view, then, deterrence required that the United States prepare accordingly. The United States must establish—through its military preponderance—that the Soviets could not gain from an attack: They could neither disarm the United States nor cow it into submission. Their arguments suffered, however, as guides to force deployment owing to (a) specious targeting distinctions, (b) fragile command and control, (c) the neglect of Soviet goals, (d) illusionary escalation "firebreaks," and (e) the lack of an "end game."

Specious Targeting Distinctions

In the abstract world, strategists could treat certain weapons—ICBMs in particular—as force targets and cities as value targets. In the idealized version of AD, then, each side targeted the other's land-based missiles in a first strike, and targeted cities in retaliation. With the hard-warfighting alternative, missiles fought missiles, that is, their warheads destroyed adversary missiles in their silos. Yet the distinction between force and value targets suffered in operationalization.

Radar installations and military command centers are force targets, perhaps, but what about military bases, communication networks, and civilian government installations, including facilities that housed top Soviet officials who would presumably direct their country's war effort? Were these force targets or value targets? Attacking them, apart from the military effects, would weaken the Soviet government's control over its citizens, and its ability to provide for the needs of a population suffering enormously post-bellum. Does that not suggest that these assets held considerable value to Soviet leaders? Conversely, the Soviets could view their missiles as value targets if they believed that missiles deterred US threats to Soviet cities.

Even if targets differed in principle, they still overlapped geographically. The high-altitude detonation of Soviet warheads—meant to impair US retaliatory attacks by disrupting US radio communications—could produce widespread collateral damage (Gottfried and Blair 1988: 19). Many "military" targets—command posts and bases—were collocated with industries and cities. Attacks on hard command posts—in Moscow, for example—would have been "virtually indistinguishable from an attack aimed at the destruction of Soviet society" (Gottfried and Blair 1988: 19). Even strikes on missile sites would not spare civilians. Whereas the United States placed its missiles in rural parts of the country—including North Dakota and Arkansas—the Soviets placed their silos (presumably for convenience) closer to urban areas. Should cities survive the blast, they were susceptible nonetheless to the effects of radiation and fallout (Levi, von Hippel, and Daugherty 1987/88), and the underappreciated effects of fire (Eden 2004).[36]

The perversity of artificial target designations was painfully apparent in the Reagan administration. It planned to target Soviet leaders at the outset of a nuclear war when, in prior administrations, the lives of these leaders would be kept at risk for bargaining purposes (see Bracken 1983: 88–89). Although the Carter administration intended to keep the Soviet leadership at "risk" initially in a nuclear conflict, with the idea of giving them "more to lose" to deter the escalation to full-scale war, the Reagan administration removed these targets from the "withhold" list. The targeting shift reflected growing concerns in US strategic circles that a US capability to eviscerate the Russian populace and industrial base would not deter Soviet leaders.[37] It also stemmed from ostensible military benefits in targeting the Soviet leadership and military command. Why sever the tentacles of the monster when one could disable or kill it by removing its head?

The underlying logic begged for scrutiny. Clarity in thinking about these targets was certainly not helped by hard warfighters who (despite their ostensibly singular focus on force targets) argued for the prompt targeting of the Soviet leadership as a "value" target with the assumption that Soviet leaders covet their own lives and government above the survival of their citizenry. It was also hurt by the implicit assumption that Soviet leaders placed little to no value on the preservation of Soviet society. Even if we assume the Soviet leadership cared most about preserving the Soviet system—its security apparatus, leadership, and so forth—what was the state absent a society, that is people, industry, effective governance, and the trappings of civilization? Put simply, could the head function without a body? For that matter, did the head exist apart from the body? The Pentagon, in the Carter administration, had defined the Soviet leadership to include around 100,000 people around the country (Ball and Toth 1990: 73).

Hard-warfighting logic collapsed under the weight of such questions. By making the Soviet leadership an acceptable target, hard warfighters undermined their argument that the combatants could fight a nuclear war within constraints. What would keep residual elements, with control of Soviet weapons—following standard operating procedures—from launching a ragged spasmodic attack against the civilian US target set? Critics had consistently warned of that danger—with strong justification, it turns out. Unbeknown to US analysts at the time, the Soviets devised an insidious answer to US targeting plans. As life imitates art, they constructed a "doomsday device," of sorts, that harked back to the *Dr. Strangelove* tale (Hoffman 2009). The Soviets automated a massive nuclear response should launch-control officers lose contact with the command (and should sensors point to a US nuclear attack).

If we look at the four categories in which SIOP targets were grouped by the Cold War's end (US GAO 1991: 2), we can easily see that moving from the first to any of the next three—and especially the fourth—amounted to countervalue targeting: (1) Nuclear Forces (ballistic missiles and their launch facilities and control centers; nuclear weapons storage sites; airfields supporting nuclear-capable aircraft; ballistic missile submarine bases); (2) Leadership (command posts; key communication facilities); (3) Other Military Forces (barracks, supply depots, marshaling points; airfields not supporting nuclear forces; ammunition storage facilities; tank and vehicle storage yards); and (4) War-Supporting Industrial and Economic Factories (industrial: ammunition factories, tanks and armored personnel carrier factories, petroleum refineries, railway yards and repair facilities; economic: coal, basic steel, basic aluminum, and electric power).

We could argue that "Nuclear Forces" are counterforce targets, but even that distinction takes no account of their vast number, or locations proximate to urban areas. Indeed, a counterforce strategy aimed "strictly" at nuclear-force targets would require many more times the weapons needed to destroy Soviet cities (Lortie 2001: 23). Thus, the United States could not destroy military targets without imposing significant "value" damage through errant blasts, fallout, and the destruction of collocated targets (McKinzie et al. 2001). Although warfighters touted the idea of counterforce exchanges that would "spare" urban targets, the indirect and collateral effects on cities—ensuing, for instance, from the spread of fallout from "surgical strikes" on nuclear missiles—would include the deaths of millions of Soviet and US citizens. The destruction of roads, governmental facilities, communication links, vital industries, and so forth would destroy critical economic, societal, and governmental nodes, with (cumulative) contagion effects that would, in turn, destroy society as it previously existed.

Adding targets to the US strike list, from any of the three other categories—despite a military rationale ("war supporting industry")—would enormously increase the human toll. The SIOP arguably served, then, as a blueprint for an indiscriminate war: "When you're talking about thousands of nuclear weapons—and the mass fires, fallout, and destruction that they can cause—the differences between counterforce and counter-value quickly begin to blur. Millions are going to die—instantly or over time—either way" (Lortie 2001: 26).

Yet hard warfighters focused on the benefits of reducing Soviet nuclear might—especially, hard-targeting capabilities—over the costs that US (and Soviet) hardware could inflict in civilian damage and deaths. They assumed, then, that the parties could avoid the liabilities of a punishment strategy by employing ever-more accurate weapons against the adversary's nuclear infrastructure. These costs, however, were "limited" only in a relative sense. What would keep the combatants from thus unleashing their wrath fully on enemy cities?

Fragile Command and Control

Hard warfighters assumed that the United States could employ counterforce strikes to control war outcomes. Yet the capability to control the conflict could quickly evaporate with strikes on C^3I facilities. The combatants would lose their capability to assess the damage inflicted (by either side), and with it their desire to withhold strikes on adversary (urban) targets of value. Indeed, the hard-warfighting emphasis on the "offensive" promoted an exaggerated sense of first-strike advantages (and second-strike disadvantages) that weighed toward a hair-trigger response to tactical warning of an attack.[38] The danger, then, was "that Soviet or American leaders, at the height of a crisis that seemed likely to spill over into nuclear war, might grasp at the straw of preemption precisely because preemption seemed to offer the only means by which they might protect themselves and their country" (Lebow 1987: 54). The distinction between launch-under-attack and launch-on-warning was already ambiguous (as discussed in Chapter 5), at best (Gottfried and Blair 1988: 58–59). Forces could launch in retaliation, with initial evidence of an adversary attack, of any scale, when the United States and Soviet Union were better equipped to determine the fact of an attack than to assess its full dimensions (Gottfried and Blair 1988: 85).

For that matter, the potential for restraint was only as strong as the warning and assessment capability of the *least capable* of the parties. Impoverished Soviet systems, in that regard, preordained a spasmodic Soviet response (Bracken 1983: 108) that would seemingly necessitate a US response (anticipatory or otherwise) in kind. Justifiable fears that the capability to coordinate forces for an effective counterforce attack would deteriorate quickly—and beliefs that a robust offensive could significantly degrade the opponent's retaliatory capability—could only fuel desires to lash out, to inflict maximum damage, with first indications of an enemy attack. In short order, misunderstanding and miscommunication in the fog of war would most assuredly consume the combatants—more quickly still if the parties targeted each other's command and assessment capabilities or a nuclear exchange involved some combination of strategic, theater, and tactical nuclear forces (Bracken 1983: 123, 157–158).

Operating blind, and in the dark about their own residual capabilities, the United States and Soviet Union might then have fired everything they had; the operative logic being, "use them or lose them." At that point, the parties might even have lacked the capacity to control their own forces. With communication links severed and the command disabled, military subentities might effectively have conducted wars of their own, lacking the ability to see the big picture or, by then, to change it. As Paul Bracken (1983: 99) concluded at the time, "each separated island will then face its own individual assessment problem," that is, "it will not only have to estimate the damage it has sustained, but also the damage sustained by the other separated units and the status of enemy forces." Indeed, the continued actions of any US or Soviet subunit could have fueled responsive salvos from opposing subunits.

The Neglect of Adversary Goals

Hard warfighters focused on military targets to the neglect of combatant goals that would explain how the fighting started, and thus how it would end. Were the Soviets out to destroy or neutralize the United States? If the former, why would the Soviets settle for less than the destruction of US society? After all, by initiating a strategic nuclear conflict, they had accepted the prospect of Soviet destruction. They thus viewed the stakes as high. So why not see the "job" through to the end by destroying US conventional-force capabilities and the US industrial base?

Hard warfighters sidelined questions of Soviet motivation by focusing on a troubling scenario. They posited that the Soviets might do the unthinkable if their domestic base were threatened. They feared that Soviet leaders, facing a loss in power, would launch an attack on the United States—perhaps as a "Hail Mary" ploy (a risky gambit lacking all other options) or a last-gasp effort to take the United States (and its Western allies) down with them. The Soviets presumably at that point would have nothing left to lose and perhaps something to gain (the destruction of the West).

Yet that scenario assumed Soviet leaders valued destruction, for its own sake, over their personal survival, the survival of Russian society, and, indeed, their own family members. No less problematic, the logic defied the very premises that hard warfighters so ardently pushed. Put simply, what could the Soviets *gain* from malicious destructiveness? Hard warfighters had always rejected the proposition that the Soviets would kill and destroy because it offered benefits. The fading Soviet leadership was, however, maliciously destructive; it apparently desired to inflict damage because *it could.*

Soviet leaders could have intended their attack as a last-ditch effort to save the regime. But why would they believe an all-out attack would help, rather than harm, their efforts? An attack on US nuclear forces could not save the regime; and an attack on US cities would only speed the demise of the Soviet leadership and kill all that it valued along with it. Would these leaders not be better off keeping their nuclear weapons in reserve and hoping (perhaps to the very end) that, by threatening to use them, they could forestall a change in regime? For that matter, would they not be better off fading away to plot and prepare for a return to power at some point in the future?

Of course, the Soviets might have stumbled into war. For example, they might have attacked Western Europe expecting an easy conventional victory and then found themselves in a nuclear struggle that threatened their very existence. US leaders, too, might have stumbled forward. These reasonable assumptions are difficult, however, to square with hard-warfighting rationality.

Illusory Escalation "Firebreaks"

Hard warfighters approached escalation in nuclear war by assuming the combatants would recognize clear levels or rungs of escalation and know which side could "win," at each. If the United States could win a theater-nuclear war (in Europe), for example, and no side could win a strategic nuclear exchange, a "firebreak" precluded the escalation of a nuclear war from the theater to the strategic level. Deterrence would prevail because the Soviets lacked viable nuclear options. As opposed to the softer warfighting principle of "escalation control," the United States would obtain "escalation dominance," which required the capability to defeat the adversary at a *higher* level of engagement. The goal, then, was not just to exercise "control," matching the Soviets blow-for-blow, to convince them that further escalation was deleterious. The goal, instead, was to dominate. If the United States could emerge victorious at the higher level of escalation, the Soviets would have no *rational* reason to continue fighting—or, for that matter, to go to war in the first place. The United States would "win" because the Soviets could not rationally escalate to some higher level to prevail.

For these warfighters, then, nuclear war did not amount to a race to oblivion as the parties sought revenge or wildly unleashed their wrath on the opponent. Denial strategists offered a more "bloodless" scenario: in it, weapons fought weapons, and people (cities) stood on the sidelines *relatively* unaffected. People would die—perhaps in high numbers—but killing, for its own sake, was not the point, for it served no real military (rational) purpose.

But could the conflict go as planned given the phenomenal levels of destruction envisioned and the quickly dwindling capabilities of the parties to control their military assets or determine which side was "ahead?" The challenges of ascertaining dominance went on full display when the Reagan administration, despite its hard-warfighting affinities, backed the deployment of Pershing II and cruise missiles in Europe in response to the prior Soviet deployment of the SS-20 missile. The United States was not seeking to match, let alone exceed, Soviet firepower, at least by standards of escalation dominance. The upgraded US force of 572 missiles—108 single-warhead Pershing II missiles, augmented by hundreds of cruise missiles deployed in a handful of countries—simply lacked the firepower of the hundreds of quick-strike Soviet (SS-20) missiles, with their three large-yield warheads: indeed, the cruise missiles would take hours to reach their targets.

What would it take, though, to counter these Soviet weapons if no conceivable US attack could neutralize them before they were fired? The United States could certainly hit other targets, but could dominance be established, then, by a standard other than which side ultimately suffered more military damage or human costs? Moreover, once the focus shifted to which side inflicted more damage, what conceivable firebreak could deter one or both combatants from employing the full weight of their arsenals?

Lack of an "End Game"

Hard warfighters offered a "game" without a solution. The combatants had presumably fought their hard-targeting battle until a "winner," the side that still possessed a residual force of superior weaponry, emerged from the exchange. But where would the combatants go from there? Both sides could only threaten adversary targets of "value," with whatever air-, sea-, and land-based forces that remained. For that purpose, the United States was *not* disadvantaged. For that mission, it did not require land-based missiles, with their advantages in accuracy and prompt delivery. Any nuclear weapon that could explode—eventually—near a target would do. So, where was the Soviet *coercive* advantage? Why would the Soviets assume the United States would hold back when it possessed thousands of nuclear warheads for possible use against Soviet cities?

In looking to how a nuclear war might end, hard warfighters need not have looked further than Soviet nuclear strategy. Although they dismissed the "mirror imaging" implicit in AD doctrine—the assumption that Soviet leaders, just like US leaders, valued the lives of their citizens—they engaged in mirror imaging of their own. They assumed that the Soviets would observe necessary restraint in war—escalating to their *level of advantage*—when, in the Soviet view, a limited nuclear war was a contradiction in terms. In hindsight, post–Cold War Russian annihilative military strategies in Syria, Chechnya, and Ukraine make the Cold War–era assumption of a Russian leadership respectful of limits seem more than naïve.

In sum, no set of strategists—hard warfighters among them—were immune to the challenges of developing consistent and well-grounded arguments. They focused on US weapons and Soviet targets, believing that a military resolution to a nuclear conflict was possible. Put simply, hard warfighters sought superiority in aiming to deprive the Soviets of nuclear advantages at any foreseeable level of escalation. They continued to provoke questions, then, about what military superiority meant in practice, absent the capability to deliver a disarming attack.

The Ongoing Debate: Cold War Strategies, Weapons, and Their Legacies

The Cold War survives in current debate. Old issues surface when policymakers look to strengthen nuclear offenses and defenses and offer strategies to address contemporary nuclear challenges.

The Utility of Nuclear Offenses and Defenses

Both offenses and defenses can contribute to deterrence. If a defender can convincingly counter a challenge—through offense, defense, or a combination of the two—a potential challenger has no incentive to attack.

A sports competition demonstrates the complementary effects of an offense and defense, in support of deterrence (or some other objective). A team's defense assists the team's offense by limiting the opponent's gains (point total or field position) or forcing the opponent to expend energy for paltry gains. In turn, a team's offense assists the team's defense by weakening the opponent, perhaps by creating and exploiting vulnerabilities. Indeed, the offense might put the opponent on the defensive. The opponent, in seeking to avoid loss, surrenders its capacity to take offensive action. The basic principles apply more generally, for example, to counterterrorist operations. Fearing disruption and disclosure, terrorist groups must resort to secretiveness, maintain elaborate cell structures, establish hierarchical control of information, and engage in constant movement. Such action reduces the vulnerability of the group only by also inhibiting efficient, timely, or large-scale operations. Indeed, a terrorist group's defensive measures might follow *detectable* patterns that create openings for effective counterterrorist operations. These patterns include the "bad habits" of a key operative, the recurrent use of a particular contact person, or the employment of a decipherable code.

How—and how much—offenses and defenses each contribute to deterrence is controversial, however. Controversy occurs because offenses and defenses are interdependent and because the offense–defense issue is inextricably linked to the denial–punishment debate (see Chapter 1).

Damage Limitation (Counterforce Tactics)

Given the potential substitutability of offense and defense, many hard warfighters viewed a US counterforce capability—that is, an ability to destroy Soviet missiles in their silos—as indistinguishable in principle from a capability to defend against enemy missiles. Some warfighters thus touted US ballistic missiles that could destroy Soviet missiles in their hardened silos as effectively constituting a "boost-phase" defense. US missiles could destroy unlaunched Soviet missiles to prevent their warheads from landing on US territory, much like a missile defense might knock out incoming Soviet warheads toward the end of their flight trajectory. In fact, offenses and defenses could work synergistically: US missiles, protected from attack, could then go on the attack to prevent further strikes on the US nuclear-missile force.

On balance, counterforce targeting was more efficient than active defenses (as discussed below). Offensive missiles were cheaper to build than defensive missiles and far more likely to hit and destroy their targets (assuming the offensive-missile targets had not yet launched). Defending against Soviet missiles was a phenomenal challenge. Even a highly effective defense—say, with a 90 percent kill rate (Lebovic 2007: 91)—would still leave the United States open to damage from hundreds if not thousands of Soviet warheads. By contrast, an accurate US (land- or sea-based) missile could reliably destroy a larger number of (unlaunched) Soviet missiles—with a good rate of return if each missile housed multiple warheads. A single offensive missile armed with—say, ten multiple warheads—could conceivably destroy one hundred adversary warheads if each adversary missile contained ten warheads. That level of performance was hard to match with a hundred, or hundreds, of defensive interceptors.

Denial-based reasoning—backing nuclear offenses *and* defenses—received new life in the post–Cold War era as the United States confronted a world of smaller nuclear arsenals. Indeed, overwhelming US capabilities arguably place US adversaries at a significant offensive and defensive disadvantage. US challengers must resort to subterfuge and concealment to protect their small arsenals from attacks by a larger, more capable US nuclear force. Indeed, their efforts to protect vulnerable missiles could impair their capability to respond quickly to attacks. For instance, limiting communications between command centers and submarines or mobile missiles, to prevent disclosing their positions, also reduces their responsiveness to command authorities—and could isolate missile launchers entirely if communications links were severed by an attack. Protective imperatives could force mobile-missile operations into a smaller radius—perhaps to capitalize on concealment—which could increase their exposure to attack. Or else, defensive measures could provide telltale signals—such as a large protective force—that would cue enemy offenses to missile locations (Lieber and Press 2017: 47).

These assumptions invite controversy on technical grounds. Skeptics question whether the United States possesses the range of capabilities required to detect and destroy adversary missiles before they are launched. The debate thereby reduces to a battle of assumptions about US force effectiveness including whether or not the United States can depend on warheads with larger yields to compensate for the imprecise locating of missiles. (See Lodal et al. 2010.)

Regardless, denial strategies deserve scrutiny in another respect. They downplay the cost sensitivities of the combatants (see Chapter 3) and nonrational processes that can take hold in a nuclear conflict (see Chapter 5). By placing the focus on what the United States *could* accomplish with force, they detract from what the United States *might not* accomplish with force, that is, the probability that the weaker adversary could still launch some number of weapons against their targets. The probability of surviving warheads requires that we consider the cost sensitivities of even a superior party and how those sensitivities might change over the course of a conflict. Through their narrow focus, denial strategies thus exaggerate the US incentive to act on alleged US advantages (a) by launching a preventative (first) strike on a nuclear-armed adversary or (b) by relying on a second-strike counterforce capability to convince the adversary to concede because its bargaining position will suffer after a nuclear exchange.

Missile Defense

AD advocates viewed defensive systems meant to thwart enemy offenses a bit more favorably than offensive systems meant for defensive purposes. Unlike warfighters, however, AD advocates still maintained a lukewarm to negative attitude toward *active* defenses, that is, antiballistic systems that fire missiles at incoming warheads. The underlying issues continue to fester in the post–Cold War period.

In the Cold War Years

AD advocates certainly recognized that *passive* defenses—the hardening of missile sites, the quiet mobility of submarines, bombers kept on alert, and so forth—were essential to safeguard the US retaliatory capability needed to deter a Soviet attack. These advocates were not alone in that view. Only unilateral-disarmament proponents would reject the necessity of safeguarding military assets from attack. For AD advocates, however, *active* defenses were problematic. If a party could effectively defend its cities, it could attack or coerce its opponent with impunity. In the AD view, even the belief that a party is *acquiring* effective missile defenses could destabilize deterrence. Soviet fears of improving US defenses could trigger a Soviet offensive buildup, increasing pressure on the United States to follow suit. Indeed, the Soviets, if fearing their increasing vulnerability, might strike before US defensive improvements closed the Soviet offensive window entirely.

AD advocates voiced particular concern over urban defenses. They might concede that active (point) defenses that were meant to protect (stationary) US land-based missiles (in their hardened silos), for retaliatory attacks, could stabilize deterrence. In principle, these point defenses married easily to the basic principles of deterrence: they helped ensure that some portion of the US land-based missile force could retaliate for a Soviet attack. Indeed, defending hardened targets against the pinpoint attacks necessary to destroy these targets was a less formidable task than defending a big, vulnerable, and immobile city. To protect missiles in hardened silos, a defense could selectively target warheads that posed a direct threat to the missiles, when a warhead—one of thousands in the Soviet arsenal—that hit off target, near a city—could destroy it. AD advocates feared, however, that even hard-point defenses could feed grandiose (destabilizing) visions for a full homeland defense.

AD advocates were not quite consistent in their claims. That was clear when they argued, for instance, that a party might attack before its opponent's defenses closed an offensive window of opportunity. Why would that party effectively commit suicide "out of fear of death"? They also invited criticism when arguing that defenses are both ineffective and destabilizing. Why would the Soviets feel threatened by a system that did not work? Although advocates could (and sometimes did) argue that the *belief* that defenses might work contributed to instability, a focus on beliefs was a departure for most AD advocates. Their emphasis, instead, was on assuring retaliation with some necessary level of nuclear *capability*.[39] Yet AD advocates did not hold a monopoly on inconsistent thinking. Denial strategists, well represented in the Reagan administration, seemed unfazed by the prospect that the Soviets might attack before the "window" closed, though Soviet offensive aspirations—and aggressive intent—fueled their push for a US missile defense.[40]

Fueled by consistent and inconsistent arguments, the issue of missile defense was contentious in the late 1960s before the Nixon administration moved to deploy a hard-target defense, prior to the signing of the 1972 Antiballistic Missile Treaty, which severely limited such deployment. The Reagan administration then undercut the agreement with its own creative interpretation of the treaty. It maintained that the treaty permitted the United States to deploy a vast array of technologies to defend US missiles and cities against a Soviet attack.

In the Post–Cold War Years

Controversy did not die with the Cold War. A decade after the Cold War's end, the George W. Bush administration abandoned the treaty altogether to realize the promise of a defense. The administration spoke for policy hawks who saw opportunity for a strong US defense against the smaller arsenals of a North Korea, or Iran.

A layered defense—positioned in theater, and on US territory—could target ballistic missiles in their boost phase and warheads through their terminal phase. Any such defense could gain effectiveness if a US offensive first thinned the adversary's arsenal. An effective defense would permit the United States to overcome the traditional challenges of extending deterrence to US allies. (See Chapter 3.) If fully protected, the United States need not worry about trading Honolulu for Seoul or Los Angeles for Tokyo.

Yet, the relationship to offense that made defenses attractive to US proponents made them problematic to US adversaries, and thus for efforts to control nuclear competition. US defensive aspirations complicated arms-control negotiations with Russia and provoked a compensatory expansion in the size of the Chinese nuclear force. Indeed, US hawkish defense advocates were placed in the awkward position of challenging Russian protests that expressed the very logic that these same advocates embraced to sell their offensive and defensive plans at home. Policy hawks in the US Congress, fearing that the Obama administration might trade away US defenses (or maybe allow the Russians offsetting offensive advantages) balked at the relevant text in ratification hearings for the New START Treaty. The offending passages, in the treaty's preamble, stated, "Recognizing the existence of the interrelationship between strategic offensive arms and strategic defensive arms, that this interrelationship will become more important as strategic nuclear arms are reduced, and that current strategic-defensive arms do not undermine the viability and effectiveness of the strategic offensive arms of the Parties. . . ." For treaty defenders, the preamble stated the obvious. If offenses and defenses were not interdependent, why would the United States even seek a missile defense? For treaty critics, the text hid an insidious purpose. Presumably, once the United States had formally accepted this principle, it would legitimize Russian claims that defenses gave the United States an unfair offensive advantage.[41]

Despite the more "modest" ambitions of contemporary missile-defense advocates, aspirations for a robust defense never entirely abated.[42] To the contrary, the unrealized *promise* of missile-defense technologies—against a large-scale nuclear attack—gave rise to wishful thinking. The push in the Trump administration to expand US missile defenses to reduce the US vulnerability to attack—in Trump's vision, to shield the United States entirely from a nuclear attack—rested on their hypothetical capability to protect US targets. The administration formally pressed its goals then for a multilayered defense, with a vast system of land- and space-based sensors and weapons that would move the United States closer to the Reagan-era vision.[43] But proponents had still not shown that the economics of acquisition favored the defense or that defenses were effective even against a small number of attacking warheads. The accompanying danger, then, was that the adversary would simply build more missiles to overwhelm the defense which—ironically—would only increase the threat to the US homeland.

Other issues were neglected. If defenses and offenses were complementary systems, might the appearance of effective defenses—or their repositioning—exacerbate apparent threat in a tense conflict? Might the United States act incautiously or precipitously against small nuclear adversaries with false confidence in a porous defense? The intellectual focus was, instead, on the technological *potential* of a defense, not its possible consequences within a broader political, social, and psychological context. Even "in defense," thoughts of advantage rang through.

Nuclear Strategy in the Post–Cold War Period

The Cold War's end reduced the overall US concern with the US–Russian nuclear balance. US officials continued to highlight US nuclear advantages and disadvantages. Although Cold War–era principles echoed in standards used to assess the robustness of the US force, they were not embedded in explicit strategies that could direct the course of a nuclear conflict.

The US defense community transitioned warily to a post–Cold War posture. Secretary of Defense Richard Cheney's targeting review, early in the George H. W. Bush administration yielded studies and briefings that reportedly "appalled" Cheney and Joint Chiefs of Staff Chairman Colin Powell, who were "purportedly struck by the inefficiencies, excesses, and incongruities that were built into the plans" (Lebovic 2013: 185–186).[44] Although the military would drop thousands of targets from the SIOP, albeit mainly in the former Soviet republics (Nolan 1989: 31), the military believed the resulting numbers constituted a floor level of necessary capability: given shrinking Russian territory and START I arms-control limits,[45] military planners recognized that they could make do with a smaller number of weapons (Lebovic 2013: 186). The numbers further declined through agreement: with the Obama administration's New START agreement,[46] the United States and Russia each had around 1,500 deployed strategic-nuclear weapons.

The Push for Flexibility

The US military continued to resist force reductions. It feared compromising target coverage in plans that enshrined "counterforce" targeting as if it amounted to strategy.[47] But counterforce was not the Pentagon's only nod to targeting's Cold War legacy. Soft-warfighting principles survived, as well, in the notion of "flexibility." US policymakers focused, accordingly, on acquiring options, preventative and otherwise, to address a world of diverse (even mercurial) threat. The Bush administration's 2002 NPR integrated conventional-strike capabilities into nuclear war–planning—a change that would involve arming some ICBMs and SLBMs with conventional warheads. The 2018 NPR, released early in the Trump administration, followed its predecessors in calling for the development of new US nuclear capabilities to support "tailored deterrence" strategies, "designed to communicate the costs of aggression to potential adversaries, taking into consideration how they uniquely calculate costs and risks."[48]

Flexibility seems like an inherent good. A larger set of options increases our chances of finding a useful solution if only by mixing and matching alternatives. But the focus on flexibility—and force "diversity" to enable it (see Payne 2016)—effectively turns means into goals. *Creating options* becomes the end when policymakers have no good options or lack the capacity (information and insights) to determine how (or even whether) available instruments can serve policy objectives. The push for flexibility is problematic, then, in various respects.

First, the flexibility argument wrongly assumes that adversary threats require tailored solutions. Payne (2011: 24), for instance, connects the two when concluding that "in each contingency, the details of leadership, personality, time, place, stakes, culture, ideology, religion, and communication can shape the credibility opponents attribute to US deterrence threats and the steps we might take to strengthen that credibility and, thus, deterrence." The linkage presumes, then, that solutions require finding a "weapon option" that correctly fits a specifically defined problem.

Problem definitions are variable, however, which gives officials room to define problems to allow for alternative approaches and intervention points. In seeking option *variety*, then, we can easily understate the utility of option *substitutability* for coping with threats. Big can substitute for small, imprecise can substitute for precise, and the reverse—in part because problems allow for solutions at multiple intervention points, which could better suit available capabilities. Just as Cold War–era warfighters claimed that quick, accurate attacks could effectively substitute for massive, indiscriminate targeting, the US military relied on the same logic, in reverse, to compensate for counterforce limits in the early post-war years. Either could reduce the Soviet military threat. Yet even this assumes that adversaries pursue but a single goal. Substitutability is also made possible because adversaries pursue a *variety of objectives*, including those based on widely shared—and predictable—aversions. Thus, targeting that is meant to *punish* an opponent can substitute for targeting meant to *deny* an opponent its gains. Would Adolph Hitler, for example, have considered his expansionist objectives viable had he known the price he would pay in Germany's destruction?

Second, the flexibility argument assumes too readily that US leaders can obtain the knowledge necessary to tailor threats, or attacks, to counter the complex and opaque influences on leader behavior. As Payne (2011: 24) concludes, "With enough serious analysis and smart policy choices, we can and must establish deterrence strategies and related force requirements that can be adapted to diverse opponents and contexts." But the logic here is contradictory. It assumes that analysis can deliver the proper recipe for deterring leaders who require special treatment *because* their reasoning is nontransparent and unpredictable—then, in a fast-moving crisis, when US officials are least able to separate useful information from noise. It assumes further that the recipe is found in some mix of physical ingredients—that is, the right weapons against the right targets. Such thinking takes no account of

(a) the potentially decisive role in a crisis instead of US promises of cooperation or assurances (for instance, not to capitalize on adversary weaknesses); (b) rival agendas within the opposing country's leadership; (c) adversary goals, such as status and prestige, that lack a clear material basis; (d) the disjuncture between general adversary goals and the tangible (targetable) assets that serve those objectives, and (e) the inherent bluntness of nuclear weapons despite their accuracy, speed, evasiveness, and varying destructiveness. That Payne skirts specifics—in arguing, instead, for good weapons, in large quantities—is telling.

Third, the flexibility argument assumes that military organizations will provide US leaders with appropriately "tailored" options when advance planning favors the general, practical, and routine. Standardized categories, technical criteria, and organizational considerations will drive force movements, target selection, the use or pairing of certain weapons, the timing and sequencing of attacks, and so forth. Policy problems will effectively reduce, then, to preexisting option packages. US leaders, themselves, are unlikely to consider options beyond a standard set of "tools"—such as leadership targeting—to address the "peculiar" threat posed by a given adversary.

Finally, the flexibility argument assumes that more options will increase the US payoff from an attack. Yet nuclear options invite significant costs despite the "tailoring." The consequences of "unintended" damage and death—or a misapplication of a "solution" to the problem—might influence the adversary's reaction more than the message that the United States intended to send with its "tailored" attack. That was true, for example, of the Trump administration's plan (under the NPR) to integrate nuclear and conventional weapons in war planning. We can doubt its necessity given other US conventional assets. More troublesome, "such integration actually lowers the sharp distinction between conventional and nuclear weapons—a distinction that has existed since the Eisenhower Administration."[49] It risks an errant US nuclear response; it also increases the chances that an adversary will assume the worst in combat and escalate in anticipation of a US nuclear attack.

Illustrating the problem: the nuclear-armed cruise missiles that the NPR proposed for US submarines (albeit canceled by the Biden administration) could also carry conventional warheads. Perhaps an adversary would not assume that incoming missiles are nuclear-armed even in a heated conventional-military confrontation. But what if they saw other indications that the United States was preparing to go nuclear? Or what if the adversaries had already exchanged nuclear fire (perhaps on a distant battlefield)? If the United States launched the conventional variant of these missiles, then, would an adversary not now assume that these weapons are nuclear armed (see Acton 2020)—and respond accordingly?

Integration of that sort offers *plausible* deterrence benefits: if the United States is prepared to go nuclear, as conveyed by the use of weapons that *could* carry nuclear warheads, adversaries might choose to forgo their conventional or nuclear options. Still, the NPR failed to acknowledge the relevant trade-offs behind the integration plan.

The Allure of Matching

Soft-warfighting principles, for their part, seemed to undergird the concept of "matching," when Russia—under the leadership of Vladimir Putin—emerged anew as a key US nemesis. The United States placed low-yield warheads on US submarine-launched ballistic missiles to offset Russia's employment of low-yield warheads. Echoing Cold War–era arguments, the NPR justified acquiring these weapons to *match* Russian employment at the (low) level at which it might occur.

The review did not embed matching, however, in the elaborate intellectual structures of the Cold War period. With a nod neither to escalation dominance nor escalation control, the review suggested only that low-yield weapons would fill *some niche* between the extremes of conventional and high-yield nuclear weapons. If deterrence failed, these weapons would somehow help the United States to "strive to end any conflict at the lowest level of damage possible and on the best achievable terms for the United States." The review left unsaid why current US conventional or nuclear forces were not up to the task and why the United States had to mimic Soviet employment by responding in kind. After all, deterrence is arguably enhanced if the United States has no option but to respond to a nuclear attack, of any scale, with higher-yield weapons. Instead, it reduced escalation to a matter of matching weapon types and capabilities, thereby ignoring factors that might lead the United States or Russia to increase or decrease the intensity or scope of a conflict. These factors include the party's (a) efforts to obtain or to neutralize a military advantage, (b) attempts to signal resoluteness to press an opponent to back down, (c) failure to appreciate red lines stemming from the opponent's goals, values, or concerns, (d) belief that they will lose at the current level of conflict, (e) desensitization to the costs of war or desire for retribution or revenge, and (f) inability to recognize the precise level or direction of the conflict in the "fog of war." Indeed, President Biden appreciated the escalatory influence of (at least some of) these factors when warning President Putin against using a nuclear weapon in the battle for Ukraine: "The mistakes get made, the miscalculation could occur, no one could be sure what would happen and it could end in Armageddon."[50]

Thus, if the United States could *match* the Russians blow for blow on the battlefield, or beyond, would the Russians necessarily consider or appreciate US "restraint"? For instance, what would the Russians make of a missile, with a low-yield warhead, launched from a US Trident submarine? Would they assume that the missile carried a low-yield (<10 kilotons) or a higher-yield (≈100 kilotons) warhead?[51] If they await the blast to make the call, might they assume the low-yield warhead was a "dud" or, regardless, whether it was the first salvo of a more punishing US strike to come? Indeed, would the Russians focus on individual warhead yields or the aggregate level of destruction—in kilotons (or megatons) but also in military, political, economic, or psychological effect? The same applies to the United States, as it surveys the scene of destruction. It might choose, then, to up the ante—with a higher-yield response—whether possessing or lacking lower-yield alternatives.

On this point, Bernard Brodie (1966: 69) lamented more than a half-century ago that discussions about introducing nuclear weapons into conflict centered on "mechanical phenomena, like the scale of hostilities reached or the rate at which territory is being yielded." His rejoinder: "One would rather expect the first consideration to be: What is the enemy up to?" Perhaps, we should take as significant, then, that Russia's official response to the US deployment of low-yield nuclear weapons was that it would lower the nuclear threshold and invite larger-scale Russian retaliation.[52] Indeed, if Russia's strategy were to "escalate to deescalate" the conflict, hoping to press the United States to back down—why should we assume that a "lesser" (more than a "greater") response would impress the Russians? After all, Russian leaders must have perceived the stakes to have been high to have employed nuclear weapons in the first place. For that matter, why should the United States play Russia's game with a nuclear response when that could push Russia to escalate? Russia must recognize that, at some point—given the inflicted human toll or the military effect—the United States *will* respond with (perhaps massive) nuclear force.

Ungrounded in the (maybe, changing) intentions of the parties, the logic of matching is infinitely malleable. For instance, to boost the case for a weapons system, US defense officials invoked the logic, *in reverse,* by assuming that adversary weapons will perform *down* to the level of US systems. They justified US missile defenses by presuming that Russia or China would commit only some of their missiles to a war in a regional theater.[53] Yet why would these countries use but a small share of their nuclear capability to address a regional problem? If either Russia or China deemed the issue *insufficiently* important to justify committing the full weight of the country's strategic nuclear force, why would they use nuclear weapons in the first place? If they deemed the issue *sufficiently* important to justify nuclear force, why would they devote only a portion of their nuclear weapons to the effort, especially given expressed Russian concerns about effective US defenses?

Yet the Pentagon insisted, with its logic, that deterrence would strengthen under the 2018 NPR.[54] Going further, Secretary of Defense James Mattis asserted that US weapon plans would raise the threshold for a nuclear war.[55] Bold claims such as these require substantial support; they cannot stand on their own. The potential costs of matching are significant if lowering the nuclear threshold, instilling the belief that the United States is lowering the nuclear threshold, or fostering the illusion that the United States requires weapons that the adversary possesses, conceding opportunities for arms control.[56]

All That Remains: A Pretense of Theory

The Cold War bestowed the concepts that now inform contemporary thinking about nuclear deterrence. These concepts are misleading, however, absent a broader understanding of how deterrence works, and how it might fail. The great nuclear debates, which once dominated the study of international politics, now receive little explicit attention in discussions of US relationships with nuclear powers, big or small. Cold War–era ideas enrich the vocabulary of debate but not its analytical rigor.

Cold War–era strategies *were* deficient. Strategists battled relentlessly over the consequences of various nuclear balances (or imbalances) and shortcomings in arms-control agreements. Although they focused on offsetting advantages, treaty loopholes and safeguards, retaliatory options, and so forth, they largely talked (or shouted) past one another. They could not settle, or meaningfully address their disputes, for they rested on unexamined assumptions about how Soviet goals might lead to conflict and how US and Soviet goals might change in conflict. The neglect begot fanciful thinking about escalation dominance and escalation control. Escalation dominance assumed the parties would act rationally, under the worst of conditions; and aspirations for escalation control were hardly more realistic. Strategists offered both concepts with the false assumption that the dueling parties would share similar views of operant firebreaks and thresholds.

Knowing the perils and pitfalls of Cold War–era strategizing, however, can help us appreciate deficiencies in thinking about contemporary nuclear advantages. These deficiencies—of the past and present—are as follows:

> **2.1:** *Strategists focus on the salient capabilities of protagonists at the expense of their less-salient objectives.*
>
> **2.2:** *Strategists focus on weapons in numbers, types, and locations to address asymmetries, without regard for their actual significance or the range of available options.*
>
> **2.3:** *Strategists tend to view conflict escalation narrowly as products of weapon attributes, deployment, and employment.*

Notwithstanding their deficiencies, Cold War strategies at least presented explicit targets for intellectual scrutiny when, by contrast, post–Cold War "strategies" persist, grow, and change in an intellectual vacuum. Thus, vestiges of the Cold War survive nonetheless in US nuclear plans. Current targeting categories—"military forces," "military and national leadership," and "war-supporting infrastructure"—still draw from Cold War terminology. To be sure, the Bush administration extended war plans, after 9/11, to include preemptive (counterforce) strikes against regional WMD targets; and war plans now permit a range of options, from large, preplanned contingencies involving hundreds of warheads to more limited strikes adapted to specific contingencies (see Kristensen 2010).[57] The changes register now in the concept of "adaptive planning." The (ironically labeled) "living SIOP" amounts ostensibly to "a real-time nuclear war plan that could respond instantaneously to war-fighting commands." As conceived, "during peacetime, the system would be capable of making automatic target changes daily" (Kristensen 1997: 24). Yet the question remains: How much should we value or fear nuclear "asymmetries" that *seem* to favor one side or the other? The fact is that such advantages might matter little given the nuclear aversions of the parties and the accompanying dangers should either party seek to capitalize on "advantages" or mitigate "disadvantages" with nuclear force.

2.4: *Policymakers and planners look to flexibility and matching in force employment to avoid hard questions pertaining to when, how, and where to use nuclear force.*

3

Nuclear "Superiority" after the Cold War

"Nuclear superiority" is a loaded term. It suggests, at the very least, that a nuclear edge yields coercive payoffs. The superior party can supposedly achieve success, when throwing its weight around, because it can better handle the consequences of war—nuclear or otherwise.

Such cavalier thinking about the utility of nuclear threats requires that we step back and pose fundamental questions. First, what constitutes nuclear superiority? Answers emerge when assessing the point—and criteria by which—a state can establish superiority over others; the consistency of the logic backing claims of nuclear superiority; and the mathematical calculations and quantitative models that purport to reveal superiority's effects. Second, does superiority matter, given the *shared* costs of a nuclear conflict? Answers follow when examining the strength, and limitations, of prohibitions—grounded in norms or traditions—constraining nuclear-weapons use. Third, does nuclear superiority offer meaningful damage-limitation capability? Answers come through exploring the obstacles to acquiring this capability. Fourth, does alleged nuclear superiority offer actual deterrence advantages? Answers are offered by exploring why China "settled" for nuclear "inferiority," whether superiority is required for "extended deterrence," and the challenges of deterring "rogue states" that are supposedly insensitive to cost.

What Is Nuclear Superiority?

For Cold War–era AD advocates, nuclear superiority was irrelevant when a country with a relatively small nuclear arsenal could inflict unacceptable damage on a numerically superior attacker. If the "inferior" country possessed sufficient capability to deter an attack, the "superior" country's excess destructive capability amounted to overkill. It served no real purpose; the superior country was not "superior" in a meaningful sense. Warfighting critics disagreed. As they saw it, a country's nuclear or conventional capabilities must be assessed in relative terms. "How much is enough capability?" is not a question answerable in the abstract. A single gun

The False Promise of Superiority. James H. Lebovic, Oxford University Press. © Oxford University Press 2023.
DOI: 10.1093/oso/9780197680865.003.0003

provides physical security against some threats but offers little defense, for instance, against an invading army.

At what point can we say, however, that a country possesses nuclear superiority over another? Various factors preclude easy answers. These factors include (a) the elusiveness of the "threshold" value for judging superiority, (b) the illogic behind superiority claims, and (c) the deficiencies of quantitative analysis meant to validate measures of superiority through their empirical effects.

The Elusive Threshold

A country's "superiority," in nuclear capability, depends implicitly on an elusive threshold value. To obtain superiority, a country needs to equal or exceed that value; otherwise, a country must accept its equality or inferiority. But should we understand that value in absolute or in relative terms? That is, can we say that possessing *X number* of weapons more than an opponent constitutes superiority? Or should we say, instead, that possessing *X times* more weapons than an opponent constitutes superiority? Either way, the value is not static, leaving analysts to confront the questions that challenged their Cold War–era predecessors. What targets do we presume the warheads will strike, with what effectiveness, and for what purposes? Then, should we assess nuclear capabilities before or after a nuclear exchange?

As we shall see, Kroenig's (2018) explicit case for US superiority ignores these questions. Yet even sophisticated arguments, which defer to the principle of damage limitation, confront the same basic issues. We thus have strong reasons to question the extent of—indeed, the fact of—actual US nuclear advantages.

Kroenig's Case for Superiority

Kroenig makes his case for US nuclear "superiority" with little deference to Cold War–era debates. His "superiority" is insensitive to considerations of force deployment, targeting, advance warning, command and control, pain acceptance, and so forth. It relies, instead, on simple math—which, as it turns out, is also simplistic. That is, it begs more questions than it answers. Should we assume that a superior party simply has more weapons than a competitor? In other words, do we set the threshold for superiority at the level of the opponent's capability? By this standard, we must recognize, then, that a one-warhead advantage counts the same as a thousand-warhead advantage. Alternatively, should we assume that some *ratio* of advantage constitutes superiority over an opponent? If so, why two rather than three or four times the opponent's capability? Then, can we ignore force *sizes* entirely? After all, possessing two nuclear weapons when an adversary has one weapon is a potentially useful advantage; possessing two-thousand additional weapons when an adversary possesses one-thousand nuclear weapons might not constitute an "advantage."

Convenient arithmetic becomes the fallback, then, to avoid difficult questions. What do we miss, for instance, when we assume a competition limited to warheads in the opposing arsenals? After all, nuclear advantages emerge from factors other than types and numbers of nuclear weapons. Precision-guided conventional weapons can now threaten land-based strategic nuclear forces or other targets once vulnerable only to nuclear weapons; in turn, anti-satellite weapons and cyber weapons can threaten the command and control of nuclear forces (Koblentz 2014). Even if we were to assume a conflict confined exclusively to a nuclear battlefield, the remaining questions challenge the analytical reliance on simplistic ratios. Can the damage inflicted by any one side limit the damage that the targeted party can inflict in return? How much damage would a party need to inflict, then, in retaliation for an attack to deter an adversary from using its nuclear weapons?[1] In more specific terms, saying that a party can land two or three blows for every blow it receives, or can land two or three blows more than the opponent can, makes no allowance for pain tolerance. Can we say with confidence, for instance, that a government would willingly accept *any* nuclear destruction on its home turf, even if it wins the battle of "blows landed"?

Without the specification of threshold values, based on discernable and meaningful criteria—backed by viable absolute *or* relative logic—superiority remains a nebulous principle that cannot yield operational standards. Superiority refers only to some *unknown* point on an absolute *or* relative continuum at which the balance shifts, slightly or decidedly, toward one of two parties.

Conceptual frailties do not stop analysts from tying superiority to various concepts and numerical standards. For Kroenig, various relativist terms are synonymous with superiority. These include a "favorable nuclear balance of power" and "military nuclear advantages." To give operational meaning to these concepts, he references a "state's expected cost of nuclear war," which he defines metaphorically (Kroenig 2018: 3–4). He assumes that, when a big and a small car race toward each other, the driver of the big car obtains the coercive advantage. It will likely win this game of "brinkmanship" because the driver of the small car has the most to lose in a crash. Simply put, size differences matter—whether measured absolutely or relatively.

Probing this vividly deceptive metaphor highlights the numerous conceptual and operational deficiencies in Kroenig's thinking about superiority. First, should we assume that the two drivers will accept even some damage? The example ignores the absolute value the drivers place on their bodies and vehicles. Under what circumstances would most drivers willingly accept a broken arm or leg? For that matter, a door scratch, for some drivers, is a burden too hard to bear; other drivers might take comfort knowing they left their "nice car" at home. Second, should we assume that the two drivers will evaluate the damage in relative terms? More directly, under what *extreme* circumstances would a driver willingly accept bodily damage knowing the driver of the other car will suffer more? Third, should we assume that the cars are otherwise unoccupied? If we do, we implicitly embrace the extreme views of Cold War–era warfighters that adversary leaders put their own lives—and the cause of government—above all else, including family or society. If we assume, instead, that the cars carry passengers, whose lives the driver values, does the driver of the large car obtain a coercive advantage knowing that, at most, a collision will kill one or two passengers in the large car but three, maybe four, passengers in the smaller car? Fourth, should we suppose that the drivers know, with certainty, the outcome of the crash? The results of collision tests on vehicles vary depending on speed but also the angle and point of impact. Some car drivers "miraculously" escape horrific crashes with eighteen-wheel vehicles; other car drivers die in collisions with motorcycles, even deer or farm animals. In these crashes, size was not the determining factor. On this point, note that automakers engineer cars—of all sizes—for greater safety (with airbags, seat restraints, metal reinforcement, and fuel-tank relocation). Conversely, people could redesign their cars for increased lethality (as we learned from the *Mad Max* movie franchise), with battering rams, tire shredders, and the like. Thus, once probing Kroenig's vague definition—through his compelling metaphor—we cannot know a priori which of the two drivers necessarily holds the advantage.

Of course, Kroenig (2018: 7) maintains only that the bigger-car driver will accept a higher "risk" of damage, not the deadly consequences of a head-on collision. His logic here *seemingly* allows him to escape the implication that a nuclear exchange is anything but costly to all parties. But a willingness to incur risk, here, is but a sleight of hand, for Kroenig requires, nonetheless, that we accept that relative damage in a nuclear war determines the relative acceptance of risks. He insists bigger-car drivers will win any battle of nerves ("resolve") because they presumably will "win" any resulting battle of actual cost. If both "drivers" regard a collision as catastrophic (and thus cost prohibitive), however, the bigger-car driver is no more inclined than the small-car driver to accept risk.

To his credit, Kroenig (2018: 25) recognizes that the risk acceptance of the parties also depends on their payoffs—the gains from a confrontation, not just the relative costs.[2] But he must also acknowledge, then, that considering payoffs further undercuts his argument. If the driver of the small car seeks vengeance, would that not swamp relative-cost considerations in assessments of risk propensities? The answer: maybe, maybe not. An answer, in the abstract, is impossible. We cannot know whether payoffs exceed costs, whether costs predict risk acceptance, or much else that Kroenig accepts, by assumption, without reference to case specifics. Here as elsewhere, specious arguments survive because the terms of reference remain remote from view, and immune from scrutiny.

For that matter, specious arguments survive through redirection. Despite the allusions to cost, risk acceptance, and payoffs, Kroenig's operational definition of superiority ultimately rests on an arbitrary threshold value. Kroenig (2013) supposes that a party acquires superiority when it possesses *one* nuclear warhead more than an opponent. Should we assume that a leader—with a one-warhead advantage—who anticipates the arrival of three-hundred enemy warheads sees the dangers in quite the same way as a leader anticipating the arrival of three enemy warheads? If we do, perhaps that undercuts the argument that superiority matters. Conversely, why should we assume leaders would knowingly accept *any* nuclear damage? Maybe, one warhead landing on national turf is one-warhead too many. Kroenig (2018: 17) seems to concede that point: "Even a single warhead detonated in the United States would be a tragedy of historic proportions." We should remember that US policymakers implied, at least, that an attack of any proportion was cost prohibitive when warning of the dangers of an Iraqi nuclear weapon before the 2003 US invasion of that country. In September 2002, National Security Advisor Condoleezza Rice memorably pronounced, in her challenge to skeptics, that the "smoking gun" evidence they sought (concerning Iraq's WMD holdings) might appear, too late, as a "mushroom cloud." That was *one* cloud—presumably, one cloud too many. President Obama *publicly* channeled that sentiment at a United Nations Security Council summit on nuclear weapons. In his words, "Just one nuclear weapon exploded in a city—be it New York or Moscow; Tokyo or Beijing; London or Paris—could kill hundreds of thousands of people. And it would badly destabilize our security, our economies, and our very way of life."[3]

Maybe we can imagine a contingency under which leaders would willingly accept that cost. Given the high stakes, let us suppose they would tolerate the potential destruction of a city, or even some number of cities for some larger good. Can we reasonably argue, however, that each warhead landing constitutes a discrete event, similar in effect to any other? The effect of any "small" set of attacks will resonate widely courtesy of economic bottlenecks; overwhelming demands on available (national and local) governance; the vertical and horizontal interconnectedness of communication, supply, transportation, and service networks; and the consequences should uncertainty and fear prevail, social norms collapse, and people be forced into survival mode. Even these considerations ignore the broad environmental effects of fallout and climate change resulting from a nuclear exchange. These potential effects, if anticipated, would certainly prove sobering for even a leader that possessed an overwhelming warhead "advantage." These effects underscore the limits of calculating winners and losers in a nuclear exchange from simple numbers of immediate fatalities.[4] In this regard, we might remember a famous quote, often attributed to Nikita Khrushchev. In his vision of nuclear war, "the survivors will envy the dead."

To be fair, simple explanations can account for seemingly complex phenomena. Kroenig establishes, for instance, that his statistical findings—showing the "superior" party achieving more favorable crisis outcomes—are insensitive to whether he employs the binary or ratio variant of his central independent variable. That is, his results do not change whether the analysis weighs a 5–1 more heavily than a 2–1 advantage over an opponent. But is this a convincing robustness test? If we buy into the relative logic, the size of the relative "advantage," as a precautionary hedge against risk, *should* influence crisis outcomes given the host of factors that will determine damage inflicted in a nuclear conflict,[5] the margins of uncertainty involved, and variance among leaders in their willingness to take risks.

The Implications of "Damage Limitation"

Even in the late Cold War period (see Chapter 2), some strategists accepted superiority as a theoretical possibility (Glaser 2014: 133), while others viewed it as a viable aspiration. Now, of course, US advantages in intelligence, surveillance, and nuclear-delivery capability give the United States a significant edge against competitors in *residual* nuclear hardware—the "balance in capability" after a nuclear exchange. Under the right conditions, the United States could significantly deplete—maybe even destroy—the adversary's nuclear forces. By contrast, countries with lesser arsenals are unlikely to engage in damage-limitation strikes against US nuclear forces: these strikes—against far-more numerous US targets—would effectively disarm the attacker. In all scenarios, then, a nuclear exchange would leave the attacker vulnerable—from a now weaker position—to US retaliation from all legs of the US triad.

To be sure, superiority, primacy, and other such terms retain meaning if the United States—under *all* conditions—could *fully* disarm its opponents with *assuredness*. But what should we conclude if the adversary could still launch some number of weapons against the United States or its allies?

Lieber and Press (2006a, 2006b) assert that the United States has approached the higher standard. Their post–Cold War claims that the United States was pursuing a first-strike advantage over Russia caused considerable consternation within Russia and US defense-policy circles: "Today the United States stands on the verge of attaining nuclear primacy vis-à-vis its plausible great power adversaries" (2006b: 7). Critics contested their assertions of US force effectiveness. They questioned the authors' assumptions about US force capabilities, the likelihood of a catastrophic Russian-force failure, and the political conditions that would permit the United States to take Russia by surprise (see Blair and Yali 2006). Presumably, the United States would not simply strike out of the blue, without a grave war threat. After all, massive Russian retaliatory capability would still reside within the margin of error in US attack calculations. If the United States allowed the conflict to build, Russia would presumably bring its forces to a high state of readiness—and might even strike first, given the logic underlying US plans to strike (that more damage limitation is better than less). Critics also asked whether the apparent effectiveness of US forces might therefore do more harm than good. It might even "backfire" on the United States if it foolishly sought to capitalize on its alleged advantage militarily or coercively: "The risks may outweigh the benefits if American pressure triggers reactive nuclear alerting and escalation in a crisis—increasing the danger of accidental, unauthorized, or hastily ordered nuclear attacks" (Blair and Yali 2006: 52).

We are hard pressed, then, to find unilateral coercive advantages in the US–Russia strategic relationship absent a US damage-limitation capability that permits the United States to avoid, *with certainty*, the nightmarish consequences of a nuclear exchange. Lieber and Press (2006b: 9) concede as much in noting that a US "counterforce strike would entail enormous risks and costs." They argue, instead, that their analysis demonstrates that "Russian (and Chinese) leaders can no longer *count on* having a survivable nuclear *deterrent* [italics added]." But, as we shall see (in Chapter 5), even if the United States could secure a high probability of a successful strike, the costs of failure, by the same standard, can weigh heavily against taking the shot. In that regard, the key words here are deceptive. "Count on" suggests that certainty is required to deter an attack, and then certainty on the part of the challenger rather than the defender; "deterrent" suggests, in turn, that the adequacy of a nuclear force is assessed by some independent standard when deterrence rests on a potential attacker's sense of the risks. Put differently, is a 50-percent chance that one Russian nuclear warhead will hit a US city a sufficient or insufficient deterrent? That is not a question we can answer, devoid of context.

Lieber and Press (2017: 38) correctly conclude that "(t)here is no deductive reason to believe that a country with a 95 percent chance of successfully destroying its enemy's nuclear force on the ground will act as cautiously as a country that has only a 10 percent chance of success"—that "whether leaders exhibit equal caution in these two very different situations cannot be deduced; it is an empirical question." But that is only one side of the issue. We also cannot conclude deductively that additional capability reduces the possessor's risks. To the contrary, capability can feed temptations or induce fears that instigate the very wars that the capability was acquired to deter.

In a similar vein, Austin Long and Brendan Rittenhouse Green (2015) hold open the possibility that the United States, with its technology advantages, can execute a disarming strike, especially against less-capable US nuclear adversaries, including China. They assume that US relative capabilities, at least, give the United States a coercive edge: "The key feature is asymmetry between the two parties and some reasonable probability of success in limiting damage by one side. The other side need only believe that, at the moment of ultimate desperation, its adversary is more willing to gamble because it has some probability of limiting retaliation" (Green and Long 2017b: 197). They admit that various factors influence a party's risk tolerance (Long and Green 2015: 66). These include fears of preemption and *perceptions* of advantage. They assume nonetheless that a priori relative damage assessments can—and likely will—work coercively to the United States' benefit. In their view, that conclusion follows logically from Cold War–era treatises which attribute risk tolerance to the balance of interest and *capabilities* (Green and Long 2017b: 196–197).

Long and Green (like Lieber and Press) exhibit an impressive understanding of the technical side of nuclear warfare, which leads them to discount the merits of calculating the balance using precombat, warhead counts.[6] But that does not bring us closer to understanding the effects of any such "balance" on the likelihood or course of a nuclear conflict.

The Logically Inconsistent Case for Superiority

The case for superiority rests on inconsistent logic. We cannot argue that superiority matters when, for instance, we also argue that the mere capability to retaliate with nuclear weapons "allows states to guarantee their own survival" (Monteiro 2014: 85).[7] Likewise, the term "superiority" loses meaning when we describe confrontations between nuclear-armed states as "standoffs" or highlight the exceptional dangers in confrontations between nuclear-armed states. Kroenig (2013: 142) does both when he states that "a standoff between nuclear-armed opponents is a nuclear crisis whether or not nuclear weapons are used, are explicitly threatened, or are the subject of the dispute, because the very existence of nuclear weapons and the possibility that they could be used have a decisive bearing on bargaining dynamics." That he deems a confrontation between nuclear powers, despite their potential nuclear asymmetries, as qualitatively distinct belies Kroenig's argument that nuclear superiority matters. If it did, why would the weaker party retain any coercive influence?

Superiority loses further significance when analysts accept that the behavior of a country defies its position in a nuclear balance. If superiority counts, for example, why did US policymakers voice concerns about US global challenges in the early Cold War period when the United States enjoyed a large (absolute and relative) margin of nuclear advantage? In turn, why should US adversaries assume they were coercively disadvantaged in those years if US policymakers were unconvinced that nuclear weapons would tip the military balance? And, given a disbelieving world, why did the United States not use its nuclear weapons in various contests, regardless of the stakes, to clear up misconceptions about US nuclear prowess? In other words, why did the United States not become a bully, exploiting its alleged superiority, and the accompanying timidity of inferior nuclear countries, to its advantage (Asal and Beardsley 2007: 143)?

The behavior of nuclear-inferior countries is no less problematic. If nuclear superiority is decisive in crises, why do inferior states confront their superiors? If we conclude that nuclear weapons helped the United States "win" the Cuban Missile Crisis, we must also ask why the Soviets placed nuclear weapons in Cuba in the first place. Should US nuclear superiority not have deterred them from threatening US national security? We could argue, plausibly, that Soviet leaders sought to close the nuclear gap surreptitiously to avoid an encounter before the Soviets could respond from a position of greater strength. But the Soviet deployment of nuclear-armed missiles in Cuba would not *decisively* alter the nuclear balance; the United States would retain "superiority" by standard measures. We must ask, then, why any state—including China and North Korea—would compete in a nuclear arms race that will guarantee a *second-place* finish. One answer is that they view it as a place to stop before mobilizing national resources for a lurch across the finish line. If that were the case, however, why would they not suppose the United States would anticipate the move and leap ahead to pad its advantage?

Similar questions persist across time, and space. How can we explain China's willingness, in 1950, to join the fight in Korea against the world's preeminent nuclear power? Why did Saddam Hussein send his forces into Kuwait, inviting a possible US response? Then, in the aftermath of the 1991 war, why did he refuse to cooperate with international inspections to determine whether he had disarmed under the terms of successive UN resolutions? Indeed, why did US policymakers—given overwhelming US nuclear superiority—even care about Iraq's potential weapon holdings?

One possible answer: nuclear superiority counts most in an actual confrontation. That is, "inferior" states somehow "select into" conflicts but then yield inevitably to nuclear realities. Yet that argument does not hold up to scrutiny. Why, then, was US nuclear superiority a nonfactor in the twenty-year war in Afghanistan that ended with the United States accepting an agreement with the Taliban, and a withdrawal plan that set the stage for a Taliban victory—or why was US nuclear superiority a nonfactor as well, decades earlier, in Hanoi's decision-making over the eight years (1965–1973) of active US combat in Vietnam? Despite nuclear realities, weaker states, it seems, have abundant reasons to believe that stronger state "advantages" do not hold. For instance, weak states "often seem to believe that international and domestic political constraints will prevent Great Powers from intervening effectively in limited wars or responding forcefully to provocations" (Wirtz 2012: 16).

Ironically, efforts by the strong to reinforce their preeminence can be counterproductive. The Johnson administration's strategy of gradually ratcheting up the bombing against North Vietnam—and not engaging fully—could easily have reinforced Hanoi's view that time was on North Vietnam's side. Hanoi could have concluded quite reasonably that, if the United States would absorb the costs necessary to win in Vietnam, it would have done so. Thus, stretching out the war to slow or prevent any US progress would test US perseverance and, for Hanoi, increase the likelihood of a favorable outcome (see Lebovic 2019: 17–63). That the United States might up the ante, by using nuclear weapons, figured "little" in Hanoi's calculations.[8]

The Deficient Quantitative Case for Superiority

The case for superiority cannot rely solely on mathematical reasoning, even when analysts imply that numbers speak for themselves. Some analysts maintained, for instance, that US preponderance in available measures of conventional capability rendered moot questions about US global preeminence (unipolar status) in the early post–Cold War period (see Wohlforth 1999). Still, depending on simple math to solve conceptual problems is a double-edged sword. If our vague sense tells us that superiority is a relative concept, we must also concede that superiority decreases *rapidly* once nuclear aspirants acquire but a small nuclear arsenal.

Figure 3.1 demonstrates this, with basic arithmetic. In the figure, the x-axis indicates the number of warheads possessed by the inferior party, the y-axis indicates the ratio of superiority held by the superior over the inferior party, and the lines display these ratios for a superior party at given force sizes (50, 100, 200, 300, 500, and 1,000 warheads). Clear from the figure is that, regardless of the superior party's holdings, its relative advantage drops dramatically by the time an adversary acquires but five warheads. Indeed, the bigger the superior party's arsenal, the more dramatic the relative decline. The point is not that these advantages are insignificant politically or militarily. The point is that if advocates recognize benefits in the magnitude of relative superiority, they must also concede that these benefits collapse once an adversary accumulates a modest (absolute) store in nuclear weaponry. For that matter, they must concede that the fact, itself, of a quick collapse in the relative advantage of the superior party might also confer a bargaining advantage.

Arms-control experts have always drawn attention to the instability challenge when contenders possess small nuclear arsenals—say, one warhead each. By unilaterally adding one, two, or three warheads to its arsenal, a party could double, triple, and quadruple its advantage relative to the opponent. Under these conditions—when a breakout from an arms-control agreement can give a state a quick and

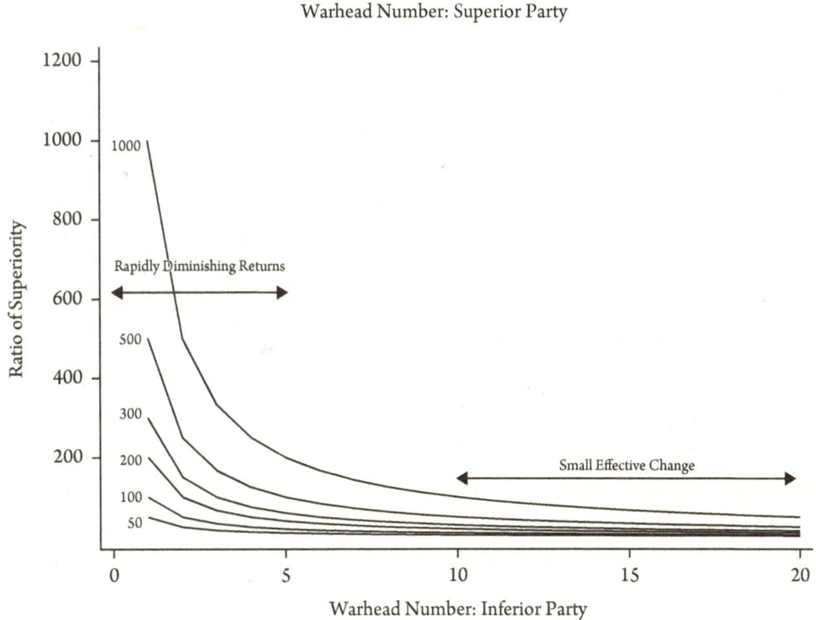

Figure 3.1 The fleeting advantage of relative superiority.

arguably decisive advantage—states are disinclined to observe such agreements, let alone negotiate them in the first place.[9] Yet the challenge, here, is from the smaller nuclear arsenal. If ratios of advantage count *for bargaining purposes*, we must recognize that, even with a large arsenal, states are vulnerable to challenges at the low end from "inferior" states choosing to build up their holdings.

Of course, one way around inconclusive mathematics is to show that these ratios of nuclear advantage have real-world impact. Accordingly, analysts ask whether superiority, as defined, translates into "winning percentages." They are not wrong to do this. That superiority, *as defined*, predicts international outcomes is a potential source of definitional validation. Yet, we must accept all such validation with extreme caution. For one thing, predictive capability constitutes a weak form of validation. It rests on circular logic: researchers affirm a hypothesis, based on questionable measures, and then use the same (thereby questionable) hypothesis test to validate the measures.[10] Tautological logic aside, validation depends on a viable model, in all respects: specification, concept operationalization, and case selection.

On these grounds, Kroenig's quantitative models prove deficient. In predicting historical conflict outcomes, with measures of US nuclear superiority, he inadequately controls for the impact of potential influences. Most problematically, Kroenig measures a country's conventional capabilities using standard Composite Index of National Capability (CINC) scores, which combine a country's global share of steel production, energy consumption, troop levels, population, and military expenditures in a single index. Analysts routinely employ the CINC measure in quantitative analyses. See also Beardsley and Asal (2009). Still, the measure's components (individually and collectively) tell us little about a country's ability to project power, win battles, apply necessary force, and persevere, if necessary, with adversity.

Take the Cuban Missile Crisis, for instance. To make a compelling case that nuclear advantages explain the crisis outcome, we must *convincingly* account for the expected outcome of a possible conventional military showdown in the Caribbean—and perhaps even a showdown in Berlin, if the Soviets staged a retaliatory attack in a more favorable venue. Yes, the United States held an "advantage" in deliverable warheads during the Cuban Missile Crisis. But the United States also held a significant advantage in a potential conventional contest anywhere in the Western Hemisphere. That the conventional capability measure is statistically insignificant in Beardsley and Asal (2009), for instance, is telling. Why should we buy their conclusion that superior nuclear capabilities influence crisis outcomes when they fail to show that conventional capabilities matter? Does it mean that countries that can build nuclear weapons should not waste resources on conventional armies?

Kroenig recognizes, to be fair, that the "resolve" of the parties could affect outcomes. As we shall see (in Chapter 6), the term invites analytical problems of its own. Kroenig hardly addresses (much less remedies) these deficiencies, however, when he reduces the resoluteness of the parties to their relative physical distance from the location of the conflict. Why should geographical proximity matter in a nuclear-armed world in which states can eviscerate one another, in a matter of minutes, regardless of where they are located? Then, why, in evaluating the stakes, should we privilege geography over topography, demography, political alignment, economic development, military basing rights, and various other factors that play into a country's strategic calculations?

Ultimately, we should recognize the models offered by Kroenig, and others, for what they are: stylized depictions. These models conform to the strictures of the data, disciplinary conventions, and requirements of statistical analysis and can thus disclose relationships between standard indicators and crisis outcomes. Yet, as currently constructed, they provide few clues to how nuclear superiority, conventional capabilities, resolve, and other factors (singularly or jointly) affect outcomes. Indeed, the crisis data set used in Kroenig's analysis treats crises as single temporal units, such that a party's resolve and capabilities remain constant throughout individual crises.

Even if we accept "resolve" as a plausible source of bargaining advantages, we must concede its complicated relationship to capabilities. Whereas leaders are presumably more resolute when they think they can accomplish their goals, they can also become resolute, perhaps, when they encounter opportunities to overcome their capability *deficiencies*. If so, weakness—under some conditions—could induce resoluteness. On this point, evidence suggests that a weaker party might seek to prolong a conflict to secure a better bargaining outcome (Pillar 1983; Wagner 2000). Its resoluteness stems directly, then, from its weakened position. The Vietnam War offers testimony in that regard: as battle conditions moved in its favor, Hanoi appeared more willing to negotiate with the United States but nonetheless adopted a hardline stance (Lebovic 2019: 53–61).

Similarly, the effects of a nuclear balance likely vary with conventional capabilities. A country with relatively weak conventional capabilities might compensate by trumpeting its nuclear might (as NATO members did, fearing that they could not halt a Soviet attack on Europe employing conventional forces alone). If a crisis resolved to that country's favor, we might conclude that nuclear superiority determined the result. Yet a somewhat more complicated interpretation is warranted. Nuclear leverage stemmed from conventional weakness: short on options, the state posed a credible nuclear threat. If so, nuclear capabilities and conventional capabilities are not *competing* explanations for the crisis outcome, as assumed when analysts pit these variables against one another in regression equations. Instead, weak conventional capabilities constitute a *necessary* condition in accounting for the influence of nuclear capabilities. Nuclear capabilities mattered more *because* conventional capabilities were weak.

Quantitative analysis serves a valuable function in the discipline. It can disclose patterns that go unrecognized in the intensive (qualitative) analysis of individual cases. But we must retain a healthy suspicion of findings that defy our understanding of decisional criteria in individual cases. These include what should be "easy" cases for verifying the superiority thesis. Although the United States *alone* possessed nuclear weapons, that fact had no discernible effect on the behavior of North Korea, North Vietnam, Iraq, and Afghanistan in the US wars with these countries.

We must also recognize that even cases that fit the apparent pattern look different under scrutiny. True, a nuclear superior—the United States—arguably "won" the Cuban Missile Crisis. Yet where is the evidence that Soviet decision-makers greeted the prospect of nuclear war with greater sobriety than did their American counterparts? After all, Kennedy supposedly calculated a 1:3 probability of a nuclear war erupting—presumably, because he thought the *Soviets* would initiate it. Rather than capitalize on his ostensible nuclear advantage, Kennedy conceded it, perhaps, by establishing a naval quarantine line around Cuba. The effect was to shift the burden of the "first move" to the Soviet Union. The status quo would prevail, then, until the Soviets challenged it by escalating or withdrawing. Why did Kennedy—backed by an allegedly superior nuclear force—not simply attack Russian vessels and dare the Soviets to retaliate?

For that matter, why was he so willing to "deal?" He ultimately agreed to a quid-pro-quo in the form of an agreement to remove US missiles from Turkey if the Soviets would remove their missiles from Cuba, despite the potential domestic political fallout. Kennedy meant the implementation delay—the months that passed before the US held up its end of the bargain—as political cover for the US concession.

Sechser and Fuhrmann (2017: 210) ultimately conclude that the Cuban Missile Crisis amounted to well less than a "nuclear victory"—a category of "hard cases" for doubting the coercive impact of superior nuclear-weapon holdings. In assessing the historical record they find, more generally, that "the weight of the evidence shows that nuclear weapons provide states with little coercive leverage. Nuclear arsenals may be good for self-defense, but they do not allow countries to dominate world politics."

Does "Superiority" Matter Given the *Shared* Costs of Nuclear-Weapons Use?

To say that a country possesses superiority requires that we ask, "Superior capability to do what?" We cannot say that a country's edge constitutes superiority unless it helps the predominant party achieve its objectives. Our valuation of superiority thus requires an understanding of the nuclear benefits that the preponderant party will receive in return—either in coercive gains, damage limitation, or cost imposition on an adversary.

Yet our net-value assessment also requires that we appreciate the costs *to the party* should it employ—or threaten to employ—nuclear force. Thus, as posed above, the question is incomplete. We should ask, "Superior capability to do what, *at what price*?" The initial question is no more useful than asking, devoid of context, "How much would you pay?" For superiority to provide a party a physical or coercive edge, then, the inferior party must believe that, for the superior party, its benefits from using, or threatening to use, nuclear weapons exceed the costs.

Even then, a central question remains. How do we calculate these costs? As we saw (in Chapter 2), AD advocates simplified their answer by offering an arbitrary standard of inflicted damage that the Soviets presumably found unacceptable. They had little to say, however, about the amount of damage that US policymakers would accept, in turn, in pursuit of US goals. At least in theory, they assumed that US policymakers for some purposes would accept national destruction.

Contradictions thus hobbled AD logic. AD advocates implied, and sometimes explicitly stated, that any nuclear damage on the United States was unacceptable. Still, they sought to deter the Soviet Union by threatening *its destruction* if attacked by the Soviets knowing that the United States, if acting on its threat, must accept its *own destruction* given the Soviet capability to retaliate. Indeed, by building up the US nuclear force, US policymakers raised the bar for the Soviets, in their quest to—at least—match US nuclear force capabilities. The inevitable consequence was that the United States sought security through force increases that reduced the security *of the US* populous in the event of an all-out US–Soviet nuclear exchange. The US forces that ostensibly reinforced deterrence, to increase US security in times of peace, would (inadvertently) make the US populous less secure (in terms of potential damage) in the event of nuclear war.

Apart from its deficiencies and arbitrariness, the AD standard was narrowly one-sided. The standard was indifferent to the US acceptance of the costs that US forces could *inflict*. Yet these costs mattered, if only in considering the international outcry that might ensue from taking the lives of innocent civilians, wherever they might live.[11] After all, costs also come in the less tangible form of a political backlash or moral discomfort stemming from violating normative prohibitions.

Of course, mainstream deterrence theorists ignore these costs.[12] As a group, they largely assume, instead, that war costs are private (inflicted on one side by another), not shared (incurred by both regardless of who first struck whom). The roots of such thinking lie in realist theory. Realists often downplay—even deny—the decisional influence of such shared costs. They insist that states do what they must to protect and promote their *own* critical interests. Still, suppositions about the shared physical costs of war—concerns about nuclear winter, the spread of radiation, and so forth—complicate lethality considerations immensely.[13] So do potential costs from violating a nuclear prohibition either in the form of a "taboo" or nonuse "tradition."

A Nuclear Taboo?

At the critical moment, would a leader "push the nuclear button," knowing the inevitable human cost on *both sides* of the conflict? In answering that question, Nina Tannewald (2007) compellingly argues that a "nuclear taboo," by which nuclear weapons became "stigmatized and delegitimized" (Tannewald 2007: 54), gradually took hold in US policy deliberations in the decades after the atomic bombings of Hiroshima and Nagasaki. Although the taboo grew slowly, that US policymakers had trepidations at all about using weapons is the important story. US leaders, who could not justify *failing to use* atomic weapons, when they promised victory against Japan in World War II, were increasingly reluctant to capitalize on US nuclear advantages. Indeed, the bomb went unused in the early postwar years, though the United States had not yet faced a chorus of criticism from NGOs, grassroots groups, and an international public seeking to end the nuclear arms race.

Believing that an "exceptional" weapon necessitated exceptional controls, President Truman seriously hoped to turn nuclear technology (and weapons) over to an international authority. He even kept nuclear weapons out of the hands of the US military.[14] His much-cited attempt at nuclear coercion—sending what were widely assumed to be nuclear-armed B-29 bombers to Britain during the 1948 Berlin Crisis—did not actually involve bombers equipped to drop nuclear bombs. Indeed, Truman could not bring himself to drop atomic payloads on North Korea or China, during the Korean War when intervening Chinese forces sent US-led forces into retreat or when the war settled eventually into a stalemate. He held steadfast despite strong US military pressure—including repeated requests from General Douglas MacArthur, the commander of the US-led United Nations force in Korea (Paul 2009: 46–47)—to employ nuclear weapons against Chinese strategic targets for potentially decisive effects.

To be sure, both the Truman and Eisenhower administrations seriously considered using nuclear weapons in Korea. Truman transferred atomic weapons to the Pacific for *possible* use against enemy targets, and the Eisenhower administration deliberated over whether to use the bomb to bring the enemy to terms (Tannenwald 2007: 125). Moreover, official hesitance in using nuclear weapons in the Korean War had an apparent strategic logic. US officials disagreed over whether North Korea offered good aerial targets; the Air Force remained unenthusiastic about employing nuclear weapons for tactical purposes, which would hurt the political case for strategic bombing, the service's core mission; and US civilian and military officials, alike, voiced concerns over depleting the still "undersized" US nuclear arsenal, potentially leaving NATO vulnerable to a Soviet attack. Using nuclear weapons might also require—or, at least, invite—an expansion of the war. Policymakers tied atomic weapons use, in debate, to extending the war to China, and some worried that targeting North Korea with nuclear weapons would provoke Russian or Chinese escalation—even undermine deterrence globally should nuclear weapons fail to yield desired effects (Tannenwald 2007: 115–154).

President Eisenhower was not opposed, in principle, to employing nuclear weapons. He viewed them as cheap and effective replacements for conventional armies. Along with Secretary of State John Foster Dulles, he explicitly sought to "conventionalize" nuclear weapons to remove psychological and political barriers to their use (Paul 2009: 54). For various reasons, however, we should not discount evidence that norms (increasingly) influenced US decisions to forgo use of the bomb.

First, evidence that casts doubt on norms mattering to deliberations of the period can support alternative conclusions. Concerns about the bomb's *provocativeness*, for example, underscore the point that US policymakers viewed atomic weapons as exceptional in their "un-usability." If not, why did US leaders not fear that a potential US conventional win in Korea would *provoke* Soviet or Chinese escalation? That policy makers engaged in exhaustive deliberations, constantly revisiting the same question, suggests, moreover, that the decision was not easy. Why so? We could imagine such hesitance were the administration considering the use of poison gas. Would the administration have shown equal hesitance in capitalizing on the US advantage in any conventional weapon of the period?

Second, US leaders stood their ground despite reasons, backed by strong and persistent military pressure, to employ nuclear weapons. During the 1958 Taiwan Straits crisis, when China shelled islands controlled by Taiwan, the US military sought authority to use nuclear weapons in any US military conflict with China, even if limiting initial strikes to Chinese airfields. Eisenhower resisted, however—preferring, instead, to first rely on US conventional forces—even though he relied on nuclear weapons to limit military spending, sought to avoid another conventional war in Asia, and believed that the United States would eventually need to use nuclear weapons if China continued its assault.[15]

Third, policymakers deferred to practical (political) constraints that had a normative basis. Practical considerations were arguably behind the heightened concerns of US leaders that using nuclear weapons would "regularize" them, cause untold numbers of civilian deaths, and provoke discomforting accusations of racial bias when Asians, again, were targeted. But do these considerations not also speak indirectly to concerns about violating a norm of nuclear nonuse, whether US policymakers had internalized that norm or not?

Fourth, norms mattered to policymakers even when they chose not to employ the language of "right" and "wrong." Policymakers tend not to sell their preferences in internal deliberations in the language of "good" and "bad." The reasons are partly psychological. We expect a person who thinks the first use of nuclear weapons unacceptable on its face to assume that civilized majorities in the world share that view. The reasons are also political. Rejecting the option on its face as wrong, even evil, would lack persuasive appeal. That the use of nuclear weapons was even up for discussion indicates that *some* policymakers deemed them a credible military option. Those who rejected the nuclear option as wrong were already convinced. To change minds, nuclear opponents required arguments short of insinuations that proponents were somehow immoral or complicit in evil.

Fifth, the *absence* of evidence of a nuclear debate is itself evidence of growing resistance linked to strengthening norms. Indeed, the *decreasing* attention, in private debate and public pronouncements, to nuclear weapons as usable policy instruments *makes* nuclear weapons less usable policy instruments: "later decisions to refrain from nuclear use were based, in part, on previous decisions to desist and a desire to continue the practice" (Paul 2009: 2). More generally, "the longer a convention persists, the more difficult it is to kill it" (Kier and Mercer 1996: 92). With time, US policymakers effectively put nuclear weapons on the shelf; that is, they tended to look to nuclear weapons less often, as viable policy instruments. That could only harden resistance to their use.[16] By choosing to forgo nuclear weapons in favor of diplomatic, economic, or conventional-military options, the United States reinforced its *own* reluctance—and ethical aversion—to using nuclear force.

Finally, contrary evidence should not overshadow an imposing fact: although the United States possessed the most formidable weapon the world had known, successive administrations chose nonetheless to forgo using these weapons. They did so during the Korean War, largely absent fear that the Soviets would tap their limited nuclear stockpile to retaliate for a US nuclear attack (Tannenwald 2007: 128–129) and despite a prevailing belief that nuclear weapons gave the United States a decisive edge to defend against a Soviet attack on the (main) European front. Even the Eisenhower administration revealed reticence to employ the bomb, as conveyed backhandedly by a legendary event. In July 1953, Secretary of State John Foster Dulles allegedly passed the message to China, via the Indian Prime Minister, that the United States would employ nuclear weapons if the current talks failed. Indeed, Eisenhower gave his approval in July 1953 to using nuclear weapons should China break the armistice (Tannenwald 2007: 148). By that point, however, the parties had effectively settled the armistice terms. Then, the message came only as a threat potentially to "expand the war."[17]

By the end of the Cold War, a prohibition on nuclear-weapons use was arguably implanted. It seems telling, then, that the George H. W. Bush administration only "hinted" to Saddam Hussein that the United States would employ nuclear weapons if Iraq used biological or chemical weapons in the 1991 Gulf War. In Secretary of Defense Dick Cheney's words, "It should be clear to Saddam Hussein that we have a wide range of military capabilities that will let us respond with overwhelming force and extract a very high price should he be foolish enough to use chemical weapons on United States forces." Once war began, Cheney repeated the threat in somewhat different terms: "were Saddam Hussein foolish enough to use weapons of mass destruction, the US response would be absolutely overwhelming and devastating."[18] We can reasonably conclude that Cheney threatened nuclear retaliation. Yet the verbal circumspection, here, is revealing. Cheney did not explicitly threaten nuclear retaliation. We need to infer that from the context and rely upon our own assumptions about the gravity of a chemical weapons attack, as judged by the administration. We still might wonder if Cheney meant any, and all, adversary uses of chemical weapons. We could certainly construct scenarios where US nuclear retaliation to a chemical-weapons attack would constitute a gross overreaction. No less telling is that Cheney felt it necessary to send the message. Despite US nuclear superiority over Iraq—by any conceivable standard—the fact of US superiority did not speak for itself. Cheney had to do the talking—and then in veiled terms that gave the administration wiggle room. Whereas his furtiveness arguably served to reinforce the gravity of the underlying threat, it also conveyed US inhibitions in threatening to employ nuclear weapons in retaliating even for a grave threat.

Over a decade later, in 2003, the George W. Bush administration used Iraq's alleged possession of weapons of mass destruction to justify the invasion of Iraq. Still, the United States prepared for war with no intent of attacking Iraq's illicit weapon sites—preemptively or preventatively—with nuclear weapons. Even under these dire circumstances, US policymakers sought to rely on US conventional power to bring down the Iraqi regime, ostensibly to allay a potent WMD threat. The potential reason: nuclear weapons were becoming effectively "unusable," as Secretary of State Dulles, a half-century earlier, feared they would (Tannenwald 2007: 150–151).

The "usability" of nuclear weapons was dealt a further blow when, in 2013, the Obama administration issued nuclear weapons employment guidance that tied US nuclear weapons use explicitly to the laws of armed conflict. These laws establish the *distinction* between civilian and military targets,[19] and require military *necessity* in the use of force, the *minimization* of collateral civilian deaths and destruction, and *proportionality* to military gains in the inflicting of harm to civilians and their property. The laws arguably impose a significant constraint, then, on the use of nuclear weapons given their inherent destructiveness. That the administration would formally concede its nuclear options—and that the Trump administration would not reverse course in the 2018 NPR—were perhaps important indications of changing global norms pertaining to nuclear weapons employment. Indeed, we could read the change to indicate that global conditions had *already* constricted these administrations' latitude for choice.

Whatever the strength of formal guidance and prohibitions, we must acknowledge that the US failure to employ, or even threaten to employ, nuclear weapons against nuclear and nonnuclear adversaries is strong evidence of a nuclear taboo that limits the advantages of ostensible nuclear "superiority." US adversaries must believe that the United States will employ these weapons if they are to confer coercive influence.

A Tradition of Nonuse?

Yet we must acknowledge limits to the prohibition on nuclear weapons employment. For four main reasons, discernable hesitance to use nuclear weapons might not reflect a taboo at all.

Fading Memories

The impact of the Hiroshima bombing has faded from public consciousness with ever-smaller numbers of people alive who experienced that war. The war is now the subject of memorials and commemoration that testify to the sacrifice and "heroism" of those who fought it and risked, or gave, their own lives for the greater good. It is appropriated, then, to celebrate the patriotism and sacrifice of those who fought *on the winning side* and not generally to remember the evils of war, much less to remember the indiscriminate killing on the Pacific front, from a US (nuclear and incendiary) bombing campaign that helped bring the war to a close.

In response to public opinion surveys, people are far more likely to rate, as momentous, the effects of the atomic bombing on "recall" than through "remembrance," that is, they rate its effect far higher when questioned specifically about the bombing than if required to volunteer the bombing when asked to list critical events.[20] Unsurprisingly, opinion polls continue to reveal—multiple decades removed from the events—high levels of US public support for the atomic bombing of Japan. Respondents tilt more negatively when asked whether they generally "approve" or "disapprove" of the decision.[21] Yet when people are explicitly asked if the decision to drop the bomb was morally wrong[22]—indeed, if they are asked whether the United States owes Japan an *apology*—public opinion turns decidedly negative.[23] Americans apparently have difficulty separating the rightness of the cause for war with Japan from the rightness of US wartime actions against that country.

The Cold War "nuclear threat" has also faded from view and, along with it, the specter of a cataclysmic nuclear war that could consume humanity. Although we can, in principle, separate laws of "consequence" and laws of "appropriateness"—rules based on the negative returns, in a cost-benefit analysis, from those based on the "rightness" or "wrongness" of behavior—the two bind together in our thinking. Our judgments about the evils of using nuclear weapons are most certainly effected by an awareness of their potential effects.

Substantial Wiggle Room

Evidence supporting a nuclear taboo also exposes its weakness. The Obama administration's efforts to align nuclear weapons policy with the laws of armed conflict paradoxically demonstrate this. True, the administration accepted principles that ostensibly limit the uses of nuclear force. But the administration still accepted, as legal uses of force, options—of large scale and indiscriminateness—that remain in US targeting plans.[24]

Notably, the administration—like its predecessors—still accepted the long-established principles of nuclear deterrence that peace resides in a threat of overwhelming destructiveness in the event of nuclear war (Lewis and Sagan 2016: 63). These principles were accepted despite prohibitions (with the standing of customary international law) on the conduct of "reprisal attacks" against civilians (Sagan and

Weiner 2021: 153–160). Like its predecessors, the administration also chose not to renounce the "first use" of nuclear weapons as a general principle. Indeed, in tying US nuclear weapons policy to the laws of armed conflict, it nonetheless omitted reference to military "necessity," a principle that proscribes the use of nuclear force when conventional force would suffice (Lewis and Sagan 2016: 68–69).

Obama's changing guidance arguably speaks to a disinclination to employ nuclear weapons and to confine nuclear strikes, if necessary, to select military targets—as does the reiteration of these principles in the 2018 NPR of the Trump administration. But history tells us that abstract principles prove infinitely malleable. The Nixon administration proscribed the targeting of population centers per se, but planners still found abundant justifications for hitting urban areas by "reducing" cities to nonhuman assets, such as industries, found in and around cities. The Obama administration appeared to straddle the same line in stating that "the United States will not *intentionally* target civilian populations or civilian objects [italics added]."[25] The term "intentionally" creates allowances for inadvertent destruction, but it also absolves the attacker of responsibility—perhaps, through creative target designations—for inflicting such destruction. An often-heard refrain in war—"the strike was not intended to kill civilians"—has dual meanings. It refers to the accidental but also the collateral deaths that constitute the accepted price for accomplishing a given military objective.

The challenge of devising meaningful, workable standards presumably confounds the interpreting of legal constraints on permissible civilian damage. By what standards do we assess the "minimization" of civilian casualties? Is it the damage that nuclear weapons could *potentially* inflict if greater in power or if arriving in larger numbers? Even a "small" nuclear explosion could produce a humanitarian disaster. "Proportionality" is no less problematic a concept given its ties to the slippery notions of "military advantage" and "military necessity." After all, Cold War–era analysts disagreed over whether, and how much, attacks of varying scale offered a military advantage or would yield a useful effect. The dispute rested on competing views of the efficacy of employing nuclear force and, more fundamentally, on diverging views of the rationality of the participants, their control over their nuclear forces, and the dynamics that could overtake the participants in a nuclear war. In the end, what US officials deem proportional will undoubtedly reflect their assumptions about their own intentions and those of a nuclear adversary. One implication is clear: if "the contribution of a nuclear attack is to limit nuclear damage to the U.S. population or to maintain the United States as a viable society after a nuclear war, then almost any degree of collateral damage could be deemed acceptable under the proportionality principle" (Lewis and Sagan 2016: 67).

In that context, we should reflect on the text of the 2018 NPR, which reads as follows: "If deterrence fails, the United States will strive to end any conflict at the lowest level of damage possible and on the best achievable terms for the United States, allies, and partners. U.S. nuclear policy for decades has consistently included this objective of limiting damage if deterrence fails." We might wonder about the considerations that influence assessments of the "possible" in times of great stress. We should also wonder about the implicit trade-off between limiting damage and ending conflict "on the best achievable terms" and, more specifically, whether fears that limiting damage to an adversary might yield a deficient resolution—short of offering "the best achievable terms." We should wonder, no less, whether maintaining options for a full-scale nuclear war—for the most extreme scenarios—has the effect of normalizing or sanctioning indiscriminate targeting in less extreme scenarios.

An Incomplete Prohibition

The nuclear prohibition is incomplete. Nuclear powers have publicly pledged not to use nuclear weapons but only against non-nuclear states. Although some nuclear states have pledged not to use nuclear weapons *first* in any confrontation with an adversary, such pledges are conditional. Whereas the Soviet Union had pledged not to use nuclear weapons first, when NATO relied on these weapons (if needed) to push back a Soviet attack, the Russian government—with its less formidable conventional military—now retains the first-use option. Indeed, Russian President Putin issued a veiled threat to use nuclear weapons against NATO countries for hindering his military aspirations, then in Ukraine (in his words, it would suffer "consequences that you have never encountered in your history").[26] More importantly, all nuclear-armed governments accept, as justifiable, nuclear retaliation for a nuclear attack.

Such a conditional prohibition is less easily maintained than an absolute one, for "conditionality" grants even potential violators (some) power to determine whether their behavior is right or wrong. Even Cold War–era hawks who thought the United States should prepare to fight (and win) a nuclear war maintained that the underwriting principles were consistent with just war doctrine (Gray and Payne 1980). Conditionality thus reduces the potential offensiveness of the behavior by instilling doubt over whether it constitutes an allowable exception. If two parties can interpret the same act differently, we can question whether the prohibition has been violated but also whether it existed in the first place. As Elizabeth Kier and Jonathan Mercer (1996: 93–94) observe, precedent—the constraining effect of past rules and behavior—rests on "conspicuousness" which "means a simplicity, a clarity, an absence of discretion that make the violation of the convention unambiguous."[27] The implications are straightforward.

> Simple and unambiguous conventions are better than complex and nuanced ones. A convention against using nuclear weapons in combat is better than one permitting occasional use of these weapons in specific circumstances. "Never" is better than "sometimes," because it is clear,

allows no discretion, and can in time acquire a symbolism that strengthens the convention. (Kier and Mercer 1996: 94)

Conditionality is even more damaging to a prohibition's standing as a "taboo." If parties believe that some uses of nuclear weapons are justifiable, even legitimate, attributing nonuse to a "taboo" is problematic. A taboo is not situational. Cannibalism—a frequently cited example of a universal taboo (Quester 2006)—is thoroughly rejected by virtually all societies. Even those who resort to cannibalism for survival remain stigmatized for their actions. We *judge* the participants even if we struggle to understand what *we* would do under similar circumstances. The prohibition—that is, the "taboo"—is so strong that we viscerally reject comparisons to seemingly similar social actions. We routinely give blood for transfusions. We harvest human organs: eyes, skin, ligaments, hearts, lungs, livers—and even hands and faces—for reuse by other humans. We still cannot fathom the consumption of human flesh for nutritional purposes. Indeed, as Tannenwald (2007: 367) recognizes, "a taboo is an absolute prohibition that is not necessarily disrupted by a violation, and does not permit reciprocal behavior in response to violation." A taboo, so understood, would prohibit the resort to nuclear arms even in response to a nuclear attack. Moreover, a taboo—unlike a normative prohibition—would not seem to erode gradually or fragmentally, as was the case, for instance, with the strategic bombing of cities in World War II (Gibbons and Lieber 2019).

Thus, in allowing for qualifications and exceptions—or thinking about nuclear weapons in conventional terms—we must acknowledge that the nuclear "prohibition" falls short of a taboo.[28] For that matter, as long as society—publics as well as elites—fail to embrace the prohibition, we cannot correctly deem it a "taboo." Experimental evidence suggests that the general US public supports the US use of nuclear weapons—for their greater military effectiveness—in various non-nuclear contingencies (see Press, Sagan, and Valentino 2013; Sagan and Valentino 2017). For them, it seems, the use of nuclear weapons is neither abhorrent nor unthinkable.

A Better Label

The apparent taboo against nuclear use can reasonably be understood more as a "tradition" of nonuse (Paul 2009: 25–37). The fear in violating a tradition is setting "precedents" that could provoke further uses of nuclear weaponry. Viewed accordingly, "the most important reason for nuclear non-use is the expectation of leaders that using nuclear weapons today will result in other states being more likely to use them tomorrow, potentially against one's own country or that of allies" (Gibbons and Lieber 2019: 34). Although a tradition might rest in normative influences, it can draw strength, as well, from the "logic of consequences" (Press, Sagan, and Valentino 2013: 189). If so, existing practices survive because abandoning them could prove costly.

Thus, the notion of a tradition takes on additional significance when viewed as established beliefs about the likely consequences of the first (post-war) use of nuclear weapons. Leaders might fear the reprobation of allies, and more neutral leaders and publics, should they violate a tradition of nonuse, which has held now for more than three-quarters of a century. They might also fear that any such action will convince other states that the threat of a nuclear attack is real (maybe imminent) and requires the precipitous, even preventative, use of nuclear weapons against potential threats. Conversely, they might fear that others will conclude that nuclear weapons *are* useful for limited military purposes. If used initially, without catastrophic social or economic effects, they might wonder whether the *illusion* of a nuclear war limited to surgical nuclear strikes will invite catastrophe.

We might conclude, then, that a tradition of nonuse gains strength through uncertainty. After all, most who write about the tradition tie its influence to concerns about the long-term consequences of nuclear weapons employment (see, e.g., Pauly 2015: 442). In that sense, the tradition fits comfortably with rational theory. Yet a tradition can also reflect the customs and beliefs that societies transmit from one generation to the next. It can thus rely on untested assumptions that here pertain to whether parties can control a nuclear conflict, whether nuclear conflicts will beget other nuclear conflicts, whether precise attacks that limit collateral damage are possible, and so forth. The danger, then, in the violation of a tradition (in setting a "precedent") is that the *first use* will become the *definitive use*. In nuclear warfare, it could inspire the unfounded or premature "learning" of lessons; provoke a contagion of fears based on a new "understanding" of how nuclear temptations might spread; and spark precipitous or preemptive weapons employment, with the assumption that those "who hesitate are lost."

The main point nevertheless deserves emphasis. Whether by reason of norms or tradition, states have raised the bar when defining conditions that would require or trigger a nuclear response. Nuclear weapons thus became "weapons of last resort," reserved for extreme contingencies. Why else would the United States commit its resources and efforts to diplomacy, or incur the human and material costs of conventional conflict, rather than push its nuclear weight around? Such behavior provides an answer. Social and political constraints offset expectations of gain from the threatened or actual use of nuclear weapons. Whether these constraints directly affect US conduct, an adversary's beliefs that these constraints hold could limit the coercive influence of US nuclear-employment threats.

Superiority Meets "Damage Limitation"

What if one nuclear-armed state could conceivably prevent another from inflicting unacceptable damage, with nuclear weapons to spare? Must we then concede its "superiority"? After all, it would seem to have "nothing" to lose in a nuclear confrontation. To the contrary, it would apparently benefit from a nuclear exchange by vanquishing a nuclear adversary while retaining a residual force to obtain coercive benefits. Yet that "damage-limitation" capability offers illusionary rewards. The nuclear "inferior" retains options; and the nuclear "superior" must confront a probabilistic challenge.

Damage Limitation Is Not Necessarily Damage Prevention

Analysts who refer to a damage-limitation capability typically suggest that a state can reduce its damage to acceptable levels in a nuclear exchange. In the early 1960s, the McNamara-era Defense Department thus flirted with damage limitation as a US nuclear option; it soured on the concept when the numbers of Soviet systems that would survive a US attack made a mockery of the term. Now, however, US capabilities—pitted against relatively small adversary arsenals—make damage-limitation a potential reality. A party that possesses this capability can protect itself from harm and, with residual nuclear might, could coerce an opponent into submission. Perhaps, then, it could even coerce the adversary into relinquishing its nuclear capabilities without firing a shot.

Unconditional superiority requires such a damage-limitation capability. That is, it requires that the superior party have the capability (a) to reduce the damage it incurs to *acceptable* levels if it strikes first or second or (b) to always get in the first, disarming, blow, given available intelligence, weapons capabilities, proximate basing, and so forth. Yet what if the stronger party can only *limit* the damage that an adversary can inflict? Obviously, less damage to one's country is better than more damage, all things being equal. The stability of deterrence—and assessments of "superiority"—partly depend, however, on whether the lesser (absolute or relative) levels of damage are acceptable to the stronger party. We cannot obtain answers from pure math, or physics. The answers lie in the messy world of politics, societal values, and individual psychology.

Physics plays a part. A big bully can physically dominate a smaller person. The bully might even push the smaller person around—acquiring concessions without a fight. The smaller person knows that, with a size disadvantage, fighting is a losing cause. Yet, by resorting to coercion—a threat to impose punishment—the bully implicitly confesses that the costs of fighting are such that the bully would rather *not* fight. After all, the bully could have punched rather than pushed or talked. Even Hitler recognized benefits in negotiating rather than fighting to achieve his goals. He did so when he obtained the Sudetenland from Czechoslovakia more "cheaply," through bargaining, not fighting, in his dealings with British Prime Minister Neville Chamberlain.[29]

What happens, however, if the smaller person stands their ground, and maybe fights back? They might do some damage—perhaps even land a lucky punch. They might even break the bully's nose. Who is to say whether the bully is willing to pay that small price for a victory? If the "weaker" party is just that—not the helpless or hopeless party—the fact of usable or meaningful superiority, or capability to limit damage, is in question. Even the lesser punishment that the smaller person can inflict might deter the bully from fighting. The balance of costs alone cannot decide the outcome if one side is more willing to absorb pain given the stakes at issue. The weaker party need not emerge without a scratch from a potential fight. It need only inflict some unacceptable level of hurt on the stronger party.

The weaker party can further capitalize on uncertainty. The stronger party might exaggerate the harm that the weaker party can inflict if the weaker party flexes its muscles. After all, fears that the Soviet Union, despite its gross military inferiority, was winning the arms race fueled infamous strategic scares of the Cold War period. In the mid-1950s, fears of a "bomber gap" afflicted the US strategic community. The ill-founded concern was that the Soviet Union had produced long-range bombers in qualities and quantities that would render US aircraft vulnerable in Europe and threaten the US homeland. Likewise, the Soviet Union appeared to grab the lead in missile technology, in 1957, with the launch of the Sputnik satellite. John Kennedy rode the resulting "missile gap" scare to the White House. The United States, with an enormous head start, anxiously built up its arsenal out of concern for US *inferiority*. Conversely, in the 1970s and 1980s, Soviet leaders feared that the United States had acquired a nuclear knockout punch precisely at the time US hawks argued that the Soviets had acquired, or would soon acquire, a disarming first-strike capability (Green and Long 2017a). Then, in 2003, the Bush administration went to war with Iraq ostensibly to neutralize its nuclear-weapons program and destroy its alleged stocks of biological and chemical weapons based on an exaggerated view (boosted by US intelligence) of the Iraqi weapons' threat. Of course, Iraq ultimately suffered the consequences; but the point remains that uncertainty about a country's capabilities could compensate for weakness under different circumstances.

Yet even viewing the problem as a "contest of wills," between two identifiable opponents, oversimplifies the problem. Personifying the combatants—endowing states with human qualities—is a useful heuristic tool. It is also potentially dangerous, for it assumes that the combatants control their own fate. We should not assume that the weaker party will retain the flexibility to avert catastrophe as required for the stronger party to win without firing a shot. By predelegating launch authority to subordinates, adopting a launch-on-warning strategy, and so forth, leaders can multiply the military effectiveness of their forces and increase their probability of use. For that reason, US presidents in the 1950s and 1960s approved the predelegation of US nuclear-launch authority under defined circumstances.[30] Ironically, states with the smallest nuclear arsenals, which play to US damage-limitation capabilities, are arguably the most tempted to limit their own flexibility, in time of war, to ensure the "effective" use of their weaponry.[31]

These facts can override strict military considerations. Savvy leaders, who possess an ostensibly superior force, must still appreciate their own inability to control crisis outcomes. With strong incentives not to test their "damage-limitation" capabilities, then, they concede a coercive advantage.

Statistical Uncertainty in "Damage Limitation"

Analysts can benefit from evaluating damage limitation in probabilistic terms and acknowledging, nevertheless, that uncertainty clouds even these assessments. The United States can rely on its larger numbers of warheads, weapon accuracy and responsiveness, and vast intelligence and reconnaissance network to detect and identify targets and assess damage for additional strikes. But the damage that the United States can inflict is still best understood as a probabilistic result. The probabilities reflect the physical challenges of disarming even a grossly "inferior" state that, on defense, can employ concealment, redundancy, hardening, and evasiveness. The probabilities also reflect the "political and strategic context" (Lieber and Press 2017: 27). For instance, the dominant party might withhold some of its nuclear might out of concern for inflicting civilian casualties.[32] The probabilities become more favorable to the inferior state as the superior state withholds more of its nuclear capability from the fighting.

A small probability that even a small number of the inferior state's warheads would reach their targets gives the weaker party coercive influence. As Glaser and Fetter (2005: 102) observe, "even relatively little credibility is sufficient when the costs of retaliation are so large." For one thing, the threat of a US attack increases the possibility of an accidental weapons launch "simply because, under tense and demanding conditions, there would be more individuals who were capable of launching a nuclear attack" (Glaser and Fetter 2005: 121). Even with the evisceration of its missile force, the weaker party possesses potential delivery options. "If the United States does not gain control of these weapons, the adversary might be able to deliver them via unconventional means, or the government might lose control of them, creating the possibility that the weapons will fall into the hands of terrorists" (Glaser and Fetter 2005: 106).

An inferior country can realize benefits—deterrence and otherwise—then, if it has some probability of retaining a small number of nuclear warheads after enduring an attack. Figures 3.2a and 3.2b show this to be a realistic expectation even for a grossly outmatched country. In these graphs, the *x*-axis represents the number of warheads possessed by the inferior party, the *y*-axis displays the probability that at least one (3.2a) or three (3.2b) warheads will survive the attack, and the five lines indicate the relative effectiveness of the attack, that is, they display the predicted results of attacks that could destroy 25, 50, 75, 90, and 95 percent of the inferior party's warheads, respectively.[33] These kill probabilities reflect the combined effects of a full complement of influences including the inferior party's defensive capabilities (and possible predelegation of launch authority) and the superior party's target-detection capabilities, weapon reliability and accuracy, attack timing (whether the attack came in waves, phases, or all at once), warhead fratricide, retargeting capabilities, and attacks on adversary command, control, and intelligence (C^3I) capabilities.

The figure reveals a significant probability that even a grossly "inferior" adversary will retain the capability to respond to an attack. Figure 3.2a shows that, with roughly a six-warhead arsenal, the inferior party has around a fifty-fifty chance, from an attack of 75-percent effectiveness, of retaining at least one warhead to deliver against the adversary. The probability that at least one warhead will survive an attack increases dramatically for that same six-warhead arsenal when on the receiving end of a less-effective (damage-limiting) attack. In fact, with an attack of 50-percent effectiveness, an arsenal of five warheads stands around an 80-percent chance of surviving the attack to deliver at least one warhead against the attacker. More sobering still, Figure 3.2b shows that that same attack produces over a one-in-three chance of leaving the adversary with at least three warheads for a retaliatory strike.

Further implications of these probabilistic simulations are revealed in Figures 3.2c and 3.2d. They restructure the prior graphs so that the *x*-axis presents the problem from the superior party's perspective, where the lines show the survival probabilities for inferior-party (pre-attack) arsenals of five, ten, fifteen, twenty, twenty-five, and fifty warheads. Here, the question is, "What effectiveness is required of a damage-limiting attack to degrade the inferior party's capability to respond?" Clear, now, is that attack outcomes are far more sensitive to attack effectiveness than to the size of the inferior-party's arsenal. Note the dramatic downturn in the survival probability of a single warhead as the attack approaches 90-percent effectiveness. Still, a point bears repeating. The superior party must accept a sizable risk that at least one warhead will survive even a highly effective attack. Indeed, with an arsenal of around fifteen warheads, roughly an even chance exists that at least one warhead will survive, and a one-in three chance exists that at least three warheads will survive, to threaten (and potentially harm) the attacker.

Figure 3.3 provides a final look at the competition between the two arsenals. It retains the *x* and *y*-axis dimensions from Figure 3.2. The distance between the lines in each set, however, now reveal the change in survival probabilities of the designated number of warheads (twenty or more, ten or more, or five or more) with an increase in the inferior party's arsenal size from fifty to one hundred warheads, for the given superior-party (offensive) kill ratios, as displayed along the *x*-axis. The message is clear. By doubling the size of a relatively small arsenal, the inferior party has greatly increased the challenge for the superior party. The size of that arsenal after an attack is highly sensitive—only at the extreme—to the superior party's kill ratios. Absent an extremely effective offensive campaign, the inferior party—by modestly building its force size—can greatly increase its chances that some number of warheads (varying by kill ratio) will survive to retaliate for an attack.

Take, for instance, the point at which the probability equals .50 that the given number of warheads will survive the attack. If the inferior party increases its force

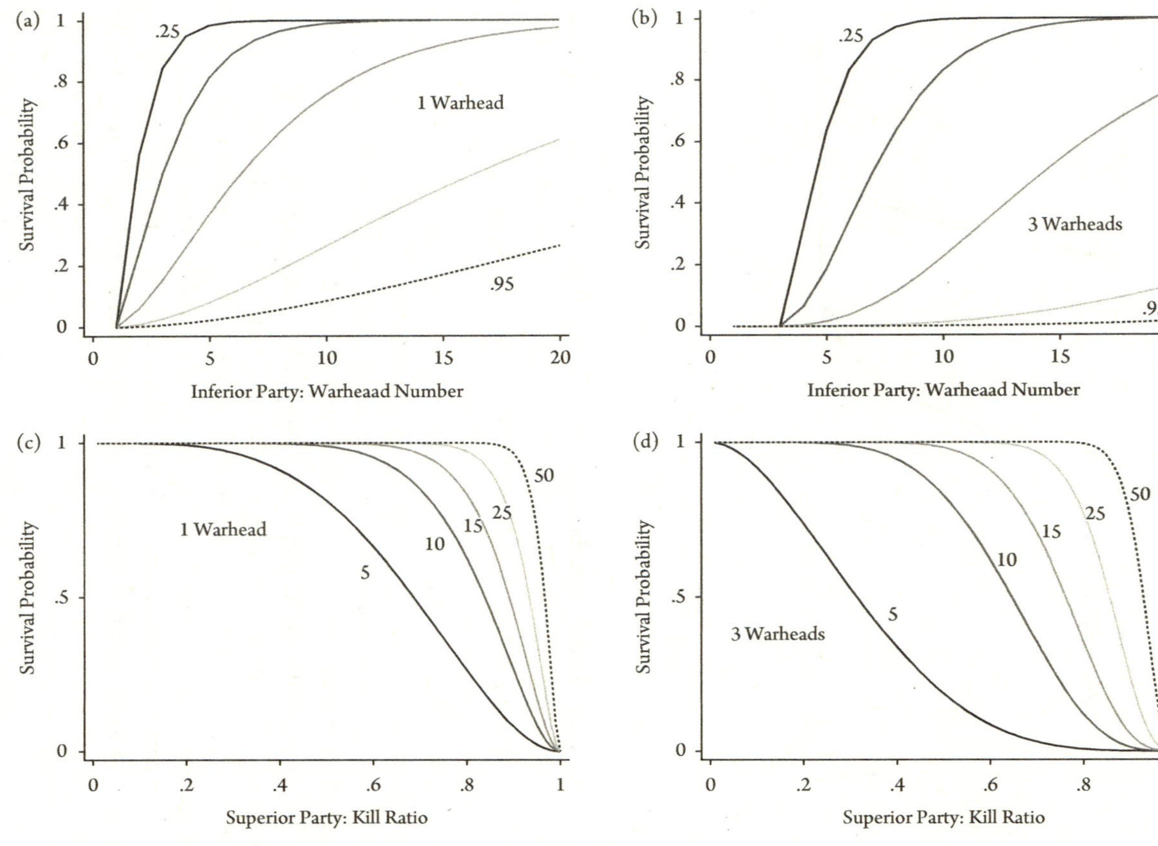

Figure 3.2 The Inferior Party's capability with small force numbers.

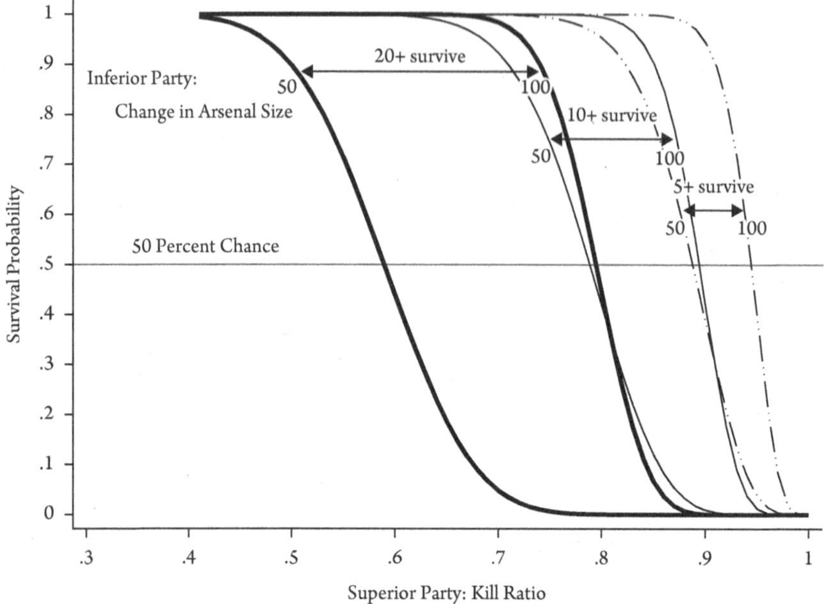

Figure 3.3 The Inferior Party's arsenal size versus the Superior Party's kill ratio.

from fifty to one hundred warheads, it can secure a 50-percent chance that (a) at least twenty of its warheads will survive an attack if the superior-party's offensive effectiveness improves from a kill ratio of roughly .60 to .80, (b) at least ten warheads will survive an attack if the superior party's offensive effectiveness improves from a kill ratio of roughly .80 to .90; and (c) at least five warheads survive the attack if the superior party's offensive effectiveness improves from a kill ratio of roughly .90 to .95. Of course, the more modest the retaliatory ambition of the inferior party, the more effective the offensive attack that the party can endure.

The main point requires emphasis. Unless its kill ratios approach .90, the superior party's offensive has virtually *no* effect on the survival probabilities of a modestly sized retaliatory force—say, around ten warheads—regardless of the size of the inferior country's initial force (fifty versus one hundred warheads). That is, the superior party requires an extremely high effectiveness level (kill ratio) to prevent the inferior party from launching a significant number of warheads in retaliation. If the superior party seeks assuredly to disarm its adversary, even these effectiveness levels might prove insufficient.

True, increasing the size of the inferior party's arsenal offers rapidly diminishing returns, should the superior party's offensive capabilities improve beyond a .90 kill ratio. Whether offenses promise 93, 94, 95, or 96 percent effectiveness might seem not to matter. It is easy to dismiss the effects of a 1-percent marginal change, especially with the uncertainties surrounding these statistics. But these small differences are quite significant. The accompanying probabilities that at least five of one hundred warheads could survive the assault is .71 for an offense with 93 percent effectiveness but .56, .38, and .21 for offenses with 94, 95, and 96 percent effectiveness, respectively. A 4-percent difference in offensive effectiveness translates into a *.50 drop* in the probability that at least five warheads will survive the attack. Still, these levels make for a demanding standard of effectiveness that will still not *solve* the attacker's problem. Even with 94-percent effectiveness, over a 50-percent chance remains that at least five warheads will survive the attack.

Admittedly, these numbers simplify the strategic problem in numerous respects. First, attacks on adversary missiles are not "independent" events. By attacking a command center, blinding a radar, destroying communication links, and so forth, an attacker can neutralize many enemy missiles; and failures in the inferior party's launch capabilities—caused, for example, by a critical bottleneck—might disable multiple missiles. Second, a superior party—like the United States—possesses abundant technological advantages that would help it track, locate, identify, and ultimately destroy adversary missiles. These advantages produce synergistic effects that will significantly boost the overall performance of attack systems. Third, a superior party might reap advantages from missile defenses. They can work in synergy with offenses to boost kill ratios—hypothetically, at least—to reduce residual adversary capability to some "acceptable" level. Finally, these very capabilities could force an adversary to rely on defensive measures—for mobile missiles, for example—that increase their exposure to attack. They might curtail communications with the command to avoid detection, which might leave them unable to respond to an attack. Mobile missiles might employ camouflage from local topography, or operate less conspicuously on certain road networks, but only by constricting their operating radius, easing the detection challenge for the offense. They might rehearse deployment patterns, to thwart an offensive, but thereby cue likely missile locations in the event of a conflict.

Certainly, these factors could combine to increase the effectiveness of a US offensive and thereby reduce the damage the inferior party could inflict in return. In that sense, the United States could reap a significant payoff. After all, "there is no magical threshold beyond which the ability to limit damage in a nuclear war ceases to matter" (Kroenig 2017: 200). We must remember, however, that all such effectiveness estimates are speculative, which matters given the high effectiveness rates required to reduce the retaliatory threat. Thus, the main issue remains: the concept of damage limitation is deceptive. It overstates the capability of a superior state and understates the capabilities—coercive and otherwise—of an inferior party. Ironically, the superior party's actions—borne of confidence in its capabilities—could induce offsetting actions, including growing its arsenal, that further weaken the superior party's position.

Does Deterrence Require Superiority?

The case for establishing superiority assumes, obviously, that it gives the possessor clear political or military advantages. But evidence and logic bely claims of any such advantage. Useful perspective, in that regard, emerges when examining China's acceptance of its nuclear "inferiority"; the overstated challenges of "extending deterrence" to US allies; and the exaggerated dangers posed by fledgling nuclear states that are supposedly unbound by the rules (and restraints) of the past.

Settling for Nuclear "Inferiority": The Case of China

Given the limited benefits of superiority, competitors can viably choose not to compete in *relative terms* with the superior nuclear state. China serves as an example, in that regard, for it currently defines its nuclear security around the requirements of a limited but viable retaliatory force. Its approach implicitly challenges the assumption that the United States obtains military or coercive advantages from an allegedly superior arsenal. The Chinese position assumes, instead, that the US–China nuclear balance is stable. Given the relatively low stakes for the United States in any conflict that would pit the United States against China, and the damage that nuclear weapons could inflict on US targets, only a small, survivable Chinese force is required to deter the United States from employing its nuclear weapons, even coercively, against China.

China's nuclear strategy is fittingly termed "assured retaliation," a phrase that faintly echoes the formal doctrinal position of the McNamara-led Defense Department.[34] The somewhat less menacing undertone of "retaliation," compared to "destruction," speaks revealingly to China's security objectives. That is, China seeks to retain some minimum level of capability that, after a US first strike on the Chinese nuclear infrastructure, falls short of assuring the destruction of the US adversary. Expert dissensus exists on the precise labeling of Chinese nuclear strategy—whether it edges closer to "minimum deterrence," "assured retaliation," or some other alternative (Bin 2007: 4). Although that is itself a source of uncertainty that could deter a US first strike (on this, see Chase 2013; Riqiang 2013: 579–580), the labels speak nonetheless to the finite purposes and capabilities of the Chinese force. In current Chinese parlance, the Chinese force is both "lean and effective" (Heginbotham et al. 2017: 20).

China's nuclear deployment preferences, and its willingness to tolerate a nuclear arsenal—undersized relative to US weapon holdings—might seem surprising given China's economic prowess, large and growing military budget, and global political ambitions.[35] The United States and Russia each possess warhead stockpiles maybe dozens of times greater in size than the current Chinese stockpile. Rather than close (or reverse) the "gap" with a dedicated (Soviet-style) push to build up the Chinese force, China has accepted its nuclear "inferiority" relative to the United States in numbers of warheads, varieties of basing modes, and damage-limitation capability.

To be sure, China *is* currently engaged in the unprecedented buildup, modernization, and diversification of the country's nuclear forces. It has moved with considerable rapidity—and *visibility*—to construct—now over one hundred—silos for ICBMs in each of three missile fields.[36] Even that development understates the nature and rate of China's force improvements. China is closing the gap with the United States (and Russia) in warhead totals; improving their accuracy, range, and penetrability; and growing a nuclear force that can deliver its payloads by land, sea, and air—perhaps to constitute more than a "nascent 'nuclear triad' " (Office of the Secretary of Defense [OSD] 2021: VIII). Indeed, these nuclear developments acquire additional menace viewed in the broader context of China's major conventional force improvements, aggressive moves in the South China Sea and toward Taiwan, and push internationally for recognition as a global power. Yet China largely depends, for deterrence, on its still undersized, land-based missile force of silo-based and road-mobile (MIRV-capable) missiles.

China has focused its concerns on maintaining sufficient retaliatory capability, as US offensive and defensive capabilities improve, not on conducting damage-limitation strikes. Given the heavy US investment in a less vulnerable SLBM force, China recognizes the inefficacy of disarming strikes against the US nuclear force (Denmark and Talmadge 2021). Although the increasing size of the silo-based force is a worrisome development, in a Cold War context—if taken to signal China's preparedness for a first strike (given the vulnerability of fixed-site missiles)—the relative size of the force, and its vulnerability, suggest that even the expanded silo-based force is intended to reinforce deterrence. The additional silos could absorb more US attack capability and complicate US planning which could give China added assurance to adhere to its declared "no-first-use" policy. Indeed, some of China's new silos are possible decoys (reminiscent of the shell-game strategy once considered by Cold War–era US force planners to reduce the vulnerability of the US land-based missile force). US planners would have to prepare their attack not knowing which silos contained missiles—preparing, then, as if all the silos contained missiles. Even this assumes that China interprets "no-first-use" to proscribe a "launch-on-warning" posture (as discussed in Chapter 5). That strategy, which the US Defense Department fears China is moving to adopt (OSD 2021: VIII), would require that China fire some number of its (presumably more vulnerable) weapons against US targets on warning of a US attack, which—of course—would invite an unbridled US response.

Presumably, China's large land mass, and its lag in technologies for countering US anti-submarine warfare capabilities, will have China relying on its land-based missiles into the foreseeable future. Thus, China's mobile missile force is central to the rough calculations employed by Chinese decision-makers to scale the Chinese force. Reflecting early American thinking about a "minimum-deterrence" force—the presumption that even one surviving US submarine was sufficient to deter the Soviet Union—China seeks to get some relatively small number of warheads—not hundreds or thousands—on US targets in retaliation. In force planning, China is especially concerned about growing US capabilities in missile defense—both at the mid-course stage, with the deployment of US interceptors in Alaska, and in theater, with US defense systems in the Pacific ostensibly meant to thwart a North Korean missile attack. China (like Russia) fears that the US integration and upgrading of these defenses will neutralize the capability to respond to a US attack. That would open the door to US preemption and coercion and weaken the constraints on US conventional options in the region. In response, China has sought evasive hypersonic glide vehicles, with their unpredictable flight trajectories, has likely pegged the size of its force to the number of US (defensive) interceptors, and has relied more generally on assuring the survivability of sufficient warhead numbers, through platform diversity, to achieve penetrability.

Although any rule of thumb owes to underlying calculating assumptions, for instance whether or how it accounts for defensive system reliability, numbers of interceptors on target, the effectiveness of defensive countermeasures, and so forth, the main threat to nuclear deterrence from China's perspective lies less in the relative balance and more in conditions that might prompt nuclear-weapons use. China exhibits some fear that the United States might engineer the logic of the stability–instability paradox to its advantage. If, in fact, nuclear deterrence is secure, the United States might employ a full range of conventional options, including conventional weapons meant to destroy China's nuclear forces, with the belief that China would not then launch its missiles.

Admittedly, the dilemma—so conceived—turns the traditional paradox on its head. Cold War–era theorists recognized that the stability of the nuclear balance opened the protagonists to conflict at "lesser" levels (though what constituted a lesser level was a matter of opinion). But the concern now is that the United States might try to nibble away at China's nuclear might under the cover of a "conventional" strike. China's response is to maintain some ambiguity about how it would respond to such an attack, that is, whether it would treat a US conventional strike like a US nuclear strike on those missiles. What this means, however, is that China remains committed to *nuclear* deterrence. In seeking to deter US transgressions with a numerically "inferior" force, China undercuts claims that superiority—in quantitative and qualitative terms—gives the United States a practical edge.

Of course, China might someday try to match—or might even match—US nuclear capability. But what would that say, then, about the practical, or dissuasive, utility of US nuclear "superiority"? For that matter, might it say only that China sought to close a gap that fed a false US sense of military or coercive advantage?

The Challenges of Extending Deterrence

Even if the United States could solve a "basic deterrence" problem—that is, dissuading adversaries (rogue state or otherwise) from striking US home territory—can the United States deter nuclear strikes on a US ally? The end of the Cold War did not reduce interest in the underlying issue of "extended deterrence." Stretching the deterrent commitment to include other states—that is, placing them under a country's "nuclear umbrella"—arguably imposes tough deterrence demands. A challenger has good reason to doubt the credibility of the accompanying deterrent threat. Why would a country invite horrific devastation against its own territory and citizenry to retaliate for enemy attacks on some *third* country?

Nuclear "superiority" arguably offers the possessor additional security to address the challenge. Even if falling short of a damage-limitation capability, a large quantitative and qualitative edge might help convince an adversary that its gains from attacking a US ally with nuclear weapons involve significant risks of a devastating US response. If anything, a larger edge is arguably required to assure allies more than to dissuade adversaries (on this, see Tertrais 2010: 8). US policymakers "knew that without the assurances provided by a robust nuclear umbrella, countries such as Germany, Japan, South Korea, and a host of others might deploy their own nuclear weapons, a development that would be inimical to America's strategic interest" (Gavin 2020: 61).

We have reason to doubt whether the capability to deter attacks against the United States is sufficient to deter attacks on US allies or to reassure them. US policymakers have been more equivocal about their likely response to a nuclear attack abroad than to an attack on the US homeland. Although the United States remained firm in its Cold War–era nuclear commitment to Europe—building up the NATO nuclear stockpile and dedicating some US nuclear forces to the European theater—the United States repeatedly fed concerns about whether it would seek to confine nuclear weapons use to Europe (in hopes of avoiding Soviet retaliation against the US homeland). In the 1960s, French President Charles De Gaulle posed a reasonable question. In a nuclear encounter with the Soviet Union, "Would the United States trade New York for Paris?" De Gaulle thought he knew the answer, perhaps because he implicitly asked another, "Would France trade Paris for New York?" France responded by constructing its own nuclear deterrent force and withdrawing French forces from NATO's integrated military command.

The question, as posed, suggests an uncertainty that could undermine the credibility of extended deterrence, as President Kennedy himself recognized. He asked, if a close ally could feel that way, might Khrushchev also question US firmness?[37] Perhaps, such questioning is inevitable. At the very least, it persists. It remains a nettlesome issue in US relationships with nuclear-armed, post–Cold War challengers. But should it?

Whether US extended deterrence challenges grew or abated with the Cold War's end remains unclear for two main reasons. First, the answer depends on whether the United States holds a decisive nuclear advantage over its post–Cold War nuclear adversaries. Although growing US offensive and defensive capabilities, against smaller adversary arsenals, better position the United States to limit damage to the US homeland in a nuclear exchange with China or North Korea than with Cold War–era Russia, US policymakers might still refrain from retaliating to a nuclear attack abroad, fearing that the targeted country might launch one or more warheads against a US city. Second, the answer depends on whether US stakes in foreign conflicts—in which the US nuclear "umbrella" is at issue—declined with the Cold War's end. Because US policymakers tend not to view China, North Korea, or Russia (despite its capabilities) as posing a Cold War–level, existential threat to the United States, they might not believe that activating the US deterrent threat—for an ally—is worth the potential cost.

Thus, those who doubt the robustness of extended deterrence presume that US adversaries will question both the US capability to deter a nuclear attack on an ally and the US willingness to accept nuclear retaliation against the US homeland as the price of retaliating with nuclear weapons to any such attack. In this, they assume an implacable US nemesis, an adversary that will take high risks or absorb high costs. They conclude—too easily, then—that the US retaliatory threat is not up to requirements.

But is that conclusion based on reasonable assumptions about the adversary's intentions? After all, skeptics assume that extended deterrence will fail with an attack on a US ally that spared the United States. Such an attack strongly suggests, however, that the adversary prefers to keep the United States out of the fight. Indeed, if either China or North Korea does not view the United States as an essential party to the disputes with Taiwan and South Korea, respectively, they might not risk a nuclear attack, whatever the target, that might pull US nuclear forces into the conflict.

To be sure, US adversaries might still question the US willingness to intervene given the adversary's retaliatory threat. However, the robustness of nuclear deterrence lies not in the probability of a response but in the potential enormity of its cost. That is, the costs that the United States could inflict on a small nuclear power like North Korea *more than compensate* for the doubts North Korea might have about the US deterrence commitment to South Korea. North Korea's "risk," here, is a function of the likelihood of the United States imposing costs and North Korea's willingness to tolerate them. Thus, a lower probability of US action is offset by the higher cost potentially inflicted. Even a very small probability that the United States would retaliate could deter an adversary when the *costs* are prohibitive of "guessing wrong."

In the film *Dirty Harry*, Clint Eastwood's character recognized the favorable mathematics when he pointed a .44 Magnum at a "punk" who thought the gun might be out of bullets. His legendary question—"Do you feel lucky?"—sent a convincing message. The enormity of the potential cost to the target, if the gun was loaded, should weigh far more heavily in the target's calculations than (even strong) suspicions that the gun was empty.

These costs achieve importance because the adversary's risk of a US response is non-negligible. The probability of a US retaliatory response would only increase if large numbers of US troops were to die in the attack or if US leaders were to believe that the attack fundamentally threatened US security. After all, US leaders would need no longer to wonder whether the enemy would use its nuclear weapons for aggressive purposes. They might now believe that they must respond to what amounts to an *existential* threat to US security.

The "Rogue-State" Threat: Are Inferior States Necessarily Insensitive to Cost?

The options available to inferior nuclear states are useful to keep in mind as hawkish analysts trumpet fears of unprovoked nuclear attacks on the United States, or its allies, by small nuclear contenders. By their reckoning, the leaders of North Korea or Iran might attack the United States when capable of landing but perhaps only a single nuclear warhead on US territory. At the very least, these leaders could subject the United States to nuclear blackmail. These concerns drive the case for a US damage-limitation capability. It would presumably allow the United States to engage in preventative strikes to neutralize the nuclear threat. But US policymakers must weigh the benefits from acting on alleged US nuclear capability advantages against the attending risks and costs. Any such assessment must give considerable weight to the unlikelihood that even an allegedly risk-prone adversary would strike out with nuclear weaponry—unless preemptively or in retaliation—given a possible US nuclear attack.

With fears of such a risk-prone or cost-indifferent adversary, however, the term "rogue state" gained currency in the George W. Bush administration. US officials voiced concerns that leaders of such states would pay an unfathomable price to accomplish their destructive goals. The perceived threat from these states grew dramatically in the wake of the 9/11 attacks on the World Trade Center in New York and the Pentagon outside Washington, D.C. The concern was that outlaw states would use their weapons or pass nuclear technology to terrorist groups, whatever the consequences. Preventative action thus became a US strategy of choice. Once US forces excised the Saddam Hussein regime from Iraq in 2003, and NATO countries backed local forces in overthrowing Libya's Muammar Qaddafi in 2011, US nuclear angst came to center on the Iranian and North Korean nuclear programs.[38]

Although Iran's nuclear goals remained suspect (even when limited by negotiated agreement in the Obama administration), North Korea's leader, Kim Jong Un, emerged as the poster boy for those who feared nuclear-armed and irrational state leaders. Reclusive, unfashionable, power hungry, and sadistic, he seemed perfectly cast for the role of a leader with irrational or, at least, indecipherable goals. Either presented a grave challenge to US security: "If you do not know what he wants, how can you deter him?" Often lost, however, in the accompanying policy debate is that, by all accounts, the North Korean leader seeks to *hold on to power*, and that (like his predecessors) he employs means, though brutal, that serve that objective. Nothing in the history of a Saddam Hussein, Muammar Qaddafi, or Kim Jong Un suggests that they would concede power they so ardently sought, savored, and defended by opening their countries to US nuclear retaliation.[39]

True, roguish leaders might try to have it both ways, escaping justice by using terror groups surreptitiously to do their bidding. But that would require that these leaders do what they seem incapable of doing—surrendering their power (and ability to control their own fate) to another group. Yes, North Korea repeatedly—even recklessly—escalated the conflict with the West, in pursuit of its nuclear ambitions. Yet, despite North Korean bluster, missile launches, and nuclear tests, and nonnuclear military options, the 1953 armistice agreement that ended the Korean War has held, now, for almost seven decades. That fact alone suggests that North Korea's leaders do not pursue their goals at any, and all, cost—rational or not. Even irrational people do not typically place their lives at risk. In fact, irrational people often unjustifiably fear self-injury or death. Rather than wandering out into traffic, their homes become fortresses to protect against the real or imagined dangers of the world.

Certainly, conditions can reach the point where a power-hungry party willingly commits suicide. Hitler ultimately did. His grand vision in shatters and his ignominious fate secure, exiting the world on his own terms seemed to him the viable alternative. Yet Hitler was a true exception. Even Saddam Hussein and Muammar Qaddafi did not choose death over dishonor and a loss in privilege. They preferred a life on the run to capture, and death. Hussein was found living in a cramped, covered hole in the ground, and Qaddafi was flushed out of a drainage pipe. When captured, Hussein fell back on his claim to power. He announced to his captors, as he emerged from the hole, dirty and disheveled, "I am Saddam Hussein the president of Iraq and I am willing to negotiate."[40] Qaddafi pleaded for his life, appealing unsuccessfully to the humanity of his captives. Both former leaders did what they thought they must to survive.

The point is simple: the exception—no matter how glaring—is not the rule. Indeed, by most accounts—including its own—the North Korean leadership looks to nuclear weapons to ensure regime survival. These weapons potentially provide an effective "veto" against foreign-imposed regime change. On this score, we should remember that Qaddafi gave up his veto, when he transparently conceded his WMD stocks and production capabilities. His effort to normalize relations with the West—and avoid the fate of the Saddam Hussein regime—ironically paved the way for the Western-supported operation that overthrew him. What lesson do you think that the world's ruthless leaders learned when they saw the widely circulated videos of Qaddafi and Saddam Hussein, in their last earthly moments?

Even those who make much of the adversary's irrationality sometimes engage in reasoning that belies their claims. That was certainly true when the Trump administration proclaimed that it would not accept a North Korean capability to deliver a nuclear warhead against the United States. In stating repeatedly that "all options are on the table," the administration indicated it preferred war—perhaps, a disarming nuclear strike against North Korea—rather than open US security to the whims, impulses, and drives of North Korea's mercurial leadership. Indeed, the National Security Council, under the direction of National Security Advisor H. R. McMaster considered a "bloody nose" strategy, among the options.[41] The idea was to inflict pain on critical targets to send the clear message that the United States would not allow North Korea to obtain intercontinental nuclear-delivery status.[42]

The proposed strategy failed a logical consistency test. The administration "constructed" a foe that was selectively irrational, that is, irrational only when it served the administration's policies.[43] It supposed that *irrational* North Korea's leaders would undertake a nuclear-suicide mission against the United States but would *rationally* back down with a US show of force. Would a North Korean leader, seeking—despite the cost—to inflict destructive damage on the United States, change his plans due to a bloody nose? The (implied) affirmative response certainly invited skepticism. Would a North Korean leadership dedicated to US destruction, for its own sake, seem more inclined, instead, to up the ante—maybe attack a neighbor—than to concede to the US threat?[44]

We gain much, then, by placing the "rogue-state" threat in historical perspective. We should remember that the aspirations and actions of the Russian and Chinese governments gave rise to the same overriding concerns. US policymakers feared what these governments might do: first, if acquiring a nuclear-delivery capability; then, in the case of the Soviets, if their internal power of control were to be threatened. They now voice these same concerns, again, in assuming that we cannot rely on the sound judgment of US Cold War–era adversaries. We should not forget "in 1964, when the PRC tested its first nuclear device, China was perhaps the most 'rogue' state in modern history" (Gavin 2009/10: 15).

Conclusions

By various criteria, the United States possesses nuclear capability advantages over all nuclear contenders. But the convenient standards that scholars and strategists employ to determine superiority, much like Cold War–era standards of advantage, rely on specious logic—here, presented in the form of recurrent perils and pitfalls.

We should question the utility of analyses that rely on measures derived from simple math, ignoring the targets that warheads will strike, their effectiveness and purpose, and the complexities of assessing outcomes before and after a nuclear exchange. Yet even the most rigorous analysis suffers if it downplays the potential retaliatory destructiveness of small nuclear arsenals and the reticence of ostensibly "superior" powers to employ nuclear force.

> **3.1:** *Simple math is insufficient for assessing relative nuclear advantages given the complexities of determining outcomes, with different force sizes, usages, and capabilities, over the course of a nuclear conflict.*
>
> **3.2:** *A focus on relative capacity to inflict damage ignores the relative cost acceptance and shared aversions of the conflicting parties.*

Some contemporary scholars flirt with the notion that the United States possesses a "damage-limitation" capability, against some competitors. That capability exists clearly if the United States can assuredly destroy an adversary's nuclear forces without fear of retaliation. That capability overrides the constraints of *mutual* deterrence, however, only should US decisionmakers believe the United States (and its threatened allies) are immunized fully from nuclear retaliation.

> **3.3:** *Superiority, primacy, and other such terms override deterrence constraints only if the advantaged party believes it could fully disarm its opponent with assuredness.*

Apart from deterrence concerns, the historical record of the Cold War and post–Cold War period reveals great reluctance by nuclear-armed states to employ nuclear weapons coercively against adversaries, despite their nuclear holdings or the conditions of conflict. The United States, for one, has consistently acted as if its nuclear arsenal conferred no definitive military advantage. Moreover, the record provides scant evidence that the targets of conventional US military action—North Korea, Vietnam, Iraq, and Afghanistan—acted out of concern that the United States would use nuclear weapons. Adversaries have initiated conflict and, in some cases, pushed the US military to its limits without fearing the nuclear consequences.

> **3.4:** *The historical record is unconvincing that alleged US nuclear advantages gave the United States a coercive advantage over US nuclear or nonnuclear adversaries.*

Evidence abounds that the nonuse of nuclear weapons has rendered nuclear weapons unusable, except under extreme conditions. That the United States and its key competitors continue to modernize their nuclear arsenal, of course, suggests that existing prohibitions fall short of a taboo. But even a "tradition" of nonuse, which invokes fears of unknown effects with the breaching of the nuclear threshold in conflict, significantly constrains the "casual" use of nuclear weapons.

> **3.5:** *US administrations, through their rhetoric, action, and inaction, have strengthened a "tradition" of nuclear nonuse which constrains behavior through fear of bucking the tradition.*

The good news, then, is that nuclear deterrence is more robust than hawkish critics appreciate. China did not seek to match US nuclear forces, in numbers and capabilities, in that the country's "inferior" capabilities nonetheless provided necessary deterrence benefits. Even North Korea has sought to avoid the high costs of war in seeking nuclear capability sufficient to deter a US nuclear attack and US efforts to impose regime change.

> **3.6:** *States with relatively small nuclear arsenals can still obtain deterrence benefits from even the small chance that they could explode a single warhead on the adversary's territory.*
>
> **3.7:** *The practical impact of nuclear "superiority" is impugned by evidence that rivals settle for ostensible inferiority.*
>
> **3.8:** *The enormous risk to a US adversary should it use nuclear weapons against a US ally weigh decisively in the adversary's cost-benefit analysis.*
>
> **3.9:** *The behavior, and incentives, of even "rogue-state" leaders suggest that they reject the potentially devastating costs of attacking the United States or its allies with nuclear weapons.*

The bad news, nevertheless, is that the dangers in military confrontations between nuclear-armed states are great. Deterrence is fragile if the "superior" party considers shedding its reserve to capitalize on its alleged capability advantages or if the "inferior" party contemplates striking first, fearing that a better-armed opponent will do just that. In other words, deterrence will likely prevail among nuclear-armed states because the parties' intentions serve as a stabilizing influence. But threats, and counterthreats, can spark harmful conflict dynamics. The parties might, then, act out of fear or temptation, with consequences they both would otherwise have sought to avoid.

PART II

COERCIVE TACTICS

Boosting Credibility to Signal a US Willingness to Act on the US "Nuclear Advantage"

How might US leaders seek to compensate should adversaries doubt the fact, or utility, of a US nuclear advantage? Put differently, how might these leaders convince a disbelieving opponent that the advantages are "real" or, at least, that US leaders *believe* they are real and might use nuclear force to avoid the costs of inaction? A plausible answer: They can boost the credibility of the US threat to employ nuclear force. Through appropriate tactics, US leaders can supposedly convince adversaries (and allies) that the United States will follow through on its threats because US nuclear advantages increase the US payoff from employing nuclear force.

Credibility became central to US Cold War–era strategic thinking with the growth in Soviet nuclear retaliatory capability. The United States had to convince the Soviets that, should it transgress with nuclear force, the United States had both the "will" and the "capability" to respond. Credibility is potentially fragile, however, because it rests on perceptions. That is, the relationship between credibility and power is like that of credit to cash. Our creditworthiness reflects our liquid and material assets (our cash, stocks, bonds, business holdings, and property) but also various intangibles, such as our payment history, earning potential, general reputation, and stake in our community. A state's credibility in international politics arguably owes to similar factors, which include a state's *perceived* adherence to commitments, relative risk acceptance, apparent resolve, and reputation for decisive action, as discussed in the following chapters.

Deterrence analysts tend to focus, however, on how "defenders" seek to boost their credibility, not how "challengers" process and interpret the defender's actions. In much of the relevant prescriptive literature, the "challenger's" expectations—as shaped deleteriously by efforts of the defender to boost its credibility—fail to receive due consideration.

4

Commitment

So how might US leaders boost their "credibility"—and the credibility of US nuclear threats—for coercive effect? Following works in international politics theory, and a long-held "wisdom" of policy practitioners, leaders can stake their turf with a firm and visible "commitment" to uphold the status quo if challenged.

The term "commitment" is sometimes used to suggest a willingness to act on a promise or threat, that is, to persevere or reveal intensity of effort.[1] So understood, the term can take the form of a verb—as in "I am committed"—to describe a strong expression of intent. Commitment is routinely employed in the deterrence literature, however, strictly as a noun—as in "I made a commitment." There, it highlights whether a defender communicates, with clarity, that threats to certain interests will prove costly to a challenger.

Through word and deed, leaders can thereby bolster their credibility to act, that is, their *apparent* capacity and willingness to meet potential challenges. A well-designed commitment—words or deeds—will presumably leave opponents with no reason for doubt, or experimentation. Neither words nor deeds are necessarily better than the other for that purpose: both involve risks that are frequently downplayed or ignored. Policymakers *might* reduce these risks with ambiguous commitments, which arguably bring risks of their own.

The Value of Explicit Commitments

Commitments are strengthened through word and deed. Yet which more than the other? Verbal and physical commitments each offer distinct benefits.

Verbal Commitments

Words allow a sender to convey messages quickly to deflect potential challenges. Contrary to a prevailing wisdom, however, they are not always cheap; and they can work effectively in the form of a verbal "red line."

The False Promise of Superiority. James H. Lebovic, Oxford University Press.
DOI: 10.1093/oso/9780197680865.003.0004

Are Words Necessarily Cheap?

Words have advantages. Making a speech—or sending out a "tweet"—is simply quicker and easier than sending an aircraft carrier halfway around the world to deliver a warning. Words also permit the sender to respond to unanticipated circumstances without necessarily binding the government by investing resources in a particular course of action. Yet leaders can still make firm verbal commitments. They can make a speech, or issue a statement, to set the stage for concrete policy steps to follow, whether economic sanctions, the withdrawal from a treaty, or the deployment or use of military forces.

Words here permit leaders to commit to a *general* course—as when the public is demanding action or agencies of government, or allies and adversaries abroad, desire signals to guide their own policies—though the policy specifics remain elusive or contentious. Soon after the 9/11 attacks, President Bush delivered a powerful speech outlining the US "global war on terrorism" before Congress. The speech was necessary because the United States had no standing plan to address that kind of threat yet recognized the imperative of committing the United States to a decisive, broad-based response to an unprecedented attack against civilians on US territory.

Of course, if saying something is *too* easy, a challenger might doubt that the defender possesses the will or the capability to back their talk with action. Likewise, a challenger might dismiss words if they seem no more than pointless bluster. A case in point: Donald Trump's musing to reporters in 2019, before a meeting with Pakistan's Prime Minister, that "I have plans on Afghanistan that if I wanted to win that war, Afghanistan would be wiped off the face of the earth, it would be gone, it would be over in literally ten days."[2] Should the challenger dismiss such talk as "cheap" (easy to say, but unlikely to do), the defender would be at a disadvantage.

But we should not assume that actions speak louder than words in the making of effective commitments. A second case in point: Israeli Prime Minister Benjamin Netanyahu's speech, in 2012, before the UN General Assembly. Netanyahu punctuated his address by drawing a literal red line on a cartoon graphic of a bomb to underscore his point: Israel would not permit the Iranian nuclear program to progress to some final stage of readiness.[3]

If leaders say something clearly, or often enough, they can create impactful realities. They can alter expectations, tie their reputations to following through on those commitments, and convince themselves, other leaders, and the broader public that living up to those commitments is the *only* course (even the "right" thing to do). These possibilities weigh against the alleged "cheapness" of a defender's talk about its commitment. Threats are not entirely cheap if, through firmness or repetition, they increase the risks for a challenger. That was the philosophy behind consistent public deference to member Article 5 obligations under the North Atlantic Treaty Alliance: an "armed attack against one" member "shall be considered an attack against them all." The less cited portion of the same article, however, reads as follows. Members are obligated to "assist the Party or Parties so attacked by taking forthwith, individually and in concert with the other Parties, such action as it deems necessary, including the use of armed force, to restore and maintain the security of the North Atlantic area." Obviously, "as it deems necessary" allows members considerable wiggle room to decide their course of action. For that reason, US leaders constantly reiterated the US pledge to stand by NATO allies if attacked. This is also why Donald Trump's failure to restate the US Article 5 commitment at the May 2017 NATO meeting in Brussels was significant, and why Trump's belated commitment to the principle in a public statement the following month fell short.[4] Other members wanted more than a reluctant general commitment; they wanted to hear *words* that spoke to a strong and unerring—indeed, automatic—commitment.

US allies recognized, obviously, that words tip priorities and precede plans and action. To be sure, US policymakers have backed their NATO pronouncements by sustaining a permanent alliance structure, holding frequent meetings of high- and low-level officials, stationing troops and deploying equipment in Europe, maintaining forward military bases, conducting joint military exercises, and deploying nuclear weapons in theater to blunt a potential Russian attack. Yet the "wrong" words, here, suggested to US allies and adversaries that the US commitment to NATO had softened.

Identifying Red Lines

The term "red line" is of relatively recent vintage: its relevant use dates back less than a half-century (Tetrais 2014: 7). It nevertheless became a go-to term in the coercive diplomatic vocabulary for delineating the bounds of acceptable behavior. Through word (and action), a defender can set a red line, which—when crossed by a challenger—will trigger potential consequences.

Red lines are possibly advantageous in removing a challenger's residual uncertainty. A potential challenger might test an uncertain commitment rather than forgo lucrative opportunities for gain. The challenger might guess wrong in doubting the strength of the commitment, resulting in conflict. Or the challenger might guess right, and the defender might concede ground, with a whimper, not a fight. A successful probe, of that sort, might embolden the challenger. Heartened by its gains, it might push forward—there, or elsewhere—at the risk of stumbling into a major conflict. For just that reason, the defender might hold firm. It might fear that, by failing to rise to the challenge, challengers might doubt the commitment of the defender to protect its more vital interests. Contemporary evidence might support such concerns. Historians will likely debate whether US decisions to pull out of the Syrian and Afghanistan conflicts helped convince Russian President Vladimir Putin that he could subsequently send his military forces into Ukraine, in 2022, uncontested by the United States.

By establishing a red line—bright and shiny—a party leaves no doubt, however, where its interests lie. The underlying philosophy is simple: the clearer, the better. Effective red lines are not arbitrary, imprecise, incomplete, or unverifiable. Instead, they are salient and easily identified. Crossing them will yield uncontestable signs of a violation, that is, visible and indisputable evidence of a transgression. Red lines thus proscribe surreptitious or circuitous crossings. Challengers cannot capitalize on a critical omission, or ambiguity, that gives wiggle room to offending behavior that can undercut the prohibition. (On these characteristics, see Altman and Miller, 2017; and Altman, 2021.)

Effective red lines draw, then, from both the material and nonmaterial worlds. They can rely on "focal points," that is, socially constructed or palpable environmental features around which the expectations and beliefs of the competing parties converge. Viewed accordingly, the notion of a red line fits comfortably with Schelling's (1960: 67–77) work on tacit bargaining. Schelling addressed the challenge that ensues in bargaining from the absence of a mathematical (equilibrium) solution. Lacking a payoff offering the best return to each party, in light of the other party's strategy, the parties take cues from a "lumpy" reality: seemingly obvious or conspicuous attributes of the social or material world that beckon, as meaningful, significant, and maybe even equitable solutions, to both parties. These features include geographical attributes such as rivers and mountain ranges that seem to constitute a "natural" boundary; a meaningful temporal separation, such as the beginning of the new year and then at the stroke of midnight; a qualitative rather than quantitative distinction; splitting the difference between bargaining positions, rather than a 51/49 solution (even if a cut down the middle advantages one party more than the other); or an absolute prohibition rather than exclusions that allow for exceptions. Indeed, as we shall see (in Chapter 8), a challenge in combating nuclear proliferation is disagreement over permissible and impermissible nuclear activities given the absence of a clear threshold that signals that a country constitutes a proliferation threat.

Physical Commitments

Leaders can reinforce the status quo by saying they will defend it. But, by "saying," not "doing," they might fuel suspicions that the commitment, or the capability to defend it, is weak.

Indeed, talking is potentially counterproductive. Governments generally understand, for instance, that territorial land grabs violate international law and will encounter forcible resistance. Thus, stating a commitment to use force in defense of one's territory is hardly necessary. If anything, it could weaken the apparent commitment. By acknowledging illegitimate claims, the defender gives these claims strength and credibility: it concedes that these claims are worthy of discussion and standing as issues. If the challenger has no chance of succeeding or no legitimate claim, there is no issue—and no reason for a verbal commitment.

Talking also sacrifices "plausible deniability." A US president can send an aircraft carrier to hostile shores as a warning to a foreign leader while preserving the ruse that the fleet was sent to handle a potential emergency or any number of possible contingencies. If its purpose remains inexplicit, the foreign leader is not placed on the spot, pressed to respond forcefully to save face. The impact of the presidential action is felt nonetheless, for the foreign leader likely knows why those forces are there. Likewise, the president is not trapped by the commitment (as discussed below): the president can preserve the option of redeploying the force, should it fail to achieve its intended objective. Of course, a firm presidential statement, tying the force to an explicit demand, could add heft to a military demonstration. But, for the reasons above, a president might demure: Words, here, might prove *costly*.

Spelling out a commitment—making it explicit when potential challengers understand it in its implicit form—can thus make matters worse. Israeli leaders certainly thought so, in adopting a "bomb in the basement" strategy. They have not publicly admitted that Israel possesses nuclear weapons, much less hundreds of warheads and a robust delivery capability, to act on the country's commitment to address existential threats. Israel's possession of these weapons has spurred acrimony, with accusations of Western hypocrisy for giving Israel a proliferation "pass" and refusing to concede similar "nuclear rights" to Iran. The question, though, is whether—despite the open Israeli "secret"—political conditions would improve if Israel declared its nuclear holdings. Likely not.

Deeds hold advantages over words beyond the deficiencies of articulating a commitment. By acting, instead, the defender can strengthen its position in various ways.

First, the defender can increase its stakes, reinforcing the importance of the relevant interests. Despite its dependence on nuclear forces, the Eisenhower administration placed US conventional forces on the front lines of a potential East–West conflict in Europe. There, they would serve as a "tripwire," meant to reinforce the political significance of those borders. These troops were too few to hold the line militarily; they were perfectly capable of fighting, *and dying*, in the initial Soviet assault to impose "private costs" on the United States. A Soviet attack on Western Europe would certainly violate international law and threaten the security of Europe. But now it would also kill Americans—which would undoubtedly place the United States—politically and emotionally—squarely in the conflict.

Second, the defender can increase the costs to those who challenge the status quo. A challenger cannot suffer the illusion that it can escape without consequences or easily accomplish its objectives. *Routine* actions that reinforce the status quo speak loudly in this context. By patrolling its borders, or keeping forces on alert, a defending country signals that its commitment is "real"—it will back up its implicit threat to defend national territory. The accompanying NATO message, then, to invading Soviet troops was that *they* would suffer a toll should they attempt to seize Western European territory.

Third, the defender can tighten the link between the challenge and retaliation, perhaps by making retaliation virtually automatic. Automaticity serves to "tie the hands" of a leader (Fearon 1997), strengthening deterrence by making a strong response to challenges more likely. The undersized NATO force—its potential inability to blunt a full-scale Soviet conventional attack—was the NATO message. The conflict would "go nuclear" sooner, not later. Any such message gains strength when the act of signaling also reduces obstacles to acting on the threat. These obstacles include the costs of responding to the threat. By mobilizing forces and closing the distance to adversary troops, the defender "sinks costs" (Fearon 1997). With forces in place, the defender can ease its physical burden—and respond more rapidly—should the adversary attack. For just that reason, counterinsurgency opponents in the traditional US army tried to limit its unconventional capabilities. They feared that, *because* it possessed these capabilities, the service would be required to engage in future Iraq- or Afghanistan-style nation-building operations.[5]

Fourth, the defender can increase its *own* costs should it seek *to avoid* conflict. By placing the country on a war footing, leaders who fail to rise to a challenge risk charges that they exaggerated the threat, backed down in the face of threat, or wasted an opportunity to take decisive action. Such thinking informs the scholarly identification of "audience costs" (Fearon 1994). The underlying principle is simple. In arousing (public) sentiments and creating expectations, leaders essentially paint themselves into a corner. Commitments, made publicly, thus assume an importance beyond the interests (and related cost-benefit analyses) that undergird them. Astute leaders can manipulate these costs to their advantage by making commitments that limit their own negotiating flexibility. Presumably, an adversary will recognize the limited bargaining space for what it is—and compromise, to make a deal.[6]

Whether these leaders sought only to engineer conditions to obtain a bargaining advantage, they must still live with the positive or *negative* consequences of their strategy. Indeed, *that* is the point of the bargaining strategy. Barack Obama's failure to follow through on his red line in Syria—when, in 2012, he warned its government not to employ chemical weapons—backhandedly makes the point. He opened himself to charges that he undercut US credibility in Syria, and beyond, when he failed to respond forcefully, then, to evidence that the Syrian government had used chemical weapons. Critics could ask, "Will allies and adversaries know the difference, then, between a viable and a nonviable US commitment?"

Of course, a party might justify its failure to act by pointing to unique or unusual circumstances that militated against acting. But will others buy that explanation? Although Obama received a Russian-backed pledge from the Syrian government to allow the removal and destruction of Syrian chemical weapons stocks, the concession did little to quell hawkish criticism.

Fifth, the defender can hold its ground to prevent the challenger from seizing a coercive advantage. Through an affront or assault, the challenger might otherwise establish that it is willing and capable of challenging the defender's commitment. The challenger might also gain ground physically, which means that returning to the status quo ante—as the defender will demand—would require an open concession. The "defender" would now encounter a difficult compellence problem (as discussed below). The challenger might even hope that its actions will create a "new normal" against which further transgressions are compared, and discounted. Each additional transgression—think Russia's 2014 assault on Crimea or China increasing its air and naval activities closer to Taiwan—will lack the urgency of the first. That Hitler violated the Versailles Treaty by militarizing the Rhineland, without resistance from other treaty parties, hurt their position to deflect further challenges. That is the strategy behind "salami slicing"—nibbling away at an understanding, proposal, or agreement, little bits at a time. Once the United States tacitly accepted North Korea's standing as a nuclear power, able to deliver a weapon *somehow* against a target (e.g., by ship), it could not easily establish red lines (including the North Korean acquisition of a nuclear intercontinental delivery capability) that, from the US perspective, constituted an unacceptable development.

Thus, physical commitments offer versatile and potentially effective means for strengthening the defender's commitments. Through its actions, then, a defender can boost the credibility of its threat to employ force, when words might prove provocative or cheap.

The Liabilities of Explicit Commitments

Despite the reputed benefits, both verbal and physical commitments have an unpleasant downside. They are often ineffective, even counterproductive; they are permissive by implication; and they can bind the committing party, to its eventual regret.

The potency of red lines—and firm commitments more generally—is frequently exaggerated. True, observers often blame diplomatic failure on perceived softness or weakness. Whereas Donald Trump, as president, appeared to blame every US challenge in the world on the weakness of his predecessor(s), and was hardly a pillar of strength in his dealings with US adversaries, Trump is not alone in highlighting the virtues of strength and determination. Observers often attribute the onset of the Cuban Missile Crisis, for example, to Khrushchev's perception (in their early meeting in Vienna) that the young John Kennedy was weak. Conversely, they frequently attribute Kennedy's success in the crisis to the president's willingness to stand up to Khrushchev. We now know, of course, that Kennedy's response in the crisis did not depend entirely on tough talk and action. Kennedy offered Khrushchev concrete concessions in return for pulling Soviet missiles from Cuba.

He also took measures to head off a military confrontation (as discussed in later chapters).

The mythology surrounding the benefits of "brinkmanship" persists. But even seemingly firm commitments often prove to be deficient signaling devices, for various reasons: (a) challenges are often unanticipated; (b) challengers view commitments conditionally; (c) policymakers speak without clarity and precision; (d) commitments are lost to contradictory signals; (e) challengers might view commitments as provocations; (f) commitments can foreclose acceptable exits; (g) commitments can trap the defender; (h) commitments can imply concessions; and (i) commitments can require challengers to make unacceptable concessions.

Challenges Are Often Unanticipated

Challengers and defenders often share different understandings of a situation. If realists are correct that the intentions of states are opaque, states cannot always know how and where their interests will be challenged. How often is a foreign leader's saber-rattling—preceding a military offensive—dismissed initially as bluff and bluster, or as theatrics meant to assuage criticism or bolster support at home? We should remember the dismissive reaction of the Israeli leadership to Egyptian military activities in the prelude to the 1973 Middle East War (Shlaim 1976), and Ukrainian President Volodymyr Zelensky's concerns that President Biden was overhyping the war threat in early 2022 with his claims that Russian forces were about to invade Ukraine.[7] If leaders can misjudge the actions of a key adversary, why should we suppose that US leaders perform better in addressing a vast array of potential threats on multiple global fronts? They might understand the extent of their commitments, and their limitations, then, only when these commitments are challenged.

Iraq's 1990 invasion of Kuwait serves as a prominent case in point. April Glaspie, then US ambassador to Iraq, is often vilified for her supposedly weak message to Saddam Hussein concerning his dispute with Kuwait—in a 1990 meeting, shortly before he seized control of that country. The Iraqi president sought the meeting, looking presumably for evidence that the United States would not derail his plans. He found evidence in Glaspie's (subsequently maligned) response: "we have no opinion on the Arab–Arab conflicts, like your border disagreement with Kuwait" (quoted in Stein 1992: 152). Perhaps Glaspie was insufficiently suspicious of Saddam's motives in seeking a meeting with the ambassador. After all, Saddam's designs on Kuwait were long-standing (he would soon claim it as Iraq's 19th province). Only in hindsight can we surmise, however, that she gave Saddam a green light—*from his perspective*—to move the Iraqi border to Saudi Arabia. Glaspie, like most of the US policy establishment, was blindsided by the Iraqi offensive.

But we need not look back that far in history. US policymakers did not respond to the Russian occupation of Crimea, in 2014, anticipating future Russian efforts to consume all of Ukraine.

Challengers View Commitments Conditionally

Defenders make commitments based on an understanding of underlying conditions that challengers might not share. Thus, challengers might dismiss claims that a foreign "audience" will rally for reasons of necessity, honor, or revenge, in response to a challenge, believing perhaps that any such response will come too late to matter (on this, see Reiter and Poast 2021). Note, for example, that Vladimir Putin supposed (falsely), it seems, that NATO countries would divide in response to his invasion of Ukraine (as discussed in Chapter 8).

Indeed, a challenger might expect that, over time, the defender would downgrade the interests behind the initial commitment as the defender came to understand (and experience) the actual costs of a military burden. Dan Reiter and Allan Stam (2002) regard that as a proclivity of democracies, like the United States. The Johnson administration committed early to uphold the sovereignty of South Vietnam. Changing US stakes and the rising political and material costs of fighting the Vietnam War nonetheless weakened the US commitment. Likewise, the Reagan and Clinton administrations exited Lebanon (1983) and Somalia (1994), respectively, when the going got tough for US forces in these countries.[8] Indeed, the lengthy—but eventually complete—US withdrawal from Afghanistan in 2021 hauntingly echoes Vietnam in its final chapter. As in Vietnam, the United States exited the conflict hardly optimistic that the host government could carry the burden (Lebovic 2019).

Although US adversaries might not read these episodes to impugn US steadfastness (see Chapter 6), some anecdotal evidence suggests they *might*. In 1991, Saddam Hussein—drawing from the last two cases—figured that history would repeat itself, and that time was his ally in a military confrontation with the United States. He knew the toll that Iraqi forces had withstood, and exacted in turn on Iranian forces, in their lengthy war of the prior decade. He did not believe that the United States would tolerate bloodletting on that scale.[9]

Policymakers Speak without Clarity and Precision

Clarity and precision are often casualties when US leaders articulate a commitment. The reasons are well appreciated: policymakers speak extemporaneously; they opt for generalities to preserve their policy options; and they respond to events based on current information and current understandings of a situation.

For critics, at least, President Obama offered a tutorial, then, on how *not* to lay down a red line when he warned the Syrian government, in 2012, about its use of chemical weapons in the ongoing Syrian civil war. Obama's awkwardly phrased red line left much to the imagination, and potential adversary exploitation: when "we start seeing a whole bunch of chemical weapons moving around or being utilized . . . if we start seeing movement on the chemical weapons front or the use of chemical weapons." The result "would be enormous consequences."

Terms like "whole bunch," "movement," "enormous consequences," and even "chemical weapons" were strikingly open-ended. In chemical weapons units, what constitutes a "bunch"? Then, what constitutes weapons "use"? Will the administration judge it by Syrian actions (the quantities involved) or by their effects (the innocence or numbers of victims)? Does movement mean preparation for use, or physical movement—indeed, does the prohibition apply to chlorine given its dual military-civilian use? Moreover, if the United States were to conclude that Syria had crossed the red line, who would right the wrong, and who would pay the price? The Syrian government or military? Would the response involve denying the Syrian military its chemical-weapons capability, punishing the Syrian government to change its cost-benefit calculus, or—at the far end—the decapitation of the Syrian government? For that matter, would it produce any reaction at all? Obama promised that it "would change my calculus; that would change my equation." Even with a chemical attack, then, his math might not support US intervention. Such lack in clarity left the Syrian government with significant wiggle room. When the government took actions that appeared to challenge the red line, it might have correctly understood the administration's lack of commitment to its words; but it might also have misunderstood the prohibition.[10]

US leaders are not uniquely culpable in this regard. That was apparent when Russian President Vladimir Putin pushed back, in 2021, against increasing NATO support (military missions, exercises, aid, training, overflights, and naval operations) for nations bordering Russia that sought to defend against its territorial encroachments. In Putin's words, "We regularly express our concerns about this, we talk about red lines, but of course we understand that our partners are very idiosyncratic and—how to put it mildly—they treat all our warnings and conversations about red lines very superficially."[11] Undoubtedly, these words were given punch and urgency by Russia's military maneuvers which included force buildups, unusual troop movements, and drills along the perimeter of Ukraine.[12] Yet red lines require loudness and clarity for impact. They fade and invite challenges, by comparison, when communicated "mildly," "regularly" in response to consistent encroachments, and deferentially in acknowledging the challenger's understanding (or "misunderstanding") of the situation.

Given Russia's eventual invasion of Ukraine, we might conclude that Putin's red line existed. If so, however, Putin arguably buried it by feeding NATO's misunderstanding with his soft, circuitous, and equivocal articulation—indeed, its shifting location (as discussed below) in the prelude to the invasion.[13]

Commitments Are Lost to Contradictory Signals

Although leaders imagine they speak for the nation, the voice of any single individual—even a US president—competes with a cacophony of intended and unintended signals. Then, outsiders might not assume that the leader's position will triumph in internal debate—or, as regards the Trump administration, that the president's voice is the most authoritative on matters of policy. How others will interpret words is hard to predict given the influence on policy of numerous interacting variables. Whether and how words matter depends specifically on whether they are (a) uttered by concordant voices, outnumbering the discordant ones; (b) backed by actual political influence; (c) consistent with political realities or principles long observed by prior administrations; (d) voiced privately but not publicly, or publicly but not privately; (e) reinforced by noticeable actions; (f) believed to apply only under restrictive conditions or in a specific context; and (g) contradicted by the same speaker on other occasions, whatever the intent.

To be sure, government leaders make much of the need to send clear messages. Yet they spend too little time thinking about whether the intended signal will stand above the noise or other statements and actions by those same leaders. We do not need to dig deep into the Cold War to detect these signaling problems. Donald Trump, in accusing Obama of a weak response to Syria's chemical weapons use, implicitly established his own red line in 2017, when the United States employed cruise missiles to crater the runways of the airbase from which the Syrian military had launched a sarin attack that killed scores of civilians. Should the Bashar al-Assad regime have been impressed that the United States imposed costs on that government for having flagrantly violated the Chemical Weapons Treaty and committed crimes against humanity? Or, instead, that US retaliation was *limited* to facilities that the United States tied directly to the attack—and then that the damage was reparable in relatively short order?

The mixed messaging continued. When the administration joined Britain and France in retaliating for a second chemical attack in April 2018, the damage was more severe and widespread. Nevertheless, might the Assad regime have expected a more punishing blow—that targeted more than Syrian chemical weapons capabilities—considering that it crossed the implicit red line, set by the 2017 retaliatory attack? Certainly, Trump diminished the persuasive weight of the second retaliatory strike by failing to respond, the same year, to over half a dozen prior Syrian chemical weapons attacks and by elevating expectations. He announced that Syria will pay a "big price" for the attack and (via Twitter) that "nice and new and 'smart' " missiles would be coming soon to Syria.[14] For that matter, did he not undermine the intended message entirely by his earlier announcement at a campaign rally that "We'll be coming out of Syria, like, very soon," his order to the military to begin preparations for an exit, and his freezing of hundreds of millions in dollars earmarked for stabilization activities in the country? All told, might his signal of "strength" have simply reinforced the message that the Trump administration, with an ISIS defeat, no longer considered this "our war" and wanted to extricate the United States from the fighting?[15] By pronouncing "mission accomplished!" (via Tweet, again) with the completion of the retaliatory attack, he might well have punctuated that point.[16]

The challenges of consistent signaling did not end, of course, with the Trump administration. The Biden administration had to navigate these challenges, even before confronting Russia over Ukraine, in sending what amounted to a message that the US red line for responding to Iranian militia attacks in Iraq and Syria had changed subtly from that of the prior administration. Whereas the Trump administration enforced a red line that necessitated a US military response when militia attacks inflicted American casualties, the Biden administration initially responded to such attacks whether or not they inflicted such casualties.[17] In lowering the bar on retaliation, the Biden administration also muddied the nature and location of the red line. The line now stood in some vague middle place between actions that inflicted casualties and any, and all, militia attacks. Questions would inevitably remain concerning what kinds of Iranian actions would bring a US response, and what form it would take. For instance, would the United States respond to Iranian attacks against non-US targets? If so, what would constitute an offending target? That US officials announced the policy shift could strengthen the message by requiring that the United States stand by its stated "commitment." But a firm pronouncement, followed by equivocal or ambiguous action, could do more harm than good for deterrence purposes.

The challenges of signaling obviously increased enormously when the United States confronted Russia over its 2022 invasion of Ukraine. The Biden administration recognized the virtues of drawing clear red lines—both by ruling out direct combat support for the Ukraine military and by reinforcing the US commitment to defend the territorial integrity of any, and all, members of the NATO alliance. The question, however, was whether that seemingly clear red line would fade as NATO sought to boost its indirect support for Ukraine or battlefield conditions changed in Russia's disfavor. Biden understood that establishing a "no-fly-zone" over Ukraine territory would require that NATO fighters fire on intruding Russian helicopters and planes—in clear violation of the stated red line—but, at some point in the conflict, Russia could view NATO supplies of military material or actionable intelligence ("active real-time targeting") to the Ukrainian military as such a violation, more so if either effort placed NATO personnel in Ukraine. For that matter, the red line could fade if confrontations, encroachments, and near misses, on land, in the air, or on the sea, elsewhere in Europe, stretched the definition of "permissible" behavior; if the borders of Belarus and Ukraine became the new front lines of a potential East–West military confrontation; if Russia came to regard supply centers, bases, transit points, or staging areas in adjacent NATO countries as unacceptably complicit in Ukraine's war defense; if Ukraine went on the offensive and started hitting military targets in Russia;[18] or if Russia viewed its exposure to punishing economic sanctions as an attack on Russia or its government and thus an "act of war" (of more gravity than implied by Putin's claim that sanctions are "akin to declaring war").[19]

Other red lines were obscured by the complexities of the issues. Even as NATO members declared that "all options are on the table" should Russia use chemical weapons in Ukraine, they necessarily equivocated in recognizing the liabilities of a broad commitment that made no distinction by type, quantity, or purpose. As British Prime Minister Boris Johnson cautioned, "It's important to recognize that there are all sorts of ways in which these things could be used, from the use of tear gas which is effectively a riot control measure all the way through to utterly devastating lethal chemical weapons systems . . . so I don't think it's helpful to be too binary about the situation because these are highly nuanced."[20] Effective red lines are inherently binary: you either cross them or you don't.

Commitments Can Foreclose Acceptable Exits

In times of conflict, a party might play to its opponent's worst fears by putting the opponent in a "no-win" situation that leaves it "no way out" of the conflict. If a commitment is intended for influence—that is, to get an opponent to change its goals or ways—the party must convince the opponent that it is better off respecting (rather than ignoring or defying) the commitment. If, however, the party promises to deliver the "worst" to an opponent, whatever it chooses to do, the commitment is ineffective or counterproductive. For instance, when in 1998 President Bill Clinton signed the Iraq Liberation Act—elevating regime change in Iraq to US government policy—the United States lost potential leverage over the Iraq government by threatening to impose the "ultimate sanction" on Saddam Hussein—whether or not he opened his country to international inspectors. Attempting to harness available leverage, then, domestic opponents sometimes extend brutal dictators a "golden parachute" (or "golden retirement"), allowing them to flee the country or to avoid prosecution for acts committed while they were in power (Escribà-Folch and Krcmaric 2017: 562).

To his credit, Joe Biden, in his early management of the conflict in Ukraine, seemed conscious of the dangers of leaving Putin without an "off ramp." He initially resisted demands to hold Russian troops—and, by implication, their leader—accountable for "war crimes" in that country. He conceded, when questioned, that Russian troops intentionally targeted civilian areas. But, when asked if they were committing "war crimes," he responded, "We are following it very closely. It's early to say that."[21] He apparently recognized the consequences of squeezing Putin to the point where he had nothing left to lose by continuing his fight. He relented, however, with the mounting Russian-inflicted carnage in Ukraine, calling Putin a "war criminal"[22]—later, a "butcher"[23]—before advocating what amounted to regime change in Russia,[24] and condemning Russian war conduct as "genocide."[25]

Before invading Ukraine, Vladimir Putin showed even less appreciation for the dangers of foreclosing the adversary's options when he issued his set of demands. He insisted on signed assurances—"precise legal, judicial guarantees"—that NATO would not expand to include Ukraine and Georgia but would also restrain its military activities adjacent to Russia.[26] He claimed, in fact, that Russia's red lines were violated by the flying of strategic bombers near Russia,[27] before Russian officials followed with a list of demands: NATO must abandon its military infrastructure in Eastern Europe, NATO must forgo military exercises in Eastern Europe and the former Soviet republics, and the United States must forswear bilateral military cooperation with the former republics. In effect, Russia set a red line that amounted to NATO's acceptance of a Russian "sphere of influence" that encompassed Eastern Europe and the former Soviet republics, denying sovereignty to countries in the region.

Of course, NATO members could justifiably wonder whether Russia had drawn an actual red line or, instead, had publicly committed to demands that NATO would surely reject, as a pretext for a Russian invasion.[28] They would have to act, and prepare, accordingly.

Challengers Might View Commitments as Provocations

The identity of the "challenger" and "defender" is frequently a matter of opinion. Challengers might see themselves as the aggrieved party, especially when the contested interest is a subject of longstanding dispute or is linked somehow to other issues the challenger judges of vital concern.

We would likely agree that stealing another's property is a criminal act. But what if the supplies in a grocery store were the only food source available to the penniless and homeless after some natural catastrophe? If storeowners chose to defend their food supply (or to charge monopoly prices for it), we might well see *them* as the challengers. If we believe that everyone has a right to eat, or that a need to survive trumps all other interests, the immediate defender—here, the grocery owner—has *challenged* a foundational societal principle, leaving the poor in the community to act in their own defense.

It is not far-fetched to think, then, that aggressive leaders would similarly see forces conspiring to deprive them of their "rights." For instance, Russian leaders could easily have viewed the Cuban Missile Crisis as a US attempt to deprive the Soviets of "equal rights." After all, the United States had deployed intermediate-range weapons in NATO countries that could hit targets in Eastern Europe and Russia. Thus, Russia sought a quid pro quo in the form of removing US missiles from Turkey in exchange for removing Russian missiles from Cuba. Three decades later, Saddam Hussein justified his stance vis-à-vis Kuwait, in his meeting with the US ambassador, by denouncing efforts to make his people "relinquish their rights" (quoted in Stein 1992: 178). Likewise, Vladimir Putin's forces invaded Ukraine on various defensive pretenses. They included combating a Nazi regime in Ukraine bent on genocide against Russian ethnic groups in the country, a preventative war to thwart Ukrainian aspirations to acquire nuclear weapons, and countering aggressive NATO moves along the Russian border that jeopardized Russian security. In his words, "We didn't come to the US or UK borders, no, they came to ours."[29] As Russia lost ground in Ukraine, Putin thus unleashed a barrage of invectives: "The purpose of the West is to weaken, divide and ultimately destroy our country"—"cultivating hatred toward Russia" and employing Ukraine as an "anti-Russian beachhead."[30]

Certainly, we can dismiss Putin's justifications as rhetoric meant to sway domestic and foreign audiences. But Putin might still possess a strong sense of grievance, rooted in Russian (or personal) victimhood, whether a result of NATO's geographical encroachments, perfidious Russian leaders who surrendered rightfully Russian territory, or a "cancel culture" that negated and rejected Russian culture, personalities, and achievements.[31]

We thus see how easily ambiguity hampers efforts to distinguish the defending from the challenging party—and thus how easily ostensible challengers might see themselves as targets of illegitimate demands, even extortion or blackmail. From their perspective, the demanding party seeks to deprive them of what is actually "theirs"—a well-deserved reputation, a fair share of something of value, or the means to pursue their legitimate goals (livelihood). They will hold the line to retain what they should rightfully possess.[32] Appreciating this dynamic affords another perspective on Saddam Hussein's fateful decision to hold the line in Kuwait in 1991. He *was the defender*, not the challenger, despite his seizure of that country. As Janice Stein (1992: 148) concludes, "he was unstoppable because of the strategic judgment he made, late in 1989, that the United States was determined to undermine his regime through economic sabotage and covert action. Once he developed a strong image of an enemy bent on his destruction, he was almost immune to any evidence that challenged the conclusion he had drawn." How could he stop? The status quo ante was an untenable place; returning to it now—backing down in the face of threats—could only make his position substantially worse.

Whether or not a party sees itself as the "challenger," it might still read provocations into defensive actions. It can easily exaggerate the decisional discretion available to the opponent; misunderstand the stakes from the opponent's perspective; show far too much concern for its own reputation and too little concern for the opponent's sensitivities; or fail to appreciate that words and actions meant to reassure or convey restraint can threaten an opponent who fears the worst ("we will only attack if . . ."). The potential for untoward reactiveness only increases as a conflict intensifies. The worsening of conflict can increase the parties' fears, stakes, and activity levels, even as it reduces the parties' capability to process information and compresses the time frame for intelligence gathering and processing. Leaders, under stress, might succumb to the various frailties of human decision-making: overconfidence, a failure to recognize trade-offs (the negative consequences of preferred actions), tunnel vision, a tendency to fit new information to existing beliefs (rather than use information to inform beliefs), wishful thinking, avoidance behaviors (as leaders shrink before disquieting intellectual challenges), and dysfunctional group dynamics (by which deference to preserve group harmony induces premature consensus).[33]

Commitments Can Require Challengers to Make Unacceptable Concessions

Commitments, by their nature, vary in effectiveness, and their potential negative impact. Notable, in this regard, is the difference between compellence and deterrence commitments. Whereas deterrence relies on threats to keep a party from doing something, compellence relies on threats to induce a party to do something.

We might think these two concepts offer a distinction without a real difference—another example of academics creating typologies for their own sake. Patrick Morgan (2003: 3) suggests some sympathy for that position when he concludes that perhaps "we should put less emphasis on the distinction between deterrence and compellence and instead treat them as interrelated components of *coercive diplomacy*, the use of force or threat of force by a state (or other actor) to get its own way." Doing so imposes significant analytical costs, however, should compellence present exceptional challenges. Owing enormously to the seminal contribution of Thomas Schelling (1966), international politics scholars widely assume that the challenges of compellence exceed those of deterrence. (See also Snyder and Diesing 1977: 25.) They typically highlight three challenges.

First, targets are reticent to make concessions to compellence demands, which are necessarily "public." That is, it requires the target visibly to concede ground. Consequently, the resulting political costs—in damage to the target's reputation, bargaining power, prestige, and honor—add to the target's material costs from a concession. A deterrence demand, by contrast, offers the target the politically palatable recourse of *not acting* to placate the coercive party: it can pretend, for instance, that it had no intention of acting in the first place.

Second, targets are loath to relinquish what they already possess, as required to satisfy a compellence demand. Prospect theory leads us to conclude that parties will take greater risks, and incur greater costs, to retain possessions—a reputation for firmness, for example—or to recover losses than to obtain new benefits (Levy 1997; McDermott 2001). If compellence demands place a party in the "domain of losses," then, we should expect it to react by accepting these higher costs and risks.

Third, targets more easily appreciate deterrence demands than compellence demands. Deploying forces along a disputed border sends a clear deterrence message, much like a blazing house fire communicates unmistakably that straying close to the flames will cause burns. The message is arguably implicit in the signal. That same fire can certainly send compellence messages: bystanders might move into action, as a precaution, knowing that the fire could spread. But these signals are weaker, for they require assumptions about wind speed, the flammability of adjacent structures, the volatility of the fire, and impediments to controlling the fire should it spread. Bystanders are likely to view these various factors differently; some might flee, others might remain behind, and still others—to defend life or property—might try to douse or contain the fire.

Although obvious analytical advantages accrue, then, from viewing deterrence and compellence as distinct influences, the differences between deterrence and compellence admittedly muddy in practice.

The consequences are not entirely negative. Deterrence and compellence can usefully combine in strategy to soften the negative consequences of a compellence commitment. When the Soviets clamped down on all land traffic into Berlin in 1948 to weaken the Western hold on the city, the Truman administration responded with an around-the-clock airlift of supplies into Berlin. Likewise, the Kennedy administration established a blockade (a "quarantine line") around Cuba to prevent the transfer of military materials to the island. Both administrations successfully turned the tables on the Soviets. They sought to compel a change in Soviet behavior by transforming the compellence—tactically, at least—into a deterrence problem. By establishing a "line" in the skies (Berlin) and on the sea (Cuba), they required that the Soviets "cross" it to continue their actions: to accomplish their initial goals, the Soviets would have to initiate violence by shooting down the planes or confronting US ships at sea.[34] Given the risks of challenging the US "line," and the costs to the Soviets of the status quo, the effect—with additional US assurances (see Chapter 8)—was to *compel* a Soviet retreat. Indeed, deterrence—in the post–Cold War period no less than in the Cold War period—offers opportunities to make compellence work. As we will see (in Chapter 8), deterrence could serve as the leading edge of a compellence strategy intended to get countries like Iran from acquiring nuclear weapons.

Still, a danger remains should policymakers confuse the two and treat a difficult compellence challenge as a deterrence problem. US Cold War–era analysts wrote volumes, for example, on an "intra-war deterrence" challenge. They assumed that the United States could tailor its nuclear attacks to *deter* the Soviets from going "all-in" with their nuclear forces should Soviet forces invade Western Europe. But was this an actual "deterrence" problem? Was the United States not also seeking to *compel* the Russians to renounce their gains by exiting Western Europe?[35]

Conflating the two is likely when (as previously discussed), actors and targets do not possess the same reference points. What the actor defines as a "deterrence" problem, the target might define as a "compellence" challenge. Thus, where a party might view its actions as an "innocent" effort to preserve the status quo, its opponent might view the accompanying threats—"do this or else"—as "blackmail" (Betts 1987: 4). From the Soviet perspective, arguably, the 1948 Berlin Crisis stemmed from a US compellence campaign. US-supported efforts to absorb West Berlin into the Federal Republic of Germany (by integrating the three western sectors of the city and introducing West German currency into the combined zones) meant that the United States, not the Soviet Union, was challenging the status quo. By its actions, the United States had effectively renounced efforts to achieve a permanent settlement, reinforced the sovereignty and independence of the West German republic, and set the stage for incorporating West Germany (and Berlin) into NATO. Likewise, the Soviets could easily have read the US effort to get Soviet missiles *out of Cuba* as a compellence challenge. Whereas US leaders viewed Soviet missiles in Cuba as a threat to the status quo (that could undermine deterrence), the Soviets could view their acceptance of geographic restrictions as a capitulation to US efforts to preserve Soviet nuclear inferiority.

Commitments Can Imply Concessions

The story of Secretary of State Dean Acheson—in the opening chapter to the Korean War—comes with a potent lesson. Acheson famously left South Korea out when identifying areas of US vital interest on a map (Snyder 1961: 34). He meant to send a clear message about areas of the world that the United States deemed essential to defend against Communist encroachments. In June 1950, with an ostensible US "green light," North Korean forces streamed across the 38th parallel, which separated the two Koreas, intent on unifying the Korean peninsula. This story is deceptive, however, in historical respects: North Korea had long pushed Soviet and Chinese leaders to back an invasion; the speech did not allay their concerns that the United States might intervene; and the Truman administration, through other means, had signaled its desire to support a South Korea that could stand on its own (Matray 2002). The story is told, nevertheless, with this implicit message: if you let the adversary know what you intend to defend, you also communicate what you choose *not* to defend. In marking your turf, you concede the rest.

The principle is familiar in the practice of law, where "detail" in a permissive or restrictive provision "invites the maxim, *expresso unius est exclusio alterius* (to express one thing is to exclude the other)" (Chayes and Chayes 1995: 11). A sign that reads "No smoking in this area" implies that people *can* smoke in adjacent areas, or in areas where signs do not expressly prohibit smoking. What was the implicit message, then, when President Biden informed Russian President Vladimir Putin, in their June 2021 summit, that sixteen US sectors (as identified by the US Department of Homeland Security) were "off limits" to cyberattacks? Was it, as Biden's critics were quick to point out,[36] that Russian hackers could assail all but these sectors without fear of penalty? Likewise, what was the implication, with Russia's 2022 invasion of Ukraine, when NATO drew its defensive red line at the borders of member countries? That Russia could grab former Soviet territory in nonmember countries, including Ukraine, Moldova, and Georgia, and not trigger direct NATO intervention? Unsurprisingly, then, splits emerged among NATO members, with some expressing concern about US public messaging that, from their perspective, placed too much focus on "what NATO will not be doing" in response to further Russian aggression.[37] Thus, even firm commitments create a gray area of exclusion that could prompt probes, claims, and actions that the defender finds disconcerting.

In international politics, the temptation to capitalize on apparent permissions is potentially great when challengers harbor doubts about the defender's preferences and willingness to enforce them. North Korea certainly had motive to test the limits of the US defensive commitment: after all, Acheson left out the Korean peninsula for a reason. He obviously believed it *was* of "lesser value" to the United States than Western Europe and held too little value (a priori) to warrant inclusion within the US defense perimeter. Tellingly, Truman initially thought the Communist attack a feint: Stalin, Master of the monolithic Communist empire, had orchestrated the attack, mindful presumably of "bigger" gains than Korean unification.

Unintended permissiveness is inherent in any commitment to a red line: it signals to challengers how far they can take their challenge without resistance. That is, it tacitly gives them permission to snuggle up to the line. A less appreciated aspect of Obama's red line in Syria—much like the Trump administration's retaliatory aerial assault on a Syrian airbase—is the implicit permission it granted challengers to do everything short of crossing the line. The Obama administration was thus backhandedly expressing indifference to the hundreds of thousands of civilian deaths *through other means* in the Syrian civil war and extending "permission" to the Syrian government to continue its violent campaign. Viewed accordingly, retaliation highlighted how little the administration cared, in fact, about quelling the violence or punishing the Syrian government for its abuses.

Thinking strategically, then, the challenger might do anything, or everything, short of crossing the red line (Jervis 2013: 110). It might seek to normalize its extraordinary behavior, hoping to change the defender's thinking about what constitutes acceptable behavior. Indeed, an adversary might "embark on a deliberate strategy of gradual escalation in order to blur the line and make it politically more difficult for the defender to justify retaliation" (Tertrais 2014: 13). China has worked that strategy well to lay claim to the disputed waters of the South China Sea. It has literally "created realities" in these waters by building "Chinese" islands to justify control over adjacent waters. North Korea, of course, has used that strategy to gain acceptance of its missile and nuclear programs. Encountering US threats of retaliation should North Korea fire missiles over Japan or hit Guam, North Korea fired missiles with a steep upward trajectory (to mimic distance flight) while (mainly) staying clear of potential territorial prohibitions. For that matter, by making the aggressive testing of long-range missiles "the issue," North Korea took the heat off its nuclear program.

Commitments Can Trap the Defender

Defenders make commitments for a reason: even the United States, despite its military power, cannot be everywhere and do everything, let alone all at once. So, a defender might choose to stake its vital interests with a visible commitment. Or else, it might stake its claims more broadly, hoping to deflect challenges in "gray areas" of less vital interest. That practice opens it to another problem—the so-called commitment trap (Sagan 2000). A defender binds itself to a response a priori and remains bound to it when deterrence fails, despite the prohibitive costs.

Policymakers might count on such stickiness. They understand that commitments are "sticky" in that they create "realities" by shaping expectations, directing resources, and sending signals that are hard to retract. That was the point when US policymakers committed publicly to AD rhetoric. They wanted the Soviets to realize that US warnings of Armageddon had helped create the reality of assured destruction if the Soviets ever attacked. But leaders often feel bound to commitments for their own sake. Just as people adhere to bad contracts rather than hurt their personal credit rating,[38] leaders might believe their country's reputation is at stake, should the country fail to stand by a promise or threat—and allies and adversaries then question the country's willingness to stand by *all* commitments.

Thus, the Nixon administration worried about the loss in US credibility from a defeat in Vietnam. It expressed these concerns though the United States had established some credibility by investing substantial blood and treasure, over eight long years, in the failing war effort—and adhering to its commitment though conflict conditions had significantly changed. Sino–Soviet rivalry had intensified, the United States had opened relations with China, the United States and the Soviet Union had engaged in détente, and so forth.[39] Adhering to the US commitment seemed the appropriate course solely because of that commitment. In effect, the United States had now committed to the fight, not to a win.

Finding Answers in "Calculated Ambiguity"?

Given the counterproductivity of commitments—stemming especially from their potential provocativeness, permissiveness, and stickiness—are parties better off issuing an ambiguous commitment? With "calculated" or "strategic" ambiguity, leaders need not concede ground; instead, they can keep their options open with a cloak of uncertainty over their intended response, should deterrence fail. But does calculated ambiguity offer an actual "best-of-both worlds" solution?

Here, too, US officials are among the believers. They issue ambiguous commitments in various verbal and physical forms—words or force movements that are open to multiple interpretations—which indicate that the United States *might* act—somehow, in some way, in some place—in the future. In verbal form, officials can choose phrases from among stronger variants ("all options are on the table"), weaker variants ("we are considering our options"), or variants in between ("we haven't ruled out any options"). Contrast these phrases with Vladimir Putin's unmistakable warning to Ukraine and the West in a televised address: "If the territorial integrity of our country is threatened, *we will without doubt use all available means to protect Russia and our people—this is not a bluff.*" He then added: "*I will emphasize this again: with all the means at our disposal.*"[40]

Or else, US officials can choose from among general concepts, without specific points of reference. For instance, they can express a desire to "preserve the balance" or "respond to aggression" in a region, or an intent to adopt firm measures, without tying the United States to a specific course of action. Officials intentionally employ concepts—"actions," "measures," "consequences," "defiance," and "escalation"—that leave targets guessing about the precise meaning. In the 1991 Gulf War, the Bush administration took a position of calculated ambiguity—as we saw (in Chapter 3)—when Vice President Cheney referred to the "wide range of military capabilities" that the United States could employ should Iraq use chemical or biological weapons. The United States signaled accordingly that it *might* use nuclear weapons should Iraq resort to chemical or biological weapons use (Sagan 2000: 91–96). Three decades later, the Biden administration adopted a similar stance. When asked how the United States would respond should Russia employ chemical weapons in Ukraine, National Security Advisor Jake Sullivan answered, "The Russians would pay a severe price."[41] Hammering the same "message," President Biden observed that Russian President Putin "knows there will be severe consequences."[42]

An alleged problem for the defender, however, is that it might trade one problem—say, "stickiness"—for one of "ineffectualness" by creating uncertainty about the fact and strength of the commitment. The challenger might choose, then, to test the commitment—and maybe others, in succession, until the defender opts to respond. The accompanying danger, then, is a "worst-of-both-worlds" outcome if adversaries are permitted to draw their own conclusions about the nature and strength of the commitment.

The adversary might doubt the commitment based on simple logic. Take, for instance, the 2018 Nuclear Posture Review which left open the possibility that the United States would respond with nuclear weapons to cyberattacks on the critical US infrastructure. But how seriously should an adversary take that threat? For a host of reasons, it might discount the nuclear threat: likely uncertainty about the source of any attack; the questionable culpability of the government nominally in control of the territory from which the attack was launched; the availability to the United States of nonnuclear options; global inhibitions and prohibitions pertaining to nuclear weapons use; and, of course, the consequences of using nuclear weapons against another nuclear-armed state.[43]

The adversary could also doubt the commitment after weighing the available evidence, which is often contradictory. What was Russia's President Putin to make of the warning from Biden's National Security Advisor Jake Sullivan of a decisive US response and "catastrophic consequences" for Russia if it used nuclear weapons in Ukraine? Was the message in what was not said—indicating the Biden administration's reticence to act—or in what was implied—indicating a decision by the administration to shed its reserve, whatever the consequences? That the administration offered more specific *private* warnings to Russia would not sway Russian leaders if they interpreted the US preference for private warnings as equivocation.[44]

US messaging on how the United States will respond to a Chinese attack on Taiwan further demonstrates this. US administrations long sought to navigate the nettlesome issue with a policy of "strategic ambiguity." Yet the 1979 Taiwan Relations Act made clear that the United States would supply Taiwan with the material necessary for its defense and "shall maintain the capacity of the United States to resist any resort to force or other forms of coercion that would jeopardize the security, or social or economic system, of the people of Taiwan." Although the United States accepts the Chinese government's principle that there is but "one China," it also expects (in the words of the 1979 Act) "that the future of Taiwan will be determined by political means." Indeed, the United States has sent in its fleet in periods of high tension between China and Taiwan to suggest a willingness to back what it "expects" with force. Going further, President Biden responded in the affirmative when asked whether the United States would defend Taiwan if it was attacked by China. Going even further, he stated, "we have a commitment"—echoing his prior claims that the US commitment to the defense of Taiwan, Japan, and South Korea were tantamount to the US Article 5 commitment under the North Atlantic Treaty. The administration managed the political fallout in Asia by denying a "change in our policy."[45]

Of course, how China will interpret the available evidence is unclear. China could view the evidence to mean that (a) the United States is not showing its hand and might, or might not, actively intervene in Taiwan's defense; (b) the United States wants to deter China from attacking Taiwan but would stand back should China attack; or (c) the United States is committed to defend Taiwan with force but would prefer not to irritate China by announcing the policy. Which interpretation holds, in China's view, would depend on China's reading of the evidence, and its tolerance for uncertainty. But China's interpretation will also draw from China's sense of the situation—the specific circumstances; available military, political, and economic options; and opportunities and threats at hand. That is, China's conclusions will reflect China's desire to act given the demands of the moment; China's urgency to make one-China a governing reality; China's beliefs about US politics (and a change in administration) as they affect the strength and nature of US commitments; Taiwan's movement toward a declaration of independence; and US responses (and non-responses) to local provocations which include China's occupation of islands currently claimed by Taiwan.[46]

The point, then, is that policymakers err in thinking that they control interpretations by crafting messages, even if reinforcing them through repetition. Whereas such messaging can smooth over political differences and avoid unnecessary conflict (the US acknowledgment of "one China"), the *sender's* intended message might not override the challenger's beliefs about what evidence matters more or what exactly constitutes evidence. An allegedly ambiguous commitment might not be all that ambiguous, at least to the target—whether or not it shares the sender's view of the commitment.

Indeed, a danger of ambiguous commitments—and all efforts to obtain or preserve decisional "flexibility" (see Chapter 2)—is that they abet shirking, in the avoidance of tough decisions. Then, an ambiguous commitment could become a firm commitment ("commitment creep") as leaders attempt shortsightedly to counter doubts about the fact or strength of the commitment or to build incrementally on their prior actions. In this regard, think of the decisions, in the 1990s, to extend the Article 5 guarantee to newly independent countries of Eastern Europe. The United States and its NATO partners maintained a general commitment to protecting and promoting democracy in Europe, beyond protecting existing members against the Soviet and then Russian territorial threat. Extending the NATO guarantee to fledgling democracies in the region seemed like the natural course even if it took NATO to the Russian border. Whether or not we approve of the NATO enlargement decisions, we should ask whether the United States and its NATO allies would as enthusiastically offer the same guarantees now to the same countries, had they not joined NATO, in the aftermath of Russia's invasion of Ukraine.

Conversely, ambiguous commitments might become "lesser" commitments if reflecting or enabling (palpable) indecisiveness. The US commitment to Southeast Asian security under the 1954 Southeast Asian Treaty Organization (SEATO) illustrates through its combination of ambitious goals and transparently insufficient means, and will, to accomplish them. Take this commitment, for instance: "Each Party recognizes that aggression by means of armed attack in the treaty area against any of the Parties or against any State or territory which the Parties by unanimous agreement may hereafter designate, would endanger its own peace and safety, and agrees that *it will in that event act to meet the common danger in accordance with its constitutional processes* [italics added]." Hanoi had good reason to remain unimpressed when, years later, SEATO protection was extended to South Vietnam. The committing party might find itself unable to deliver on its commitment, having failed to prepare adequately (politically or materially) for the challenge—assuming, of course, that the party committed in good faith in the first place.

Conclusions

The case for nuclear superiority hinges in large part on its supposed coercive benefits. Given inhibitions on the use of nuclear force, however, the "superior" power might seek to bolster the credibility of a threat to employ nuclear weapons by making commitments verbally, or implicitly through physical action. In going that route, it is susceptible to the perils and pitfalls listed below.

Effective commitments can take the verbal form of a "red line." Despite the term's popularity, a challenger might dismiss a red line as "cheap talk," should it believe that the words come too easily, or take as provocative what was better left unarticulated. Thus, physical commitments offer plausible advantages over verbal commitments: they can signal intent—absent the incitement of verbal warnings—by creating conditions that make a response to challenges more likely. Yet physical commitments, too, are often ineffective and—worse—counterproductive.

> **4.1:** *Policymakers might fail to recognize that neither words nor deeds are inherently better than the other for signaling, given the strengths and the deficiencies of both.*
>
> **4.2:** *Policymakers might fail to recognize that commitments are potentially ineffective when:*
>
> *a) a defender cannot accurately gauge the scope and nature of future challenges.*
> *b) a challenger believes that the operant conditions do not hold.*
> *c) a defender speaks without clarity and precision.*
> *d) a defender's messages are lost to contradictory signals.*
>
> **4.3:** *Policymakers might fail to recognize that commitments are potentially counterproductive when:*
>
> *a) viewed by challengers as provocations.*
> *b) foreclosing acceptable exits.*
> *c) trapping the defender by limiting its future options.*
> *d) implying concessions through exclusion.*
> *e) viewed as a "challenge" or demand by the challenger which "compels" it to act.*

In principle, a clear line in the sand might tie the formidable US nuclear arsenal unmistakably to a specific set of US interests or concerns. But commitments might fail to achieve their intended effect given the challenger's reading of the evidence and situation. Even an intentionally ambiguous commitment provides (albeit maybe misleading) clues to adversaries about the strength and nature of a commitment. For that reason, "calculated ambiguity" might prompt an adversary to misjudge a commitment and to stumble into confrontations.

> **4.4:** *Policymakers might overestimate the opaqueness of their ostensibly ambiguous commitments.*
>
> **4.5:** *A vague or uncertain commitment might offer a "worst of both worlds" solution: a commitment that is too soft to impress potential challengers yet sufficiently strong to pull the defender into action to confront an actual challenge.*

Then, an ambiguous commitment might weaken if the challenger believes "ambiguity" reflects or enables indecision, which impairs a quick and effective response to transgressions. Conversely, such a commitment might strengthen incrementally when unguided by deliberative decisions.

> ***4.6:*** *Policymakers might fail to recognize that ambiguous commitments can: a) lead incrementally to firm commitments that "trap" the committing party or b) reflect or enable indecisiveness, which weakens commitments (in appearance and in execution) when challenged.*

5

Risk Manipulation

We saw (in Chapter 4) that parties might acquire the credibility necessary for deterrence through commitments. But what if the commitments prove inadequate for deterring adversary aggression? Or what if the adversary acts or probes outside of areas of commitment? For that matter, what if the coercing party—say, the United States—is effectively the "challenger" and seeks to compel adversary concessions?

To boost its credibility, the United States is not without options. It can manipulate risk. That is, it can orchestrate conditions to convince the opponent that its probability of a costly loss is unacceptably great should it continue along a given course. By taking forceful action, the United States can hope, at least, that the increasing risk will convince the opponent to back down.

As with commitments, however, risk manipulation can prove ineffective, even counterproductive, if fueling uncontrollable escalation.[1] So, can decision-makers profitably manipulate risk? Answers, here, are provided by considering: (a) the nature of risk, (b) US Cold War–era efforts to manipulate risk, (c) "latent" risk ensuing from organizational practices, (d) risk stemming from a "bad" US decision-maker, and (e) risk involved in a potential military conflict with China or North Korea.

Considering Risk

In standard expected utility calculations, decision makers evaluate options by weighting their value by their probability of occurrence. The idea is familiar to gamblers. Take a simple example. Assume they receive a dollar each time they draw a green ball randomly from a jar that contains nine red balls and one green. If required to bet a dime on each play (and the balls were replaced after each play), players would most likely break even with ten dimes to invest in the game. Obviously, the game comes with risk to the players: they *might* lose their money (after all, it is gambling). Yet a casino would not consider these favorable odds for the house. The casino's goal is not just to break even but to turn a profit. It can serve that goal by increasing the risk to the players—here, by placing additional red balls in the jar.

The False Promise of Superiority. James H. Lebovic, Oxford University Press. © Oxford University Press 2023.
DOI: 10.1093/oso/9780197680865.003.0005

In decreasing the odds of winning, the casino could still attract gamblers to play. True, some people are averse to actions that incur a small chance of a loss though taking a chance could prove lucrative. With a two-dollar payoff, and a one-in-ten chance of a successful draw, some people might still not participate. Yet some people will accept a large probability of a loss for a more meager reward. They might play the game with a 90-cent payoff for a green ball or the addition of red balls to the jar. Likewise, one person might reject a $1,000 wager even with a high probability of a profitable return, whereas another person might accept the same wager. We would view the first person as "risk averse" and the second person as "risk accepting."

Individuals handle risk differently depending on their personalities. For that reason, risk propensities are at issue when administration officials wonder openly about whether leaders—of so-called rogue states—might take high risks to achieve certain gains. Yet people who accept risk need not be gamblers at heart, enjoying the (adrenaline) rush that attracts certain personalities to rock climbing, BASE jumping, or other dangerous activities. An individual's risk propensities vary, depending on the individual's situation. We would probably act despite the high risks when the consequences of not acting are likely far worse. Jumping from the window of a five-story building is rarely the preferred course of action. For most people, it is the best choice if the apparent alternative is dying in a blazing building. In jumping, a person chooses a risky course but the less risky of the alternatives.

An individual's risk proclivities might change over time, perhaps to reflect the individual's relative wealth or stage in life. Responsible financial advisors encourage us to move from a "risk-accepting" to a "risk-averse" investment strategy as we approach retirement age, to "lock in" our security. Once we have acquired sufficient assets to maintain the lifestyle we seek in retirement, we choose not to risk a loss that could jeopardize our retirement plans. By contrast, we might accept financial risk early in our careers, when we relish achieving financial security and have decades to recoup a loss.

These principles apply more generally. Thus, if rogue-state leaders found the status quo sufficiently unappealing, or a potential gain sufficiently appealing, they might risk war, even absent a risk-prone personality. We might not know, then, whether a leader's actions reflect risk propensities—in a *basic* willingness to accept a high probability of loss—or, instead, the leader's valuation of gains and losses. The difference is critical, however, for devising policies to deter undesirable behavior. Changing the payoff of the outcome, for instance, by increasing the potential costs of achieving it might not deter a party that is generally risk-accepting. By contrast, increasing the likelihood of a US nuclear response might not deter a party that believes it can absorb those costs (at least at the low end of a nuclear conflict).

We must show due care, then, when we claim that an adversary will accept high risks *and* high costs. If an adversary accepts the costs—because it believes, for instance, that the potential gains offset the costs—the relative "risk" for the adversary declines accordingly. Put simply, if I can afford to lose $10,000 but you cannot, the risk for me is smaller than it is for you. For me, this is *not* a risky bet, because I can easily cover the loss; that is, $10,000 dollars means less to me than it does to you.

But do parties dutifully consider the risks of behaving one way or another? Perhaps not.[2] Even intelligence professionals mishandle probabilistic information, unintentionally "loading the dice" to favor one or another option. Researchers tell us, for instance, that these trained analysts might: (a) confuse the likelihood of occurrence with confidence in a prediction, based on the quality or source of the underwriting information; (b) ignore low-probability, but still costly, outcomes; (c) treat potential occurrences as competing hypotheses rather than recognize that, together, they constitute plausible outcomes with different likelihoods of occurrence;[3] (d) emphasize the most likely occurrence, which implicitly downgrades the likelihood of alternative outcomes; (e) employ non-numerical terms (for instance, "highly likely" or "likely") inconsistently to characterize probabilistic outcomes; (f) treat low probability outcomes as more probable than outcomes for which probabilities remain uncertain; (g) screen information to fit their prior assumptions about the chances of occurrence; or (h) exhibit false confidence in their own probabilistic predictions, leading them, for example, to stop acquiring additional information.[4]

Even this assessment exaggerates human decisional skills. It assumes that parties consider probabilistic information when "in many cases people ignore probabilities and act as though events were either certain or impossible" (Jervis 1979: 310). Thus, they might focus on the "payoffs"—the benefits and costs of doing one thing or another—or act based on some prior understanding of the likely effects of a course of action. If they estimate the probability of a given outcome, then, we can justifiably suspect that the estimate is driven by a policy preference. After all, exaggerating the likelihood of a preferred policy's positive outcomes and downplaying the likelihood of negative effects is a well-recognized decision bias.

We can question the basis of even seemingly neutral risk assessments such as Kennedy's alleged one-in-three estimate of the chance of war with the Soviet Union during the Cuban Missile Crisis. First, we can question the origin of the numbers. After all, the nuclear-war data set, from which to draw statistical inferences, contained zero cases—and the United States had not previously engaged in a confrontation of that magnitude with a *Khrushchev-led* Soviet Union. Second, we can question Kennedy's understanding of expected-utility calculations. As Jervis (1979: 311) observes: "It is hard to accept the notion that Kennedy thought that the costs of war combined with twice the gains that would accrue if he won (the payoff for standing firm when the odds were two-to-one that the Russians would back down) were anything like equal to the value of losing combined with twice the payoff of a compromise (the payoff for backing down when the odds are as stated)."

When we discuss risk propensities, evaluation, and manipulation, then, we must recognize that the logic of risk assessment of economists, formal theorists, and decision modelers might not apply when policymakers wrestle with demanding decisional problems. Decision makers ignore probabilities, "invent" probabilities to serve preferred options, and act unwisely given the likelihood of a positive or negative return. Prospect theorists tell us, for example, that individuals tend to: (a) exaggerate the probability of occurrence of low-probability events; (b) discount the probability of catastrophic events over which those parties have limited control; and (c) respond to probabilities based on how problems are framed. For instance, they perceive a ten-in-one-hundred chance of a negative outcome as more foreboding than a one-in-ten chance.

This brings us to a fundamental question that challenges formal considerations of risk. How do cost and uncertainty interact, then, over their respective range of values? For instance, could the magnitude of an impending loss drive our thinking about a problem, overwhelming the decisional influence of probabilistic information? Think about how people respond when they get news that they *might* suffer from a terminal illness. Some people might reject the news entirely. In their state of denial, they assume that the worst simply *cannot* happen. Others might respond in panic; for them, the extremely unlikely event is likely. Although the doctor might order additional tests, and *try* to place the patient's mind at ease, the thought of having a fatal illness might block all attempts to comfort the patient by discounting worst-case probabilities.

Likewise, a potential for disaster, apart from its actual probability, can move a person—or nation—to action. We should remember the US reaction to the 9/11 events, in this regard. In more rural parts of the country, where lucrative targets were sparse, rumors ran rampant about potential terror attacks on shopping malls and other public places. The US government also moved into action as advisors warned of US unpreparedness for horrific attacks including the introduction of toxins into the US water supply, the poisoning of dairy products, strikes on nuclear facilities or central bottlenecks in the US transportation system, or the use of biological, chemical, and even nuclear weapons (a dirty bomb) to kill thousands—maybe millions—of people. The implicit assumption was that the United States must prepare for all these potential scenarios despite the costs (indeed, impossibility) of defending across the full range of scenarios. Importantly, these fears arose from a *low-tech* attack on aircraft employing tactics that could only work *once*: terrorists threatened to use box-cutters against compliant airline passengers who assumed falsely that their lives depended on cooperating with the "hostage-takers." Indeed, the government responded to that unrepeatable mode of attack by banning all devices—metal forks, nail clippers, and so forth—that a person could use against another passenger. The reaction in policy was hardly driven by probabilistic considerations. Once the cabin doors of a plane were locked, only a creative fiction writer could envision a scenario linking a metal fork or fingernail clipper to the destruction of an aircraft. We can say that the perception of danger had crowded out all sense of realism (probability).

If that is how things work, perhaps we should predict decisional outcomes by rejecting calculations based on expected value—whereby probability and payoff are weighted equally—and accept that the payoff—here, in potential cost—could swamp the consideration of probabilities. We might recognize, for instance, that the evaluation of probabilistic information becomes increasingly suspect beyond some cost-threshold value. That is, people respond singularly to the potential costs rather than accepting or rejecting risk based on probability. In medical terms, again, what are the costs to the patient should the "worst" happen? Might they lose a limb? A vital organ? Their hearing? An eye? Their sight entirely? We might also appreciate that governments might respond mainly to cost regardless of what "people" do. Local governments (with state and federal backing) prepare for certain contingencies—for instance, with school-shooter drills—because they involve *salient* threats. But can we say that, in doing so, they prepare for the biggest threats to the life and limb of the local citizenry?

Decisional biases are important to acknowledge when assessing the viability of strategies that we mean to manipulate and control risk. To be sure, policymakers might acknowledge the consequences should a risky gambit fail rather than accept risk with the *promise* of a positive payoff. In that sense, a famous eighteenth-century quote (from Samuel Johnson) still applies: "When a man knows he is to be hanged . . . it concentrates his mind wonderfully." Yet policymakers might overreact to threats if they believe that the *costs of failing to act* are prohibitive. They might act precipitously without processing probabilistic information that begs for a cautious response to the threat.

Manipulating and Ignoring Risk in the Cold War Years

That said, the concept of "risk" little influenced the thinking of the Cold War's hard warfighters and AD advocates. For them, nuclear war was largely a technology problem. Deterrence was secure if a party could deploy forces in the right ways and quantities. By contrast, risk manipulation was central to the thinking of "soft" warfighters who looked to hedge their bets knowing that deterrence might fail. They took guidance from Thomas Schelling's (1960: 187–203) seminal notion of a "threat that leaves something to chance."

The chance, here, partly stems from uncertainty about the manipulator's intentions, that is, how far it will press to achieve its goals. But the chance behind Schelling's thinking stems as well from factors beyond the parties' influence. In Schelling's (1960: 188) words, "whether we call it 'chance,' accident, third-party influence, imperfection in the machinery of decision, or just processes that we do not entirely understand, it is an ingredient in the situation that neither we nor the party we threaten can entirely control." Military strategists subsume these various factors in the "fog of war." Risk in that context arises from uncertainty about "who did what, when, why, and how." Uncertainty owes, then, to the rapid unfolding of events, lack of real-time intelligence and communications, and breakdowns in command and control, as individuals and organizations act outside of the chain of command. Indeed, subordinates might take it upon themselves to act in explicit violation of government policy. Curtis Lemay, the head of US Strategic Air Command (until 1957),[5] colorfully insisted that he had no plans to wait for a Soviet attack (and presumably authorization) and would preempt with signs of Soviet mobilization (Kaplan 1983). When a party intensifies or extends its military operations, an adversary is frequently left not knowing whether an escalatory move was unauthorized or inadvertent or whether the opponent acted out of strength or out of weakness—that is, whether it is feigning strength to hide weakness or gambling with some big move to prevent growing weakness.

In the thick fog, however, parties act on what they know, or think they know, about their situation. Consequently, they might overreact or underreact to events, miss opportunities to pursue beneficial outcomes, contravene the behavior of allied parties, or act unintentionally to provoke further conflict, causing it to spiral out of control.

For success, then, the risk-manipulating party must accept some—perhaps substantial—risk that "anonymous" forces of war will cause the conflict to escalate toward some catastrophic end, hoping the opponent eschews the same strategy and chooses to deescalate the conflict. But the guiding logic here is paradoxical, even inconsistent. It suggests that a party can manipulate risk because it knows the point of no return while conceding that the strategy works because no one knows how far to push given the element of chance. Thus, in focusing on manipulating risk, strategists implicitly minimized the challenge from unknown risks. Indeed, the most dangerous Cold War–era confrontation—the 1962 Cuban Missile Crisis—brought dangers that were then unknown to US (and Soviet) leaders. The challenges of risk detection did not end, of course, with that crisis.

The Cuban Missile Crisis

Few, if any, decisional periods in US history have received more attention than the Cuban Missile Crisis. The focus owes to the Kennedy mystique, the literate group of administration advisors who wrote of their experiences in the aftermath of the crisis, and the seemingly heroic image that emerged from the crisis of a US president facing down an existential threat. The focus sometimes owes to myths surrounding the earlier meeting between Kennedy and Khrushchev in Vienna, where the Soviet leader supposedly assessed Kennedy as weak; implicitly, the crisis thus communicates the importance for deterrence of maintaining peace through strength. Of course, the crisis acquires considerable attention from its positive outcome. Where contemporary wars seem interminable, and end inconclusively or in defeat, the Cuban crisis offers an apparent winner and loser.

Yet the Cuban Missile Crisis backhandedly conveys just the opposite message because of the high stakes and the resulting seriousness with which Kennedy and his advisors approached their deliberations. A half-century after the momentous events, ongoing disclosures—documentary evidence and revelations from military participants—reveal in frightening respects how uninformed the decisional participants were about the potential risks, and thus how little control Kennedy and Khrushchev exercised over events. We came close—in ways that the main participants could not know—to a nuclear cataclysm.

First, the US and Soviet leader lacked control over the crisis in ways that neither leader manipulated, nor understood. Indeed, they courted disaster (Dobbs 2008a, 2008b):[6] The administration's effort to *offset* risk—by raising the alert status of US forces—unintentionally *manipulated* risk. Ironically, the administration's efforts at escalation control empowered lower-level defense officials to take actions with severe escalatory potential.

Kennedy was informed belatedly that the Navy had dropped depth charges on Soviet submarines—an ever-present threat to the US surface fleet—to force them to surface. The Navy's standard operating procedure—which was communicated (via the State Department) to Soviet officials—was not communicated to Soviet submarine officers in the vicinity of Cuba. Kennedy blanched when he learned that the Navy was using explosives to get Soviet submarines to surface. We can only imagine his response had he known—as we subsequently learned—that some of the submarines carried nuclear torpedoes (with near Hiroshima-sized destructive power) and that the crew of one of the submarines took a fateful vote, assuming they were under attack. A nuclear conflict was averted by a two-to-one vote of the senior officers in *favor* of firing the missile, with the boat's captain voting with the (losing) majority. Lacking unanimity, the submarine surfaced—creating the false appearance that the US military—and its civilian leaders—remained in control of the situation.

Elsewhere, things that could go wrong did go wrong, narrowly averting a potentially disastrous escalation of the conflict. Unbeknown to US civilian leaders at the time, a US ICBM test went off as *scheduled* over the Pacific. It was fired from Vandenberg Air Force base—one among other ICBMs that were mated with their warheads (Sagan 1993: 79–80)—though Strategic Air Command forces, at DEFCON2, were near maximum alert status. Elsewhere, an American (U-2) spy plane—on a routine air-sampling mission—mistakenly strayed into Soviet territory. (The event would lead Kennedy to famously retort, "There's always some son-of-a-bitch who doesn't get the word.") The incident prompted the scrambling of US (F-102A) interceptors, outfitted only with nuclear-tipped missiles because their base was at DEFCON 3 alert status. The fighters—in combat—had only a nuclear option. Even if not firing their missiles, we could easily create crisis scenarios in which these aircraft provoked a Soviet nuclear response (on these, see Sagan 1993: 139–140). Soon after, the Soviets shot down a US reconnaissance plane over Cuba. US decisional participants were unaware that the decision to fire at the plane had been made on the ground, expressly in defiance of Khrushchev's orders. In the informational vacuum, participants could argue for US military action. With advisors divided more on the size and discrimination of a follow-on military strike, not the wisdom of a strike—and Air Force Chief of Staff Curtis Lemay itching for an all-out strategic assault—a forceful US response was likely with future incidents (Pious 2001: 96–97).

Second, Kennedy administration officials assumed they possessed far more latitude for conventional military action than conditions on the island permitted. The administration could have made conditions decidedly worse had it chosen to invade the island, unaware that the Soviets had deployed tens of thousands of troops to the island, armed with many dozen tactical nuclear weapons. The Soviets, who were apparently more committed to Cuba's defense than US leaders recognized at the time, were positioned (at least) to eviscerate the US base at Guantanamo and strike invading US troops in the event of a US attack (Schulte 2012: 36). If the Soviets had deployed these weapons merely for symbolic purposes—that is, to reinforce the Soviet commitment to Cuban sovereignty—the Soviets logically would have advertised their presence. To the contrary, Soviet officers on the scene downed the US reconnaissance plane to prevent it from passing over the Soviet tactical missiles. The Soviets also had to know that placing the missiles in Cuba might invite their use. After all, the Cuban theater was ripe for conflict: the United States had backed the ill-fated invasion (at the Bay of Pigs) of the island in the prior year and continued its efforts to subvert and sabotage the Castro government. Castro himself lobbied the Soviets hard during the crisis to strike the United States with nuclear weapons in the event of a US attack.

Third, the parties poorly understood each other's goals in the confrontation. Neither side appreciated the red lines of the opponent—what it would and would not tolerate. Perhaps that is obvious. Although Kennedy had publicly warned the Soviets, in September, that the United States would not tolerate Soviet offensive weapons in Cuba,[7] the Soviets *had* placed missiles in Cuba without expecting they would soon have to withdraw them. The Kennedy administration flew blindly into a potential abyss. Kennedy and his advisors had no clear sense of Soviet goals—whether, for instance, the Soviets, in placing the missiles in Cuba, were intentionally inviting a US response to justify a Soviet attack elsewhere in the world. In consequence, they could not predict what the Soviets would do if the United States struck targets in Cuba (see Bell and Macdonald 2019).

Events did not bring clarity: throughout the crisis, administration officials showed little capacity to read Soviet behavior. They could track Soviet ship movements and detect progress in Soviet missile-site construction. But they struggled to interpret the evidence, which they could read either as good or bad news, and relied, then, on (worst-case) assumptions in assessing whether the Soviets had mated their missiles to warheads (Acton 2020). In this, they confronted the inevitable challenges of decision-making in crisis. Still, as we shall see (in Chapter 7), US officials were far more attentive to available options than to their potential consequences given plausible Soviet goals in the conflict.

We might take comfort believing that the military facts of the situation—US conventional superiority in and around the Caribbean and the catastrophic consequences of nuclear war—weighed heavily against a forceful resolution of the crisis. But we assume, then, that leaders fully appreciated the risk that US military actions—authorized or not—could have triggered a massive nuclear exchange. That was a likely consequence of any Soviet first use of nuclear weapons given US ill-preparedness for *limited* nuclear action.

Although history most certainly judges the administration's decision kindly, we should not judge the quality of a decision by the favorability of the result. We must not forget the two-sided nature of the problem, and the concessions—not the risk-taking—by *both* sides that kept the parties from war. Ultimately, a compromise proved possible because Kennedy remained open to a trade (the US missiles in Turkey) and because Khrushchev made a very public concession to avoid a violent resolution. That point deserves elaboration. Khrushchev conceded ground by accepting a mere US promise to remove a handful of missiles from Turkey, down the road—without a public quid pro quo—and an informal (eminently revocable) US pledge to respect Cuban sovereignty. Moreover, he did so over heated objections, and claims of betrayal, from the Soviet Union's only Western Hemispheric ally, possible costs to Soviet prestige and coercive leverage, and domestic political costs that would eventually bring an end to his leadership—just as Kennedy resisted intense pressure from his own military advisors to go in for the kill.

Beyond the Cuban Missile Crisis

In the Cold War years, US leaders sometimes engaged in risky behavior—with no intention of manipulating risk. Most troubling, however, are the occasions when US officials failed to recognize the grave risks.

A critical "non-crisis," early in the Reagan administration, demonstrates this. US–Soviet tensions had peaked again—now, two decades removed from the Cuban confrontation. Reagan had labeled the Soviet Union an "Evil Empire." The Russian military had shot down a South Korean civilian jet liner over Soviet territory. The United States had positioned Pershing II missiles in Germany proximate to Soviet command-and-control targets. Moreover, the Reagan administration was preparing for the worst. With US policy hawks warning that the Soviets would breach SALT II arms-control limits and were prepared to "win" a nuclear war, the administration (like its predecessor) pursued a warfighting strategy that would allow the United States to fight, and presumably emerge victorious, from a protracted nuclear conflict.

What better time, you might ask, to launch a mock NATO attack on the Soviet Union? In November 1983, NATO performed a five-day exercise—a war game, dubbed Able Archer 83—to simulate a nuclear attack on the Soviet Union and its Warsaw Pact allies. The exercise was designed, and implemented, for authenticity. High-level government officials participated; soldiers wore protective gear and moved nuclear weapons to prepare them for "launch"; across communication channels, military officers requested permission to destroy select Soviet cities; NATO forces went on simulated general alert; and commanders and staff operated generally with necessary furtiveness and secrecy.[8] Such realism, while the friend of dutiful preparation, is the enemy of a stable deterrence relationship. As Soviet military commanders were kept in the dark about the purpose of the NATO activities, they took compensatory action. The Soviets placed their air and ground forces in the vicinity on high alert, fearing an imminent strike.

We worry what leaders will do when "they get the 'call' at 3 o'clock in the morning," a concern made famous by a Hillary Clinton campaign ad.[9] Can we depend on a US president to make good decisions, on the fly, in a crisis? Will they possess the insight, experience, and sound judgment to make the right choices—under high stress—with the limited information available? These are good questions (and are addressed, again, below). We are not wrong to hope that good leaders—based on their personal character, instincts, and experience—make good decisions. The unappreciated lesson of Able Archer is, however, that grave dangers exist in what leaders, however competent, tend to ignore, assume away, or take for granted. Yes, we must worry about the consequences of that fateful call; but we must also worry about what might happen when that phone *doesn't* ring.

The good news, however, for Cold War–era US policymakers, was that Soviet leaders seemed loath to manipulate risk. Unlike US policymakers, the Soviets did not alert their nuclear forces for signaling purposes during the Cuban Missile Crisis and the 1973 Middle East War. Thus, the Nixon administration's decision to move US forces to DEFCON 3 status during the 1973 Middle East War apparently left the Soviet leadership confused, not impressed (Gottfried and Blair 1988: 145). To the contrary, Soviet leaders eschewed actions that might concede control of nuclear forces to those within the military command. Even in war, the Soviet command structure would remain highly centralized, betraying a cautious Soviet approach to nuclear options. Indeed, the Soviets had every reason *not* to invite or manipulate risk given Soviet vulnerabilities in the event of a US first strike. The centralization of its command and control, considerable blind spots in early warning, and overinvestment in land-based missiles created inviting targets for a US nuclear offensive that became more threatening, still, with improved US capabilities. By the early 1970s, accurate MIRVs on US ICBMS vastly increased the number of warheads the United States could place accurately on hardened Soviet targets.

Given Soviet vulnerabilities, the Soviets were unlikely participants in a proposed US game of nuclear chicken. So, the Soviet goal was not to manipulate risk but to end a nuclear war, should one occur, on the most favorable military terms possible: Soviet military doctrine stressed the benefits of seizing the offensive. In fact, the costs the Soviets might endure would weigh against the success of a risk-manipulation strategy. If the Soviets initiated a war by invading Westerrn Europe, they likely coveted the gains of war or feared the costs of the status quo. If that were the case, however, Soviet actions were more sensitive to these gains and costs than to the increased likelihood of nuclear catastrophe (which the Soviets had accepted when initially attacking Western Europe). Thus, for the Soviets, US efforts to manipulate risk, and stoke fear, might have upped the value of the territorial gains, and the costs of withdrawal. Risk manipulation, in this context, would be inherently counterproductive.

The bad news, then, was that US nuclear manipulation strategies were, as the saying goes, "too clever by half." They ingeniously allowed strategists to reinforce deterrence to counter the suicidal implications of a nuclear exchange. They nevertheless *assumed* a pliant opponent who was willing to play by the same rules. The United States, in seeking to bargain—with a nuclear warning shot, for example—could well have triggered its own demise.

Latent Risk in Conflict: The Organizational Challenge

The challenges of maintaining command and control in nuclear warfare exacerbate hidden risks. Civilian command and control is fragile, at best, once missiles start flying. Yet even with nominal civilian control, leaders must cope with rules and procedures devised by professional military organizations when planning and executing an attack. Military organizations, *as organizations*, perform by their own rules and procedures, which civilian leaders lack the access, professional knowledge, and real-world experience to appreciate, let alone critique and influence. Scott Sagan argues that military organizations—within fledgling nuclear states—will likely fall short, then, in maintaining command and control over nuclear weapons: "professional military organizations—because of common biases, inflexible routines, and parochial interests—display organizational behaviors that are likely to lead to deterrence failures and deliberate or accidental war" (Sagan 2003: 47). Sagan's (2003: 46–87) reasoning centers on two factors: organizational priorities and organizational influences.

Organizational Priorities: Preemption by Another Name?

Organizations design plans, and rules, to serve their own priorities. Accordingly, military organizations plan to go on the "offensive," seizing the initiative to accomplish set military objectives, in time of war. In the early 1950s, then, the US Strategic Air Command prioritized the buildup of the US nuclear weapons arsenal, and the capability to deliver weapons effectively against the Soviet target set, over controls on US weapons use: "In sharp contrast to weapons development and deployment, the command system underwent a slow and chaotic evolution without the semblance of a grand design" (Gottfried and Blair 1988: 49). Indeed, Bruce Blair—who once served as an Air Force launch-control officer—insists that US ICBM launch codes were once set to "00000000," a sign that concerns about weapons availability exceeded fears of an accidental launch.[10] In the language of the nuclear literature, the imperatives of "positive control"—ensuring that forces would successfully launch—trumped concerns about "negative control," ensuring against an accidental or unauthorized launch.[11]

The military initially expressed its offensive priorities with a clear preference for preemption. When preemption proved risky, it sought the "next best thing" in launch-on-warning. Although civilian leaders pushed back, the military command was well positioned to determine whether and how the United States responded to a warning of a Soviet attack.

Preemption

The distinction between prevention, preemption, and retaliation seems sharp and straightforward. With prevention, a party attacks another *before it can initiate* an attack; with preemption, a party attacks another when it is *about to initiate* an attack; with retaliation, a party attacks another *after it has initiated* an attack. The Eisenhower and Kennedy administrations seriously considered—but ultimately rejected—a preventative attack on the Soviet Union and China, respectively, before they posed a nuclear threat to the continental United States, just as administrations more recently considered preventative options to thwart the feared nuclear ambitions of Iran and North Korea. For successive US presidents, and their advisors, prevention held unsettling ethical and practical implications. They expressed discomfort with the idea of launching an unprovoked attack (see, e.g., Sagan 1989: 22) but also exhibited doubts about its ultimate effectiveness. Various administrations have recognized that an aerial assault would "set back" an incipient nuclear-weapons program, not terminate it.[12]

Preemption, by contrast, was far less controversial, in large part because it more closely resembled retaliation than prevention. In a world of perfect information, preemption is effectively retaliation. If you know with absolute certainty that another is about to strike, a preemptive and retaliatory strike only differ practically in their relative effectiveness (political, moral, and legal issues aside). Of course, we always lack absolute certainty, and must cope with uncertainty by degree. Given such uncertainty during the Cold War, US policymakers pondered whether the United States should respond militarily to "strategic warning" of Soviet attack preparations or wait, instead, for "tactical warning," from US sensors, that an attack was underway.

We might assume that tactical warning, compared to strategic warning, significantly reduces uncertainty about the fact of an impending attack. Strategic evidence, which points to a potential attack, could have a benign explanation: the adversary is mobilizing forces for political effect, engaging in a military exercise, and so forth. But an attack—as revealed through tactical warning—*is an attack*. Or is it? The answer depends on the capabilities of the early warning system, that is, its likelihood of generating false (positive) warnings of an attack, when even small degrees of uncertainty matter greatly given the grave consequences of guessing wrong. Risk here is incurred given the possibility *and negative consequences* of acting in error.

No matter. The Eisenhower administration assumed the risk of acting on strategic warning was worth taking. It placed thousands of US nuclear weapons abroad, focused in Europe. Yet prepositioning amounted to a double-edged sword: these weapons increased Western security only by increasing their vulnerability to a surprise Soviet attack. The Eisenhower administration addressed the threat by embracing preemption, hoping to at least *limit* the damage to the United States and its forces through a disarming first blow on signs of Soviet preparations to attack. The consequences of being the second to strike—especially given the administration's dependence on nuclear over conventional forces—were too unappealing.

Launch-on-Warning

Of course, the attractiveness of preemption as an option diminished with growing Soviet nuclear might. By the sixties, preemption could no longer prevent the Soviets from a devastating response whether the United States struck first or second. That meant that the United States would have to bear the consequences—in catastrophic cost—should it misread the "signals" and launch an attack based on false evidence of a Soviet attack. Although civilian leaders and military planners soured, then, on preemption, the challenges of acting under uncertainty remained in US planning via the launch on warning (LOW) and launch under attack (LUA) options. Both options gained a foothold in government in the early 1960s.[13]

In principle, the two options stand somewhere in the middle between preemption and retaliation. With LOW, US forces would launch with evidence of the pending arrival of Soviet warheads; with LUA, US forces would launch immediately with the arrival of Soviet warheads on US targets. But military planners muddied the distinction between LOW, LUA, and preemption. Because LOW, like preemption, relied on inferences drawn from available evidence, LOW "perched precariously between" the alternatives of preemption and retaliation (Blair 1993: 170). Precariously, indeed: the military used the terms LOW and LUA interchangeably. Launch under attack effectively *was* launch on warning.[14]

The distinctions eroded because the military sought to maximize *target coverage*.[15] With constraints on US force sizes, it lacked alternatives to satisfy the demanding requirements of a plan designed so that attacks occurred in proper sequence (e.g., strikes were timed to avoid the "fratricidal" effects of one warhead on another), complemented the effects of other attacks (e.g., attacks on early-warning and air-defense systems would set up attacks by penetrating bombers), or ensured the destruction of targets through redundant strikes (to meet high levels of "damage expectancy"). Tampering with the plan, as by delivering it piecemeal, would prove severely disruptive: "the subsequent implementation of one of the large attack options would probably result in a very ragged attack, the timing thrown off, redundantly targeted systems conflicting, targets left intact because their assigned weapons had been used previously, weapons left unfired, and weapons uselessly fired at assigned targets that had already been destroyed" (Terriff 1995: 111). From the military's standpoint, attacks had to unfold as planned to avoid a dangerously inefficient mess.

The US military command "solved" the vulnerability problem, then, by relying on LOW. The alternative was unacceptable. "Riding out" a nuclear attack would expose US strategic forces to "such severe command disruption and force attrition that it would have been effectively neutralized in terms of target coverage" (Blair 1993: 33). Still, the preferred option exacted a severe trade-off. The wisdom of a US launch decision was tied to the efficacy—and fallibility—of a US command and control system, designed with one overarching goal, for a single use: "carrying out a one-time, prearranged plan whose essential objective was to destroy a long list of targets" (Blair 1993: 35).

Given vulnerability concerns, military officials and planners were pressed to accept the risk of a false positive—what in statistics is known as a Type I error (falsely rejecting the null hypothesis)—to ensure the availability of the full complement of US nuclear forces to conduct US targeting plans. From their standpoint, the risk was acceptable given the consequences of a Type II error—wrongly dismissing evidence of an impending attack. With absolute certainty of an attack, as achieved from perfect knowledge of Soviet missiles in flight, LOW would amount to retaliation. But as uncertainty about the fact of an attack crept in, retaliation would edge dangerously closer to preemption in its essential features.[16]

Civilian Intervention

To be sure, civilian leaders expressed their discomfort with LOW and LUA. They wanted more control over wartime outcomes, fearing that war might start on false warning of a Soviet attack; and, if war occurred, they desired options short of an all-out exchange. They hoped that, by observing limitations, they could curb some excesses and maybe buy time in war for the Soviets to reconsider their position. Yet civilian leaders lacked the knowledge, time, and organizational presence to overcome military resistance.

Civilian leaders, policy hawks among them, expressed concerns about relying on LUA. Even the Reagan White House chose not to make LOW or LUA the US default option. Neither option allowed time to assess the degree and scope of a Soviet attack—indeed, to verify the fact of a Soviet attack. Still, the option was not foreclosed due to concerns over US ICBM vulnerability. Paul Nitze, for one, speculated that the United States might need to adopt a LOW posture if the US land force "becomes as vulnerable to a first strike as seems at least theoretically possible."[17]

Even Henry Kissinger, who strongly opposed LOW for its decisional shortfalls, saw LOW as useful for reinforcing deterrence.[18] Officially, so did the Reagan administration.[19] National Security Decision Directive 13 (NSDD 13) concluded that "we must leave Soviet planners with strong uncertainty as to how we might actually respond to such warning." Many in government (and research institutions) saw an upside if the Soviets *believed* that the United States would launch its missiles before Soviet warheads arrived.[20] NSDD 13 gave life to LOW and LUA, regardless, by establishing a need to "prevail" in a nuclear war and failing to rule out either of the two as a permissible option. NSDD 13 stated, in fact, that "it will remain our *policy* not to *rely* on launching our nuclear forces in an irrevocable matter upon warning that a Soviet missile attack has begun. . . [italics added]." Certainly, the administration could have voiced a more emphatic prohibition if it desired one. It thus indirectly sanctioned LOW (and LUA) as credible—indeed, usable—response options (across the three legs of the US triad).[21]

Lower-level commanders had the capability, and authority, to use their nuclear weapons under exceptional conditions—for instance, with a surprise Soviet nuclear attack, when contact with the president or their successor was impaired.[22] Although the decision to launch nuclear weapons was (and remains) for the most part a presidential decision, we must ask nonetheless whether a civilian leader would say "no" to a military launch request in the minutes before Soviet warheads were expected to hit their US targets. Military commanders would likely have pressed a president hard—presenting the situation, and options, perhaps in less than a minute—"pressuring him to quickly authorize retaliation while under apparent or confirmed attack" (Blair 2018: 8). A president might have heard that a limited US response would confound plans for a coordinated, large-scale US offensive; that it would compromise residual US nuclear forces by giving the Soviets an opportunity to strike US command and control centers and land-based missile sites; and that any delay would seriously impair the US command's ability to *limit* further damage to the United States. Perhaps they would have heard that a limited US nuclear response would allow the Soviets to seize the initiative *as they surely would*. Their command and control were perilously vulnerable to a US strike, and the Soviets lacked the capability to determine the full extent and scope of a US attack to appreciate its potential limitations.[23]

With major last-minute adjustments impossible to implement, civilian options would have reduced to a binary choice—do nothing or respond with the full force of some prepackaged alternative. Whatever the choice, a system designed for target coverage would likely have defaulted quickly to serve that imperative, as mounting damage, communication failures, breakdowns in command, and information deficiencies removed civilian leaders from the decisional loop. Lower-level commanders, acting within their authority, would have responded to threats, as they perceived them, or acted on instructions, as they understood them.

Thus, US civilian leaders sought an early warning system, technologically and organizationally, that would permit a quick, accurate reading of the Soviet threat and, if necessary, a rapid (presidentially authorized) US response. In that sense, the system was meant to preserve US options: the United States would not be caught off guard, forced to respond belatedly from a weak position. Through their resource allocations, equivocations, and deference, however, these leaders wedded the US strategic-nuclear force to a precipitous response posture which ironically constrained US options.

Organizational Influences

We see that risks stem from the military's preference for positive control. But risks also result when organizational "fixes" weaken negative control. Organizational theory tells us that organizational rules and practices block effective correctives and impair "learning" from mistakes.

Given the primacy of rules and practices, organizations favor solutions that minimize the extent of problems and require only incremental adjustments or some new corrective combination of subordinate tasks.[24] Indeed, organizations might reject the need for a corrective. For instance, management might blame the operator—maybe opting to retrain personnel—rather than acknowledge that existing practices make "operator error" more likely. Or the operators might keep supervisory personnel in the dark by "managing" the problem, not solving it—opting, instead, to minimize immediate distress and disturbances. Individuals within the organization are caught, then, between the organizational desire to perform effectively and to hold to the past and suppress failure.[25] We cannot assume, then, that the rational pursuit of organizational goals will prevail.

Absent a full assessment of problems, and opportunities for realistic testing, organizational "correctives" invite additional challenges. These challenges arise when new rules conflict with old rules, create uncertainty about how to act in given circumstances, divert resources and attention from other problems, or solve immediate problems by creating bigger future problems. An invisible bypass or work-around will not figure into official plans and might be recognized as a "problem" only with some unusual sequence of events ("Who could have foreseen . . . ?") or with non-routine conditions, as when US nuclear forces shift to a higher state of readiness. Unfortunately, civilian leaders might recognize the problem only after the fact—when it is perhaps too late to forestall disaster. Even under favorable circumstances, we can expect "some mismatch between prepared procedures and actual requirements" (Blair 1987: 114). Rules and procedures—new or old—cannot cover unexpected contingencies as will most certainly arise in a tense nuclear crisis: "The most stressing and potentially dangerous nuclear command problems arise when no rules cover the situation, either because no one anticipated that it could arise or because unanticipated shocks fundamentally change the operating environment in which the system is designed to work" (Bracken 1983: 12). Perversely, under these circumstances, the redundancies that planners design into systems for safety and security add to their "complex interdependence," a myriad of linkages that are not fully known or understood (Sagan 1993: 32, 39–40). They can lead to catastrophic failure when the safeguards meant to protect performance, or prevent disaster, interact with other system components.

To be sure, notorious incidents—safety violations and breakdowns in operator discipline—could compromise positive control. But informal rules, short-term fixes, and a lack of discipline—especially in hair-trigger situations—could lead to unauthorized weapons use.[26] We should not assume, then, that solutions designed to work in peacetime will abate challenges of command and control in a nuclear confrontation.

Challenges emerge, in fact, from the initial design of the system. The task of assessing and responding to a nuclear attack was distributed horizontally within the US government. Thus, the operational arm of the US nuclear force—the Strategic Air Command—was intentionally denied the task of providing warning of a nuclear attack (Bracken 1983: 21). Distributing responsibilities across the government provides institutional checks to prevent the accidental or precipitous launch of US nuclear weapons. Yet the same horizontal structure compromises control through decentralization and ambiguous command authority (Bracken 1983: 174). Horizontal systems invite miscommunication between subordinates and those "up the chain of command" as subordinates work out procedural details and solve problems—as *they define them*—to get things "to work." Misunderstanding among organizations and among the levels of command is inevitable. Barriers keep organizations from fully comprehending each other's practices; and superiors, by merit of their position, lack a grounds-eye view.

Hierarchically organized, semi-autonomous units within the system might operate semi-independently, then, with only an impaired understanding of the actions, procedures, biases, and limitations of other units. "Individuals responsible for preparing or implementing military options are likely to consult with subordinate and superior commands, but they may have little or no contact with horizontal commands that are charged with preparing or implementing other aspects of the same operation" (Lebow 1987: 77). Conditions are ripe, then, for a contagion of error in a nuclear crisis. Questionable evidence of an attack, gleaned from disparate sources, could overwhelm informational processing capabilities within the tightly coupled network: "tight coupling allows two or more errors that would ordinarily be minor and isolated to conspire together to produce a large error that propagates throughout the system."[27] The "conspiracy" is enabled by distance, despite the interconnectedness of the system: "Because these errors come from far-flung parts of the system, operators do not recognize their relationship until it is too late" (Carter 1987: 637).

The convergence of such evidence could give leaders a false sense of the existing threat in the compressed time frame in which a response is required. Paul Bracken's (1983: 53) prescient conclusions in this regard are worth quoting in full:

> While the complexity of the system has made us safer from accidental war it protects us only against the discrete, isolated failure. Multiple errors or malfunctions are a different matter altogether, because they invoke reactions by humans and involve organizational procedures and computers. The problem with compound accidents, especially those involving human behavior, is that the number of possible reactions is enormous and no design can protect against all of them. The likelihood that multiple events will lead to trouble increases when there is increased military activity. Thus, when forces are placed on alert, the complexity of the warning system may not only cease to produce redundancy; it may also amplify the mistakes.

For the same reason, the tight coupling between the *US and Soviet* warning and response systems invited catastrophe. The danger existed that each side would "misjudge" adversary military preparations in a crisis since they would produce unfamiliar routines. Each side knew how the other performed in *exercises* but that could not prepare them entirely for "the real thing," especially if either superpower sought "to mask the extent and intensity of its military preparations" (Lebow 1987: 69). Acting within an informational void, US precautionary responses—including a dispersal of weapons—could have activated Soviet precautionary responses, provoking a US response, and so forth. Moving to a higher alert status (say, DEFCON3) reduced the vulnerability of US forces, while also enhancing their capability to engage in attacks. By shifting the "prevailing balance toward positive control" (Steinbruner 1987: 540), US forces thereby constituted an inherent threat to Soviet nuclear forces (Blair 1987: 84). Each iteration thereafter—actions sparking same-side actions or enemy reactions—could magnify the prevailing sense of danger as various subunits acted, per their role or capability, and the system became ever-more tightly coupled (Bracken 1983: 56–59, 64, 72). In a nuclear crisis involving the United States and North Korea, we could easily envision a scenario whereby repositioning even US defensive assets would invite offsetting North Korean preparations, feeding such a reactive spiral. In a period of high stress, parties might fear the worst, and prepare for it. At some point, they might act on their preparations.

The challenges to crisis control only increase because the authoritarian governments, in the US adversary pool, construct these rules and procedures to serve their own parochial political interests. Because nuclear-control mechanisms reflect the structure of civil-military relations within a country (Feaver 1992/93: 174–178), Soviet leaders adopted negative over positive controls despite Soviet vulnerability to a US surprise attack (Meyer 1987: 488–491). Whereas "all states face a sharp trade-off between negative control (assurances against unauthorized use) and positive control (assurances for authorized use) over their nuclear weapons," the bias toward negative control "may be particularly strong in authoritarian states, where the threat of internal unrest to nuclear forces, strains in civil-military relations, and worries about political succession may all bias decisions toward prioritizing negative control over survivable nuclear C2 [command and control]" (Green and Long 2017b: 196). In other words, authoritarian leaders seek tight controls over nuclear weapons, as extensions of the constraints these leaders impose overall on the military, with the aims of preventing it from overthrowing the regime and preserving the government's key prerogatives.[28] So, why is this bad news, not necessarily good news?

A risk exists when leaders—fearing a loss in political control—distribute positions within the military to reward political or personal loyalty. Too few professional hands might remain on the "button" to protect against weapons use; indeed, the "wrong hands" could seize control of these weapons in the event of a military or civilian coup. The risk of a loss in control increases when a foreign conflict, itself, is instigated by domestic instability within the nuclear-armed country (the called wag-the-dog syndrome). The risk is graver still if these leaders, in developing countries, fear the vulnerability of their nuclear commands and weaponry, as by a quick strike by a geographically proximate adversary (Feaver 1992/93: 178–180).

As the chances of war increase, leaders might relax negative controls, by predelegating command authority to *unready* subordinates, to ensure positive control of the nuclear arsenal.[29] Assumptions about who is authorized to do what—under what circumstances—will likely vary across the command. They will also change as the conflict unfolds, and "local" interpretations of the rules meet local perceptions of the operant circumstances. National leaders, whatever they might think, cannot fully control the ins and outs of nuclear weapons employment: "The operational details that affect the balance between positive and negative controls . . . are too extensive, too diverse, and too remote to be regulated by executive edict" (Blair 1993: 30). Indeed, such control assumes that how nuclear-weapons employment will unfurl is knowable when no one can know how individuals and organizations—lacking crisis (positive-control) experience—will respond to the series of unanticipated events that could trigger a launch or predict exactly how procedures, shortfalls, bypasses, and command, communication, and intelligence failures will interact to affect outcomes.

Leaders might avail themselves of some response assuming *falsely* that time will eventually permit a more informed and deliberative decision, and that the action itself has not narrowed the latitude for choice sufficiently to lead to war. We should remember, in this context, that European leaders on the precipice of World War I acted with the same false notion about escalation: "They believed that it consisted of discrete, reversible steps, and they believed that high levels of military preparedness in no way prevented diplomatic efforts to resolve crisis" (Lebow 1987: 113). Even if leaders see the precipice ahead, they might still move forward. They might believe that, by delaying a decision, they will concede the upper hand in war, bargaining, or domestic politics. They might fear, for instance, that the adversary will benefit militarily from the "first move," the opponent will acquire a psychological edge when hesitation conveys indecisiveness and irresoluteness, or domestic critics will pounce with signs of government weakness.

So, we understandably hope that our leaders can handle a crisis. We might even take comfort in knowing that US nuclear weapons are entrusted to military professionals. The "near misses" of the Cold War tell us, however, that we cannot remain cavalier that checks will work in "normal" times, let alone in intense nuclear crises if leaders thrust and parry, using a variety of coercive bargaining tools. Who would have thought, back in 2007, that US air force crews would outfit nuclear-armed cruise missiles on a B-52 bomber—thinking the missiles were unarmed—and then fly the bombers to multiple US airfields, where the weapons would remain essentially unguarded for many hours?[30] We are told that, with precautionary measures and safety checks, such accidents could not happen. Is confidence warranted, then, that the impossible will not become the probable when standard procedures are relaxed or changed? Conversely, is confidence warranted that organizational practices will not prevail in a crisis that ill-suit the demands of the moment? After all, "how a political system performs in a crisis is a function of personal, group, institutional, and cultural patterns and interactions that were established long before the onset of the crisis" (Lebow 1987: 19). We might suffer the consequences when control over weapons takes a back seat to organizational imperatives and procedures.

Postscript: The Risk of Bad Decision-Making (by a Bad Decision Maker)

Of course, a well-designed organization—led by an impulsive and erratic *US leader*—might not produce better outcomes. Whereas US policy hawks had long fretted about the consequences of mentally unstable foreign leadership, many in government during the Trump administration expressed concerns that the decisional "limitations" of the incumbent US leader might lead to an unnecessary war. The concerns peaked in the final days of the administration with fears that Trump might manufacture an emergency to justify his hold on power. Chairman of the Joint Chiefs of Staff General Michael Milley even sought to reassure his Chinese counterpart that the United States had no plans to attack China, and told House Speaker Nancy Pelosi that he would intervene, as part of the chain of command, to contravene a rash presidential nuclear-launch order.[31]

That would require acting *outside* his legal authority, however. The president can consult with key civilian and military advisors to decide how to respond to a threat. Such *discretionary* consultation would most certainly occur if the president was apprised by military commanders of a pending nuclear attack on the United States. Moreover, a presidential order that came "out of the blue" would likely generate discussion and queries up the chain-of-command—and possible *extra-legal* intervention by military commanders and/or the president's civilian advisors. Yet neither the JCS Chairman, nor the Secretary of Defense, nor anyone else in government can veto a presidential launch order. Trump, to put his fiendish (deranged?) plan in motion, need only have authenticated his identity with a code (contained within a briefcase that always accompanies him), chosen from a menu (within the briefcase) of preplanned nuclear options, and transmitted the launch order to the Pentagon command center (see Woolf 2021). An ongoing crisis or emergency would likely reduce the desire, time, and opportunity for those charged with implementation to question the presidential command.[32] For these reasons, members of Congress proposed legislation—or, at least, wanted to consider legislation—to check the president's nuclear-launch authority.

The worst-case fears about Trump's behavior went unrealized. But an underlying question remains. Should the system incorporate checks on a US president, by requiring broader consultation (with the Secretary of Defense, key Congressional leaders, or others) before a launch order is permitted (Whitlark 2019)? The dilemma points again to a basic question: Can we design a system that "always" performs when it is needed and "never" performs when it is not (Green 2019)? We might focus on creating a safer system or on making a system more flexible or responsive. But, in seeking to do both, we inevitably compromise one for the other.

So, the better (albeit more complex) question is, "What risks will officials accept in system unresponsiveness to avoid the risks of an errant launch from misinformation, inadvertency, or (presidential) bad judgment?" Confounding an answer is that it is situation dependent. The answer depends, in part, on changing security conditions—and thus (fallible) judgments about the presence or absence of threat. We might accept a heightened risk of an inadvertent launch in a crisis, fearing controls that might impede a quick launch when time is of the essence. But how can we know with certainty whether the risk of being too slow to fire is greater than the risk of being too fast on the trigger? The answer is only complicated by the inherent interdependence of the two risks. Relaxing the hindrances to positive control—to facilitate a quick launch of missiles—increases the *need for negative control* given accompanying changes in the threat environment should the United States take the adversary's precautionary reactions as provocations.

The point, then, is that focusing on organizational fixes—at all levels—skirts a basic question that we cannot answer with an organizational chart. The answer requires reevaluating the fundamental trade-off—designed into the system—that favors positive (over negative) control of that portion of the US ICBM and SLBM force on launch-ready alert. As a former CIA director put it, "the system is designed for speed and decisiveness," not for "debate."[33]

Recognizing Inherent Risk: Nuclear Escalation in Post–Cold War Conflicts

Risk of a nuclear confrontation is inherent in any conflict between nuclear-armed states. US conventional confrontations with China and North Korea are hardly exceptions in that regard, though these two potential US military adversaries represent opposing ends of the risk continuum.

Confronting China

The Chinese government shares features with its Soviet predecessor and its current Russian successor: the Chinese government values centralized control and does not cavalierly brandish its nuclear weapons. China officially maintains a no-first-use policy, places a high premium on retaining (negative) control over its nuclear forces, possesses a nuclear force sufficiently robust to survive a nuclear attack, and—historically, at least—has kept its nuclear forces at low readiness (with missiles stored separately from warheads).[34] These factors militate against accidental and precipitous Chinese nuclear force use. Of note, then, are Chinese attitudes toward predelegation. China appears loath to transfer launch decisions to military officers in the field. Indeed, China generally appears sanguine about the chances of a precipitous nuclear-weapons launch in a US–China military confrontation.

For various reasons, some experts insist that a major conventional confrontation with China need not turn nuclear. They argue that China does not expect the United States to employ nuclear weapons in a confrontation and would avoid local intervention that could turn nuclear (Cunningham and Fravel 2019: 75). They maintain that a US conventional confrontation is controllable in part because neither the United States nor China could win a nuclear exchange. From their vantage point, the United States has great incentive to avert conditions—including conventional confrontations and irrevocable commitments to support allies with force—that could lead to the use of nuclear weapons. They assert further that the weapons and command structure that China maintains for fighting a conventional war is sufficiently distinct—by nature and location—from that for fighting a nuclear war. The separation allows for US strikes on the conventional infrastructure that would leave the nuclear infrastructure relatively unscathed. Inasmuch as China lacks a significant damage-limitation capability vis-à-vis the United States, it is unlikely to go nuclear, short of a direct, immediate threat to the Chinese nuclear force.

Risk, however, goes with the territory: "The generally relaxed beliefs about nuclear escalation currently espoused by China's strategic community seem unlikely to persist in a world where the outbreak of an intense conventional war would have recently proven many of this community's other working assumptions incorrect" (Talmadge 2017: 51). The danger here is multidimensional.

First, threat depends on the target's stakes. If Chinese officials had engaged in conflict believing China's vital interests were at risk, their sensitivity would increase accordingly to US conventional military activities that place those interests at risk. Would China's assessment not depend, then, on whether the conflict arose from a dispute with North Korea or a confrontation, instead, over Taiwan—and, thus, in China's view, a challenge to China's national sovereignty? If China was responding to multiple affronts, with a strong sense of grievance? Or, for that matter, if China believed it would "lose" the war if continuing to observe nuclear restraints? Indeed, with the confidence gained from acquiring second-strike retaliatory capability, "Chinese leaders might become much bolder in defending their perceived regional interests," especially "if China's aggressive behavior at the conventional level was launched in order to defend what Chinese elites sincerely believed to be the legitimate status quo" (Christensen 2012: 463).

Second, threat depends on the target's expectations and assumptions. If Chinese officials had misjudged the probability of war, or if they assumed falsely that the United States would not engage in certain conventional attacks that raise the risk of a nuclear conflict (Chase 2013: 61), they might quickly reevaluate their assessment that the United States will not seek an advantage by disabling China's nuclear force. In turn, if US leaders harbor concerns as to whether China adheres to a strict no-first-use policy,[35] China's *temptation* increases to get in the first nuclear blow, fearing an imminent US first strike. The significant US damage-limitation (but not damage-avoidance) capability vis-à-vis China only increases the US propensity to attack and China's incentive, then, to preempt that attack (even at the risk of instigating a catastrophic US nuclear response).

Third, threat is variable, as the vulnerability, expectations, and assumptions of the combatants change with the ebb and flow of a conflict. China, in peacetime, might downplay the probability of a nuclear war, but will that thinking hold when the bullets, and missiles, start flying? What might China do in a military conflict, once US military forces assault Chinese airfields, missile installations, radar facilities, command centers, and communications networks?

In threat assessment, we tend—quite naturally—to focus on the tangible aspects of the strategic problem—the factors that we can more easily comprehend and manipulate (attack). But threat lies also in important intangibles. These include the commitment that Chinese leaders maintain to their goals; their unwillingness to back down and incur humiliation and defeat; the tendencies to act based on misleading, partial, and unreliable information; the willingness of leaders and commanders to take risks or react impulsively under stress; and the sheer random element when anonymous individuals within an organization make a call with unseen or cascading consequences.

US war plans only exacerbate these dangers. US military plans for a conventional war with China have centered on taking the offensive against Chinese conventional targets. Plans have called for a quick blinding attack on Chinese military capabilities, assuming the advantage goes to the side that takes the offensive—escalating precipitously to achieve a knockout blow. Indeed, what was once called the Air-Sea Battle plan sought deliberately "to create confusion at the start of the war, making it very hard for the adversary to understand signals of restraint and declarations of limited intent" (Rovner 2017: 704). Under these conditions, pessimists ask, can we feel secure that safeguards adopted in peacetime—and not reality tested through employment in war—will prove effective? After all, China must relax safeguards in combat to permit positive control over its nuclear weapons, if only to ensure they continue to serve a deterrence function once Chinese conventional forces are engaged in combat. Indeed, a Chinese doctrine that deemphasizes nuclear-weapons use, and minimizes the threat of nuclear war (even in a conventional confrontation), invites the dangers of precipitous launch when combat realities challenge preexisting rules, practices, and conceptions.

The parties might provoke escalation even if intending to prevent it. What if Chinese officials were to believe that time was moving against them—that the US military could slowly but deliberately disable their nuclear capabilities, leaving China hostage to US terms to end the fighting? Might they not hit US forces (or targets), selectively, to bring US policymakers to their senses? Signaling "could easily be interpreted not as a demonstration of resolve or as a warning but as preparation to conduct actual nuclear missile strikes, possibly decreasing crisis stability or even triggering escalation rather than strengthening deterrence" (Chase 2013: 80). As problematic are actions mistaken for signals. Observers note, in this regard, that China arms some of its missiles interchangeably with conventional and nuclear warheads, that the United States might sink Chinese nuclear-armed submarines, mistakenly believing they are attack submarines, and that the Chinese nuclear command is not entirely independent of the conventional command (Talmadge 2017: 55).[36] Incidental or indirect attacks on Chinese "conventional" targets could lead China to misjudge US intentions and respond with nuclear force.

With US strikes on portions of the China's nuclear infrastructure, China might assume the United States is deviously seeking to nibble away at China's nuclear-response capability (Chase 2013: 69). US military planners might judge US strikes as largely benign: they overwhelmingly spare Chinese nuclear targets or center overwhelmingly on conventional Chinese targets. But any non-negligible portion of the Chinese nuclear infrastructure nearly hit, indirectly hit, or hit directly could feed negative Chinese expectations, hardened in conflict. Much depends on whether Chinese leaders believe their nuclear force is sufficiently robust and invulnerable to survive a possible US attack. A non-negligible probability exists, then, that China will launch missiles precipitously or that China will predelegate launch orders to field commanders, with the accompanying risks, to guard against a cutoff in communications with the central command.

As James Acton (2018) presciently warns, the dangers of escalation to a US–China (or US–Russia) nuclear conflict have only increased due to dual-use C³I networks, which clouds the purpose of attacks on those systems, and to fundamental interdependencies between conventional weapons and nuclear targets—on both sides of the conflict. That the same satellites are employed for conventional and nuclear missions makes them inviting targets *for* antisatellite weapons—but also makes inviting targets *of* antisatellite weapons. The underlying logic, which places any weapon in a degree of separation from any target, could present irresistible escalatory logic in an environment rich for misperception—that is, with an absence of real-time information on adversary concerns and considerations, suspicions that attacks by conventional weapons on dual-use targets is a smokescreen for the nuclear counterforce purposes of the attack, and doctrinal ambiguities and inconsistencies that suggest, to the United States, that its opponent might seize the nuclear initiative in war or escalate to test US "resolve" or avoid an impending military loss. That US officials, in war, might fear a loss in US damage-limitation capability, with an attack on US satellites, missile-defense systems, or C³I systems—*neutralizing an ostensible US nuclear advantage*—might further fuel the march toward a nuclearizing of the conflict. We should remember that the 2018 NPR indicates that cyberattacks alone could prompt US nuclear retaliation. For that matter, then, China's fear that deteriorating conditions will provoke a US nuclear response might induce the very behaviors—Chinese attacks on US C³I—that feed US apprehensions.

Thus, the "intangibles" of political pressure, organizational constraints, psychological stress, and misperception are no less central to risk despite their remoteness from view. Indeed, the very factors that comfort many experts—the Chinese emphasis on negative control—could become a liability should China feel threatened and shift to a positive-control posture. The risks of a precipitous, inadvertent, unauthorized, or accidental launch might increase—perhaps, dramatically—when China moves into unfamiliar territory. Negative control procedures are rehearsed and well understood; positive control procedures under realistic conditions are not.

We cannot know how Chinese leaders, let alone their military subordinates, might act under duress and uncertainty in a nuclear-charged environment. It would bring unexpected challenges as communication links are disrupted, commanders cannot separate good information from the bad, useful information becomes scarce perhaps due to its overwhelming quantities, and old rules designed to keep weapons from being launched poorly fit the new demands and environment. In the final analysis, those who downplay the risks in conflict—just like those who promote the manipulating of risk—understate the parties' (changing) intent and the forces that influence or negate it.

Confronting North Korea

North Korea differs substantially from China in force vulnerabilities, likely uses of nuclear weapons, and, of course, national leadership and behavior. Indeed, North Korea's mercurial and enigmatic leadership, aggressive missile and weapons testing, and vividly colorful threats make North Korea an outlier among states. For some, then, it appears unbound by the logic required for deterrence to work.

So, when might North Korea employ its nuclear weaponry? North Korea maintains that its nuclear forces protect the country's regime and territory. We can assume, then, that the probability of North Korean nuclear-weapons use would grow enormously if foreign forces pose a threat to the country's government or geographical integrity. For that reason, we can also assume that such a threat—absent North Korean war initiation—is remote.

Most experts believe, in fact, that any North Korean first use of nuclear weapons would occur at the onset, or in the conduct, of a North Korean invasion of South Korea (Bennett 2010: 26). North Korea might launch nuclear weapons to weaken the capability of allied forces to resist the offensive and, simultaneously, lend credence to the country's long-standing threat to use nuclear weapons if attacked. For these dual purposes, it can attack allied bases in Guam, Japan, and South Korea or opposition forces when mobilized for attacks (Narang and Panda 2020). Were its offensive against South Korea to fail, North Korea might employ some fraction of its nuclear arsenal—for these same military and coercive purposes—with the hope of denying victory to invading forces.[37] Whether on offense or defense, North Korea could withhold a portion of the country's ICBM force to threaten US targets, to deter an all-out US nuclear response. US leaders might hold back, then, knowing that a US nuclear strike would kill huge numbers of civilians and might still not prevent North Korea from lashing out, in retaliation, with one or more nuclear weapons. One reading of North Korean strategy suggests that the country is moving toward an "asymmetric escalation strategy" that mates conventional and nuclear capabilities to aid an offensive and constrain US nuclear options (Kang and Gibum 2017: 503).

Still, any use of nuclear weapons—by either side—would invite a substantial risk of a nuclear conflagration. At this point in the conflict, North Korea would likely be primed for follow-on attacks. After all, North Korea would have launched a conventional attack either believing that battlefield conditions had shifted in the country's favor (behind North Korea's nuclear-deterrence shield) or seeking to redirect domestic threats to the country's leadership through a diversionary war. If so, the chances are great that North Korea's leaders would have either *downplayed*—or *accepted*—the *risk* that the fighting (nuclear, or not) would spell the end of the North Korean regime. Indeed, the gloves might come off with the first use of nuclear weapons. North Korea might opt not to keep the bulk of its force in reserve. Its leaders might believe that the United States would not hold back—and fear their own country's inability to retain nuclear command and control under attack (see Cho 2009: 75).

In important ways, then, the two sides—at the precipice of an all-out nuclear war—would encounter the principles that Thomas Schelling (1960: 207–229) made famous in his notion of a "reciprocal fear of surprise attack." No party can afford to strike second when a decisive advantage goes to the initiator. Although US leaders need not fear disarmament by force in a war with North Korea, the logic still applies. That nuclear weapons offer North Korea a means to deny opponents a conventional military advantage, not just to punish them for their transgressions or to coerce concessions, only increases the likelihood that a North Korean nuclear attack would come early, and the likelihood that the United States would seek to prevent or preempt it.

Reasons exist, then, for pessimism—but also optimism—in assessing the risk of nuclear escalation in a Korean conflict. Hawkish analysts tend to take North Korea's threats at their face—quite understandably, perhaps—given North Korea's isolation from the rest of the world, the personality cult that surrounds the country's leader, and the opaqueness of the country's decision-making. They highlight uncertainty surrounding the North Korean regime, and its potential pursuit of seemingly suicidal goals, to justify US "worst-case" thinking and preparedness. By their thinking, if the adversary has no concern for costs, or defies predictability, the United States must act precipitously—even preventatively—to limit the costs the adversary can impose.

Yet we should remember the Cold War–era fear of hawks who argued that, with an impending loss of power, Soviet leaders would take their foreign adversaries down with them (see Chapter 2). Although hawks warned of the dangers of "mirror imaging"—assuming Soviet leaders shared "our values"—the mirror image, it turns out, was more accurate than the image that hawks offered. Our concern should focus, then, on conditions that make any war with North Korea more likely, since nuclear deterrence would undoubtedly remain precarious in a conventional confrontation with that country. Then, North Korea might still hold its nuclear punch in reserve, when losing the battle with foreign forces, awaiting the breaching of some "red line."

Conclusions

A party can boost the credibility of threats to employ nuclear force by manipulating risk for coercive leverage. Following Thomas Schelling's notion of a "threat that leaves something to chance," the party assumes that it can press the opponent to reduce the risk by deescalating the conflict.

Yet risk manipulation requires that the party ignite a fire to prevent an inferno—here, in the form of an all-out nuclear conflict. It presumes paradoxically that a party can achieve a favorable resolution by risking the consequences the tactic *is meant to avoid*. It rests, then, on heroic assumptions about the ability of *both* parties—given their shared reading of the situation, and its escalatory potential—to exercise the control required for the tactics to work. Where the manipulator sees restraint in its own behavior, the target might see reckless provocation or, worse, an imminent threat. It might conclude, then, that it must protect itself, by seizing the initiative and taking the offensive. At the very least, it might respond with forceful signals that send the opponent down the same path. The tactic thus invites various perils and pitfalls.

> ***5.1:*** *Policymakers might accept risks and trade-offs of precipitous action without duly considering arguments, information, and available alternatives that weigh toward a cautious response.*

Risk-manipulation tactics assume, in fact, that states are rational, coherent entities. But actions initiated by government leaders have unanticipated consequences when military organizations, charged with implementation, act by their own rules. Indeed, the interdependence between the actions of the opposing militaries could take the parties over the brink.

> ***5.2:*** *Government leaders lack familiarity with the rules and practices of military organizations that confound policy implementation and the execution of orders.*
>
> ***5.3:*** *Policymakers exaggerate the options available for the gradual or discriminate employment of nuclear force given military efforts to maximize force capabilities in conflict.*
>
> ***5.4:*** *Policymakers might fail to recognize that the interdependent actions of competing military organizations can fuel escalation toward war, even an all-out nuclear conflagration.*

Then, US leaders might act rashly out of ignorance, fear, or political self-interest. Solutions are encumbered, however, by a critical trade-off in the choice between negative and positive control.

> ***5.5:*** *Constraining the nuclear-launch authority of a US president compromises the US capability to respond as quickly, and lethally, to a nuclear attack; but conceding complete launch authority to the president could empower a poor, venal, or malign decision-maker.*

The dangers of poor decision-making increase in the fog of war, exacerbated by the contemporary interdependence between conventional and nuclear systems. Efforts to control risks in a nuclear confrontation with China or Russia are confounded by dual-use weaponry, the interdependence of a country's conventional and nuclear C^3I networks, and the potential uses of conventional weapons against nuclear targets. How any nuclear-armed party responds to an attack is highly dependent on its assumptions about the changing political and strategic context, rooted in assessments of adversary intent. Great danger exists, then, when combatants act out of fear of adversary preemption, take the offensive to reduce the costs of a nuclear exchange, or lash out to protect against seemingly inevitable defeat.

> **5.6:** *At some point, maybe early in a nuclear confrontation, the parties might stop manipulating risk and start preparing for the worst given the inherent risks of the confrontation.*

Thus, even the "superior" nuclear party is impaired when trying to boost its influence by manipulating risk. The challenge here, again, is from tactics that are ineffective, at best.

6

Resolve and Reputation

To establish its credibility to use nuclear force, a nuclear "superior" arguably has other possible options. It can bolster its apparent "resolve" to convince the opponent that the cost of its transgressions will exceed any gains. Or else it can build—and then store—credibility for future use by establishing a "reputation" for standing by commitments or taking decisive action. A party can supposedly harness both mechanisms by establishing a "reputation for resolve."

The concepts of resolve and reputation are problematic, however, when employed to understand deterrence (or compellence) relationships. Outcomes that we often attribute to resolve, or reputations, owe to other plausible influences. Indeed, efforts to establish resolve or build a reputation for action might again prove ineffective or counterproductive. The questionable contribution to coercive influence of demonstrating resolve and forging reputations becomes apparent when assessing each concept in turn.

Demonstrating Resolve

Resolve is commonly viewed as purposiveness, toughness, or resiliency. Although its usefulness for explaining bargaining outcomes might thus seem self-evident, resolve might hinder, more than assist, our understanding of how deterrence and compellence relationships work. The concept's deficiencies become apparent when examining the nature of resolve and challenges in communicating it.

The Nature of Resolve

To predict which of two conflicting parties will prevail in a confrontation, many academic and policy analysts look to the parties' relative resolve: their dedication to the effort or willingness to persevere despite challenges, setbacks, and overall costs. Conceived thusly, confrontations between states reduce to contests where victory goes to the side that "wants it more." That seems true, at least, in athletic contests.

The False Promise of Superiority. James H. Lebovic, Oxford University Press.
DOI: 10.1093/oso/9780197680865.003.0006

Fired up by an inspirational coach or player, the victor rises to the challenge at a decisive moment, perhaps snatching victory from impending defeat.[1]

The prominence of the concept in deterrence theory belies the theory's roots in realism, with its explanatory emphasis on state interests and capabilities.[2] Thus, the impact on crisis outcomes of participant resolve—as compared to the "balance of power" and "relative stakes"—remains a subject of dispute in the international politics literature. Skeptics doubt that a party—despite its supposed resolve—can realize success against an opponent with the greater capability or interest in defending or in challenging the status quo. Their reasoning applies even to an athletic contest. If a competitor acquires satisfaction from its "acts of determination"—believing, for instance, that enduring pain builds character or toughness—or if it expects acclaim from fighting through injury, then "interests," not resolve, explain its behavior. We can certainly question, then, whether "resolve" matters when a party means its displays of determination to "signal" its interests and capabilities to a doubtful adversary (as when a competitor chooses to "up its game," with a quick offensive burst, hoping to give the opponent a "dose of reality").

The term's presence in the literature—indeed, in the popular lexicon—therefore requires that we ask whether resolve has explanatory power. What should we make of Vice President Mike Pence's gesture, in an early trip to South Korea, where he articulated a high-pressure, no-negotiation policy with North Korea to convince it to abandon its nuclear and ballistic-missile programs? After venturing out from the Freedom House building, Pence stared across the demilitarized zone into North Korea because, as he put it, "I thought it was important that people on the other side of the DMZ see our resolve in my face."[3] The implication was that an appropriate facial expression would strengthen the underlying message. But, if so, could that same expression—meant presumably to convey "we really mean it"—substitute for an absence of US interest or capability backing the policy? Moreover, could an expression add weight to the message assuming that it was grounded in US interests and capability?

The issue, here, is fundamental to the US approach to other major conflicts, including, as of early 2022, the confrontation with Russia over the future of Ukraine. The Biden administration tried to signal US resolve to aid the defense of Ukraine, in the preinvasion period, in response to Russian moves portending an offensive against that country. The United States supplied Ukraine with weapons, and it warned Russia of serious economic retaliation and a fortification of NATO's position in Eastern Europe should Russia attempt a land grab. Indeed, Putin's actions abetted the US effort, by giving NATO new life and purpose: members boosted their troops levels and weapons commitments to frontline states in the conflict.[4] But the question remained, could these efforts help to deter a Russian military offensive given Russian advantages in relative interest and capabilities?

After all, Russia held considerable advantages. These advantages included Russia's military deployments along Ukraine's border, Russia's absorption of prior sanctions (imposed by the West for the Putin government's various misdeeds), and Russia's greater interest, by reason of geography and shared history, in controlling the future of Ukraine. Not only did Russia's President, Vladimir Putin, insist that NATO directly jeopardized his country's security by extending the NATO security guarantee to countries on the Russian perimeter, but he maintained that NATO actions betrayed aggressive designs on Russia. Going further, he denied the legitimacy of an independent Ukrainian state, claimed it as part of "One Russia,"[5] and dismissed Ukraine as a fictitious creation of his predecessor's failed leadership.[6]

By contrast, NATO members indicated that their interests dictated *avoiding* a military confrontation with Russia—even tolerating some level of Russian aggression. Whereas the European Union had responded with only limited economic sanctions when, in 2014, Russia seized Crimea, members diverged now in their preferred response to Russian transgressions.[7] Germany depended greatly on gas supplies from Russia and was now blocking the supply of German-made arms to Ukraine. France, in turn, saw opportunity in the dispute to push for a European Union–led security arrangement and negotiating approach.[8] For much of the citizenry of NATO's member states, defending against Russian territorial threats was not a priority: The reality of forceful efforts to overturn the territorial status quo in the region, in the last world war, was confined to distant memory or buried with the increasingly forgotten toll in the military graveyards of Europe. Indeed, the leaders of the United States, Britain, Germany, and France were distracted by domestic political needs and pressure, due in no small part to an ongoing pandemic and the accompanying challenges of economic recovery. NATO members feared, of course, that a less-than-decisive response would feed Putin's desire to roll back NATO to its Cold War–era defense line. Still, NATO representatives affirmed that the North Atlantic Treaty's Article 5 provision did not apply to Ukraine and early US joint diplomatic efforts included select NATO countries but not Ukraine nor alliance members bordering Russia.[9]

Could any show of US (or NATO) "resolve" override these salient facts conveying the limits of US (and NATO) capabilities and interest? To be sure, critics rightly charged that President Biden conceded too much in a January 2022 press conference in which he stated his belief that Russia was going to attack Ukraine ("he has to do something"), that NATO remained divided nonetheless on the appropriate response, and thus that NATO would tailor its response to the scope of a Russian landgrab, even suggesting that it might accept a "minor incursion" by Russian forces (meant perhaps to consolidate Russian holdings in Ukraine).[10] Yet, by his statements, did Biden not simply provide Putin with a clearer sense of the Western opposition's overall cost-benefit analysis (by "telling the truth")? Ultimately, Putin had to decide whether the potential costs of an offensive (including the future challenges of an occupation) offset his greater interest in controlling the future of Ukraine.

Whether resolve has independent standing and importance is a question that also arises in scholarly treatments. Resolve suffers as an *independent* explanation, for example, when scholars reduce it to the "balance of interests" or the relative stakes of the parties. Richard Betts (1987: 12) notes, for instance, that resolve plays an important part in the thinking of "risk maximizers." Yet he associates the maximizers' viewpoint with a conception of resolve rooted in interests: "If both sides have an unbearable amount to lose from going to war, then as they edge toward the brink the two sides' respective resolve to fight rather than concede flows from how much they have to lose if they back down" (Betts 1987: 14). Although resolve supposedly stands apart from interests and capabilities—which perhaps lends them credibility—the independent contribution of resolve in the formula is unclear.[11]

Similar problems afflict contemporary analyses. In "Nuclear Superiority and the Balance of Resolve: Explaining Nuclear Crisis Outcomes," Kroenig (2013) goes to great lengths to draw out "resolve" as a key explanatory concept. His attempts to define the concept fall short, however, given its indistinguishability from, and causal origins in, other factors explicitly studied in his analysis. His project motivation illustrates this. In Kroenig's (2013: 143) words:

> I demonstrate that nuclear crises are competitions in risk taking, but that nuclear superiority—defined as an advantage in the size of a state's nuclear arsenal relative to that of its opponent—increases the level of risk that a state is willing to run in a crisis. I show that states that enjoy a nuclear advantage over their opponents possess higher levels of effective resolve. More resolved states are willing to push harder in a crisis, improving their prospects of victory.

Kroenig, here, provokes more questions than he answers. Is superiority the basis of resolve? Is resolve different from risk acceptance? Indeed, Kroenig (2018: 23) claims elsewhere that the more resolved state is one "that is willing to tolerate the greatest risk of nuclear war." He is not alone in suggesting that the two concepts are interrelated—maybe indistinguishable. When Mercer (1996: 15) defines resolve as the "extent to which a state will risk war to keep its promises and upholds its threats," resolve amounts to risk acceptance.

For his part, Joshua Kertzer (2016: 3) defines resolve as a "state of firmness or steadfastness of purpose," "second-order volition," or "willpower." In his view, it is not reducible to an actor's intentions nor equivalent to an actor's capabilities. But Kertzer fudges the definitional issue with claims that resolve is "unobservable," when its observability matches that of other international-political concepts, including "intentions." All useful concepts become observable through valid operational definitions, predicated, in turn, on clear conceptual definitions. What makes a concept "unobservable," then, is its lack of conceptual clarity.

Rather than address that issue directly, Kertzer (2016: 26) opts for a "novel solution." His proposal: "rather than infer resolve via its consequences, we study resolve via its causes." In one project, Kertzer (2017) thus links resolve to individual time and risk preferences: the timeframe in which people seek outcomes and the risks that they will run to achieve them. Both, he establishes, predict a willingness to stay the course. We can reasonably ask, however, whether he is studying "resolve" *rather than* time and risk preferences. Even if resolve is related to time or risk preferences, we still do not know whether these preferences constitute defining features of resolve or, instead, its sources or consequences.

Roseanne McManus (2017), in a multimethod study, aims to rescue the concept of resolve by placing the focus on US presidential "statements of resolve." The empirical focus could assist our understanding of resolve by illuminating its concrete manifestations. The analysis lives and dies nonetheless with her definition of the key concept.[12] When she defines statements of resolve, then, as "the value that a country or leader places on a particular issue," she begs the question, "Why call it resolve?" Indeed, elsewhere in McManus' analysis, resolve amounts to a capability to act, a willingness to act, or "commitment." She defines statements of resolve, then, "as public statements which indicate that a country is committed to a position" (McManus 2017: 46). In operationalization, these statements of "resolve" amount to "conflictive behavior," which includes demands, threats, and negative assessments.

Definitional issues—here and elsewhere—obtain urgency because those who use the term *suggest* that the resolute push bigger, harder, or longer than their interests and capabilities warrant. Resolve thus amounts to an exaggerated reaction to conditions, a catalyst that spurs action, or the fuel that sustains it. The problem for analysis, however, is that we cannot define resolve without understanding its source if we are to distinguish its effects from those of other variables. Note, for instance, that Kertzer (2016: 35) posits that resolve is sometimes "situational," and depends, then, on the interests and capabilities of a party. Thus, we must ask, what specific mechanisms propel leaders (or, instead, states) to act with more vigor than they otherwise would? With dutiful probing, we might conclude that "resolve" reflects the impact of some set of sources that we can *usefully identify*.

If resolve stems from state interests or capabilities—or a leader's (or public's) aspirations, fears, expectations, or aversions—we might conclude that it is more outcome than explanation. We sacrifice explanatory and predictive power, then, by focusing on resolve at the expense of its sources. Alternatively, we might conclude that resolve reduces to one or more of its sources. That appears true of Keren Yarhi-Milo's (2018: 5) analysis in which she ties resolve to a "willingness to pay high costs and run high risks" in crises. But (as discussed in Chapter 5) a willingness to absorb costs (to achieve benefits) is tantamount to the lowering of risk. Of course, a party can do both—accept risk and absorb cost—but a willingness to accept high costs and run high risks is *potentially* a contradiction in terms. It is one thing to go up against a competitor who accepts a greater risk of war. It is quite another to challenge a competitor who believes it can *tolerate the risk* of war because it can profitably *absorb the war's costs*. Contrast a boxer who jabs from the outside to avoid getting hit versus another who fights on the inside and willingly takes the punches.

Yet, we are not out of the conceptual woods even should we carefully distinguish the effects of cost versus risk. We have come full circle if a party's ability to absorb cost stems from the party's capability—for instance, its large population base or physical size—or the party's interest—for instance, in negating the threat posed by an opponent's nuclear capabilities. If so, calling the resulting behavior a display of "resolve" simply confuses the issue by distracting from the underlying influences.

Jettisoning the term entirely, then, might yield a productive focus on other variables with well-documented effects. The additional push, attributed to resolve, could reflect an upswell in public support owing to a leader's manipulation of audience costs (Fearon 1994) or capitalization on a "rally 'round the flag" effect (Baker and Oneal 2001; Baum 2002). Or else, "resolve" might reflect the state of a bargaining relationship—whether the party is seeking to deter or compel another's actions, and then whether the party is winning or losing. As noted previously (in Chapter 5), research in behavioral economics reveals that individuals take higher risks to regain amounts lost than they would to obtain similar gains from a neutral prior position. Again, think of a gambler who bets cautiously and responsibly, but makes bigger bets—with smaller odds of winning—when hitting a losing streak. To attribute the gambler's behavior to "resolve" misconstrues the situation. Indeed, it falsely suggests that the gambler acts due to internal rather than situational forces.

At the risk of circularity, one way out of this conundrum is to assess efforts to establish a "reputation for resolve" (Dafoe and Caughey 2016: 346; Lupton 2020; Yarhi-Milo 2018: 5). If a party believes it might receive concessions in a current or future conflict, by establishing such a reputation, it will act with *firmness* in the present to acquire gains—*beyond those attributable to its interests and capabilities*. Yet the reputational form of the concept does not ease the analytical burden. Researchers must still establish that: (a) the party's current actions exceed some baseline of behavior set by the party's interests and capabilities and (b) the observed effects are not different from those attributable, for instance, to efforts to establish "credibility."[13] Even then, confusion results when parties seek reputations or apparent resoluteness as a matter of "interest."

Thus, the term "resolve"—in its widespread and varied use—obscures its conceptual and causal foundations, hindering our understanding of whether and how it affects conflict outcomes. When viewed instead, for example, as cost acceptance, we better appreciate the link to other variables but also the causal boundaries and limits. We know that leaders do not sacrifice for the sake of sacrificing. Their behavior is conditional, situational, and directed. By contrast, resolve stands implicitly as an undifferentiated, unbounded, and undirected force. We are left uncertain over whether it is asset or outcome, and why and when it varies or yields coercive benefits. So, we have a choice. We can treat resolve as a synonym for some other variable ("credibility"), depriving it of separate meaning. We can treat resolve as a *mysterious* residual category, attributing to it what we cannot otherwise explain. Or we can clearly define resolve, distinguish it from other variables, and then seek to ascertain its independent explanatory contribution—if any.

If we adopt the latter approach, we must recognize that bad operational definitions can undermine even sound conceptual ones. Despite their commendable rigor, Todd Sescher and Matthew Fuhrmann (2017: 260) fall a bit short, then, when including resolve among their "military escalation" controls in an analysis of compellence outcomes. Resolve is present, then, "if the challenger employed demonstrations of force or engaged in military mobilizations during a compellent threat episode." We can question the dichotomous treatment of the variable (it is either present or not) and its representation in absolute form rather than a ratio of the challenger's resolution over the defender's resolution. More noteworthy for present purposes is that resolve reduces, in the analysis, to various physical gestures. The dichotomous variable "is coded one if the challenger employed demonstrations of force or engaged in military mobilizations during a compellent threat episode" (Sescher and Fuhrman 2017: 260). Rather than indicating resolve, such behavior might convey its absence. A bully who makes a show of removing his jacket and rolling up his sleeves for a fight could well be looking to buy time—to avoid a fight.

This is not to minimize the challenge in separating and isolating the effect of resoluteness from other explanatory effects, as the Cuban Missile Crisis well illustrates. We can certainly debate whether the balance of capability or relative stakes favored the United States or the Soviet Union during the crisis. We only add to our explanatory misery when we attempt to introduce a "balance of resolve," apart from these capabilities and stakes. Consider the evidence.

The United States *seemed* to hold an advantageous position. First, it had strong reason to resist. The Soviet Union had moved its missiles off the US shoreline, giving the Soviets the capability to strike much of the continental United States with limited warning. Second, the United States, without question, enjoyed conventional military superiority in the Caribbean: The Soviets had to traverse vast distances by sea to contest US combat power in the region. Third, the United States enjoyed nuclear superiority on critical measures: it could deliver larger numbers of warheads quickly from Europe—or over intercontinental distances, by missile—when the Soviets still relied heavily on bombers that would take hours to arrive on target (and then on one-way flights, lacking fuel for a return). Fourth, the Kennedy administration faced extraordinary pressure from Congress to take a decisive stance. Republicans admonished the administration for failing to act sooner, and Democrats feared the political fallout from a less-than-firm response (Zeisberg 2015: 148).

The Soviets, nevertheless, had a decent hand to play. First, to say that the Soviets were disadvantaged in nuclear delivery capability against US targets is not to say that the Soviets had *no* such capability. US officials knew the Soviets, at their disposal, had hundreds of warheads and many dozens of missiles that would virtually guarantee the destruction of some number of US targets in the event of a nuclear exchange (Bell and Macdonald 2019). Second, even if the United States enjoyed conventional primacy in the Caribbean, the Soviets could opt to contest US power in Berlin or elsewhere seeking a "grand compromise" that would permit the Soviets to retain their Cuban arsenal. Third, once placing missiles in Cuba, the Soviets could establish a new status quo, which would help them overcome their nuclear disadvantage. They would have to remove the missiles—quite visibly—to satisfy US demands. Such a retraction would impose reputational costs on the Soviets and bring unwanted political scrutiny on Khrushchev back home (as it eventually did, leading to his ouster). Fourth, the Kennedy administration offered concessions that revealed the administration's uneasiness about the "fairness" of the status quo ante and trepidations about a US–Soviet military confrontation. The administration engaged the Soviets early in talk about the possibility of offering a pledge to respect Cuban sovereignty. It agreed to trade US missiles in Turkey for the Soviet withdrawal of missiles from Cuba despite the objections of the Turkish government and the political sensitivity of such a concession (hence the ruse and secrecy surrounding the deal). Moreover, it maintained open dialogue with the Soviets, through various channels, and seemed reticent to remove the Soviet missiles with force. Finally, time was on the Soviet's side. Success in keeping Soviet missiles out of Cuba would little affect the US–Soviet nuclear balance once the Soviets acquired the capability to deliver massive nuclear power against US targets by missile, bomber, and submarine. Although this potential reality could hurt the Soviet bargaining position, much would depend on the US reading of the situation, as when Robert McNamara initially expressed doubts about the practical difference should missiles hit US targets from Cuba or the Soviet Union.

So, who won the battle of resolve? If resolve draws from interest or capabilities, its contribution to the outcome of the crisis is unclear. If the term has unique explanatory worth, its contribution to the crisis outcome is even less clear. After all, President Kennedy and his advisors hardly relished the thought that the US–Soviet confrontation would turn nuclear.

To be sure, Danielle Lupton (2020) makes a strong case, backed by experimental and documentary evidence, thatperceptions of resolve matter in international conflict and, indeed, were decisive in the Cuban Missile Crisis. Yet her conclusion—Khrushchev sent missiles to Cuba upon doubting Kennedy's resolve—can take us only so far.[14] Written statements and reflections are revealing. But so are less accessible understandings and assumptions that go unrecognized or unarticulated. The alleged balance of resolve should have us scrutinizing the place, timing, and nature of Soviet action. Why would Kennedy's apparent equivocation at the Bay of Pigs—in a failed effort to overthrow Cuba's Communist government—have convinced Khrushchev that Kennedy would accept Soviet *nuclear weapons* on the island? Why did Khrushchev present Kennedy with a fait accompli, by secretly transporting missiles to the island, if he believed Kennedy lacked the resolve to block Soviet transgressions? Why did the Soviets initiate action in Cuba rather than Berlin, where they enjoyed a greater conventional and nuclear military advantage? After all, Kennedy equivocated allegedly on matters pertaining to Germany including the construction of the Berlin Wall (Lupton 2020: 94). Could Khrushchev's beliefs about US interests not explain much of the evidence attributed to resolve? Lupton (2020: 99) asserts, for instance, that Kennedy's response to the missiles in Cuba caused Khrushchev to revise his assumptions about Kennedy's *interests* in Cuba—and thus that interests "help explain how Kennedy was able to alter his reputation during the missile crisis." Does that not suggest, then, that interest considerations could also explain—maybe entirely—the initial missile deployment (regardless of whether Khrushchev used words that suggest he considered resoluteness)? Was Khrushchev's Cuban gambit designed to strengthen the Soviet global military position and, if so, what does that say about the actual importance of resolve or its causal underpinnings? Why did Kennedy's more cautious blockade approach not reinforce Khrushchev's belief that Kennedy was reluctant to employ force, nuclear or otherwise? Why did Khrushchev not read Kennedy's (unexpectedly) tough stance in Cuba as sufficiently "situational" that it would still permit Soviet action against Berlin?

Answers are not easily provided with reference to resolve alone. We potentially lose much explanatory power, then, when neglecting the intentions of the parties as they reflect their objectives, perceptions of interest and capability, and sense of the ambient threat and changing probability of war due to factors that neither leader could control.

As a first step in addressing the definitional conundrum, analysts should find it useful to ask, "From what, and where, does resolve derive?" Is it a function of ideology or belief? Does it rest in the characteristics—or personal character—of government leaders, or instead in behavioral forces or national culture, religion, and society?[15] Lacking clear and consistent answers to these questions, resolve will amount to a vague synonym for other ill-defined variables or an ambiguous "residual category"—a "trash-can" variable of sorts. It could encompass reputation concerns, the irrational effect of emotion, pure stubbornness, or any number of other influences, depending on whether individuals, governments, or states are the subjects of analysis. It might also reduce to readily available empirical indicators such as the relative employment of violence in a confrontation (see, e.g., Lupton 2018).

Given the lack of clarity, assessments of resolve could easily amount to a conflation of outcomes and explanations. Put differently, had the Soviets retained their missiles in Cuba, would we judge the balance of resolve in the crisis differently?

Communicating Resolve: Lessons for—and from—the Trump Administration

Of course, Donald Trump proved no friend of conceptual clarity, or sound theory, when he implied early in his administration that resolve dwarfed all other influences in bargaining. He seemed to act on what he preached when he assailed the North Korean nuclear program with fiery rhetoric. He claimed, for instance, that the United States was "locked and loaded" and would destroy North Korea if sufficiently provoked. He backed his rhetoric with military demonstrations. Later, basking in the glory of having secured the release of three Americans from a North Korean prison and the upcoming 2017 summit with North Korea's leader in Singapore, Trump mocked his critics who supposedly claimed he would get the United States into a nuclear war. His response, "And you know what gets you into nuclear war and you know what gets you into other wars? Weakness, weakness."[16]

Inasmuch as US capabilities and stakes had not changed from prior administrations, what had presumably changed was US *resolve*—a willingness to stare down opponents and dare them to take their best shot. Indeed, by withdrawing from the Iran deal, the administration maintained it was signaling its resolve—strength and toughness—to North Korea, to facilitate denuclearization negotiations with that country. In Trump's words, "I think it sends the right message."[17] His National Security Advisor John Bolton echoed these sentiments when arguing that by getting out of a bad deal with Iran, the US sent "a very important message to North Korea that we're not in these negotiations with them just to get a deal."[18]

Such statements, and actions, belie the many challenges in signaling resolve. Despite Trump's claims and bravado, we can question the practical impact of his words and demonstrations in a messy world of distracting and conflicting signals. We can ask, more generally, whether a determination to send a message of resolve can overcome the incentives of an opponent to remain steadfast and unyielding.

Leaders Focus on the Message They Want to Send, Not How the Adversary Will Receive It

While leaders usually think they are signaling toughness, they often do not realize that their words and behavior may be sending several conflicting messages. During the Vietnam War, for example, both the Johnson and Nixon administrations tried to use aerial bombing to show US resolve to Hanoi. Such messages required that Hanoi ignore the many other messages sent—by US domestic protests, the state of combat in South Vietnam, and, for that matter, the need *to send messages* because the administration could not impose a solution. The United States always retained the option of upping the level of violence. It could have bombed the dikes in North Vietnam and flooded its countryside or engaged in the saturation bombing of Hanoi. That it chose not to do so *was a message*.[19]

Even if Trump believed he was coming to the negotiations with North Korea with a reputation for resolve, his words and behavior sent other messages. Trump shifted his rhetorical tone dramatically with early signs that North Korea was willing to discuss the "denuclearization" of the Korean peninsula—a term which, in the North Korean diplomatic vocabulary, would bind the United States no less than North Korea. Indeed, it could place the onus of adherence on the United States, before North Korea would take concrete steps. Moreover, he took a victory lap before direct negotiations even started—touting them as a major achievement. Had he not thereby conveyed an absence of resolve? After all, he had seemingly settled for a small concession—an agreement only to talk. Might that not suggest to North Korea that it held the cards since Trump seemed desperate for a win? For that matter, when he encouraged talk that he deserved the Nobel Peace Prize for his initial efforts—and sold them publicly as a major contribution to world peace—did he not also reveal absent resolve? Did that not suggest that his goals were personal, not professional, and that he would settle then for a pretense of success, over a viable agreement? That Trump would lower the bar to get *his* deal?

Trump sent various incompatible messages about the nature of the deal he would accept. He backed out of the Iran deal, criticizing it for failing to control Iran's non-nuclear activities and for permitting Iran to build up its nuclear capability in the future. But he did not explicitly challenge terms of the deal that allowed Iran to retain its nuclear infrastructure. Might North Korea reasonably assume, then, that a "denuclearized" North Korea could still possess a robust nuclear program? For that matter, North Korea had reason to discount Trump's criticism of the Iran deal. It was one in a long line of prior agreements he had dubbed the "worst deal ever."

Befitting the rush to summitry, the resulting "deal" between Trump and Kim Jong Un amounted to an agreement that the parties would "work towards the complete denuclearization of the Korean Peninsula." That agreement left much to the imagination, and the future. Indeed, it likely left North Korea to conclude that: (a) *nuclear weapons* brought a US president to the bargaining table, (b) surrendering nuclear weapons would mean losing the country's most valuable security asset, and (c) the country need not expend its critical bargaining chip to gain US concessions. These views likely gained strength in the months following the Singapore summit. Trump heaped praise on his agreement, and the North Korean leader, despite North Korea's demands, continuing nuclear efforts, and unwillingness to take positive steps toward increasing the transparency, much less moving toward the dismantling, of the country's nuclear program. North Korea continued to restate its sole pledge to "permanently dismantle" a missile engine test site. At the same time, it called on the parties to consummate a peace agreement that would formally end the Korean conflict, and presumably end the US justification for maintaining forces in South Korea and a state of belligerence toward North Korea.[20]

The stalemate that ensued, as the months ticked by, could only reinforce the North Korean position. Indeed, in the interim, North Korea continued to build up its nuclear arsenal. Trump implicitly conceded that by not acting negatively, North Korea had essentially met US conditions and need not take positive steps, at least in the short term, to satisfy US demands. While US officials hoped North Korea would provide an accounting of its various nuclear sites as a first step toward denuclearization, Trump celebrated the absence of negative behavior: "A lot of tremendous things. But very importantly, no missile testing, no nuclear testing."[21] He fueled suspicions that he would accept the status quo as a more permanent arrangement when, at the United Nations (in September 2018), he repeatedly praised North Korea's leader for his past restraint and indicated that the United States was in no hurry to achieve nuclear dismantlement: "We're not playing the time game. If it takes two years, three years or five months, it doesn't matter. There's no nuclear testing, and there's not testing of rockets."[22]

Simply maintaining the status quo—not testing—became the implicit US requirement of the North Koreans. In a speech before his departure for the second summit with Kim in Hanoi (in February 2019), Trump again took the burden of concessions off his North Korean counterpart, by suggesting that Kim was already doing enough. In Trump's words, "I'm not in a rush," he said. "I just don't want testing. As long as there's no testing, we're happy."[23] His implicit blessing of the status quo dovetailed with his assessments of "great progress" in ongoing negotiations;[24] the value he seemed to place on just preserving his "special relationship" with Kim (constantly referencing the "beautiful letters" that Kim had written him);[25] his persistent coveting of recognition, in the form of a Nobel prize,[26] that a deal would afford; his seeming lack of concern that Kim had previously refused to meet in Pyongyang with Secretary of State Mike Pompeo who was seeking concrete North Korean steps toward disclosure and denuclearization; and his general optimism about the chances of making a deal. Sure, lower-level negotiations did not suggest a deal was close. But Trump seemed sure that a deal was within reach, and Trump was notorious for blindsiding and overruling subordinates. Most importantly, Trump seemed anxious for a deal. He was under severe pressure domestically with numerous—indeed, ever increasing—investigations by the Justice Department and Congress for misconduct associated with the 2016 election and potential obstruction of justice in the ensuing probes.[27]

With the start of talks on February 28, Trump pronounced to the press, "No rush. No rush. No rush. There's no rush, we just want to do the right deal. Chairman Kim and myself we want to do the right deal. Speed is not important, what's important is that we do the right deal."[28] The audacious itinerary said otherwise. Without prior agreement on specifics, the parties had scheduled a signing ceremony later the same day. What was Kim to conclude? Was the message not, "stick to your position and you will win out?"

Inevitably, perhaps, the summit could not close the divide. Kim pressed demands for the lifting of sanctions on the North Korean civilian economy in exchange for the permanent dismantling of the Yongbyon nuclear facility.[29] That would still leave North Korea with possibly over five-dozen nuclear warheads, a robust uranium-enrichment capability, and a vibrant missile program for delivering the warheads—which all would remain shrouded in secrecy. Trump left the summit early; a "working lunch" between the two leaders was among the casualties.

Trump demurred when asked at a post-summit news conference if his goal remained North Korea's "complete, verifiable denuclearization." His response: "I don't want to say that to you because I don't want to put myself in that position from the standpoint of negotiations—but, you know, we want a lot to be given up."[30] Soon after, he signaled his openness to some unspecified incremental (smaller) deal on the road to something bigger.[31]

We might agree that we have much to gain from pursuing modest objectives—that denuclearization will occur only through an arduous, long-term process—and that talking is better than fighting or provocative missile testing. We might also agree that Donald Trump is an atypical case—as a master of inconsistency—that offers few insights into how most leaders craft their messages. The issue here, however, is whether Trump's own conduct fed expectations that undercut his apparent resolve. He came into the negotiations with a track record of threats and a willingness to walk away from "bad deals." His apparent resolve—manufactured for effect (at home and abroad)—got lost in communication. The reason: Trump had a strong incentive to concede ground, to a determined opponent, to get a deal. In the months that followed, Trump's tolerance for (indeed, excusing of) North Korea's short-range ballistic missile tests, criticism of "ridiculous and expensive" joint military exercises (pairing US and South Korean troops), and harangues against South Korea for paying inadequately for its own defense could only reinforce a North Korean sense that Trump would not hold firm.[32]

To be sure, then, we can read Trump's behavior as a primer on how *not* to sell resoluteness. If so, we concede that the effort could succeed in the hands of an able practitioner. We must ask nonetheless whether efforts to appear resolute compete at a great disadvantage with the noise and distraction of the international political environment.

Leaders View Signals with Their Own Assumptions about Adversary Intent

Cognitive theorists (as discussed in Chapter 5) offer a simple but important wisdom: what *we* think is inherently more salient to us than what *others* think and do. It thus significantly shapes our view of the facts—indeed, what we think *are the facts*. That parties *impose* their interpretations on the words and deeds of others inevitably compromises the effectiveness of any signal of resolve.

Many US presidents have obsessed nonetheless over the global effects should the United States fail to act decisively or stick to its guns. In Iraq and Vietnam, for example, they feared the United States would lose credibility and acquire a reputation for irresoluteness for not having followed through on a commitment. World leaders see events, however, through a prism of their own assumptions about how governments are predisposed to behave. Consequently, the US departure from

Vietnam did not have the effect that US presidents had feared. Other countries drew their *own* conclusions about the meaning of the US exit.

Whatever message US leaders seek to convey, then, North Korean assumptions about US thinking will drive the North Korean approach to negotiations. Even serious North Korean negotiating offers will reflect the minimum giveaways that North Korean leaders *assume* US leaders might accept. Not surprisingly, then, North Korea's recent bargaining concessions centered on the Yongbyon site—the biggest and most visible nuclear complex in North Korea—though its relative importance to North Korea's nuclear ambitions has waned. That the country's leaders would regard the partial or full dismantlement of the complex as a bargaining chip is understandable: the United States has traditionally made concessions in exchange for constraints on the facility. Under the 1994 Agreed Framework, negotiated by the Clinton administration, North Korea committed to shutting down its main reactor at Yongbyon—a source of plutonium—and sealing the fuel at the site (for the promised construction of two light-water reactors).[33] The facility took center stage again, in 2008, when North Korea blew up the 60-foot cooling tower for a reactor at the facility as a concession to the Bush administration. That did not prevent North Korea, years later, from equipping the reactor with a new, river-cooling system[34]—or from continuing to enrich uranium at Yongbyon, and elsewhere.[35] From North Korea's standpoint, production at the Yongbyon complex had become the essential basis of any deal. Trump, himself, encouraged that thinking. When, in September 2018, North Korea announced its willingness to permanently dismantle the (now aging) Yongbyon nuclear complex for unspecified US concessions, Trump reacted with assorted accolades—"very exciting," "great responses," and "tremendous progress." The country's leaders were understandably surprised in Hanoi, then, when Trump rejected the same deal.[36]

North Korean negotiating offers will also reflect what North Korean leaders *assume* they can afford to accept given accompanying vulnerability concerns. Policymakers focus on signals of resolve as if they override concerns about "trust" when the parties must cooperate—to some (varying) extent—to produce desired results (Kydd 2007).[37] Any such cooperation creates openings for an opponent who "cheats" to capitalize on the resulting opportunities for gain. Placing a country's fate in the hands of another country—the United States, for example—requires quite a bit of faith that its leadership will keep its word. The nuclear agreement with Iran (as discussed in Chapter 8) required its leaders to trust that the other parties would stick to the deal. Iran's leaders did not assume the United States would discourage trade and investment with Iran through a lingering threat to re-impose sanctions.[38] Or that the United States would reverse itself and re-impose sanctions once Iran had dismantled or frozen its nuclear capabilities. They were wrong. Presumably, North Korea took note.[39]

With a US reputation for reneging on deals, could North Korea believe that any reliable deal is possible, short of capitulation to US demands? After all, the United

States did not just withdraw from the Iran agreement; it violated it (ACA 2018). For North Korea's leaders to agree to total denuclearization, they must trust, for example, that the United States will stick to the deal so that it will not use any acquired information subsequently to attack North Korea's remaining nuclear facilities. They must also believe the United States will not exploit the country's exposure to try to push for a new regime. A regime that insists on absolute control and defines national security in the most generous terms is unlikely to welcome measures that include providing outsiders access to sites that *might* house illicit weaponry and production materials, identifying key security and intelligence personnel, providing requested documentation, making program officials available for interrogation, disclosing decisional structures governing weapons use, and satisfying observers that conventional military programs lack nonconventional applications. Information is power. Such disclosures constitute useful intelligence for plying the political weaknesses of the North Korean regime and exploiting its military vulnerabilities. The North Korean regime is certainly loath to accept that, much less the vulnerabilities that would result from jettisoning the country's nuclear capabilities.

Consider North Korea's heated response to John Bolton's (pre-Singapore) invocation of the "Libyan model" for North Korea. Bolton was apparently referring to former Libyan leader Muammar Qaddafi's willingness to give up his weapons of mass destruction—openly and completely—to placate the West.[40] But North Korea seemed more impressed by the consequences. A few short years later, the United States, Britain, and France took up arms to remove Qaddafi from power. North Korean security fears were unlikely reduced by Trump's pre-Singapore suggestion that the Libyan model (as North Korea understands it) "would take place if we don't make a deal."[41] If North Korea views denuclearization as a trick to make North Korea vulnerable, US resolve, no matter how it is signaled, will not produce a deal. Even if its leaders feel secure, North Korea might seek to "lock in" U.S. commitments, with a significant US down payment, before taking irreversible steps toward denuclearization. Such a requirement will only increase the difficulty of getting a deal.

Thus, when attributing potential outcomes to apparent resolve, we must recognize that behavior is always subject to multiple interpretations and that these interpretations vary with the *predispositions* of the interpreting party.[42] We must also recognize the limits of coercion in overcoming such predispositions. Absent the capability or willingness to impose a resolution, parties must depend, at some level, on the adversary's sense of the benefits of cooperation.[43]

Unbridled Resolve: The Madman Theory

Some US leaders presume implicitly that a strategy of resoluteness can overcome limits to capability and interest. They thus flirt with a strategy that originated in the works of Cold War–era bargaining theorists. They argue that a rational person—deferent to calculations of benefits and costs—could prevail in bargaining by feigning irrationality.[44] The prohibitive costs to the "madman," should it follow through on the threat, would normally undermine the threat's credibility. Here, however, threatening actions tantamount to suicide increases the credibility of the threat, for it highlights the irrationality of the threatening party: "You would have to be mad to threaten something like that." The madman "wins," presumably, because no sane person would fool with a madman.

Whereas "brinkmanship" assumes that the stronger of two parties will take a conflict to the brink to realize potential gains, the madman theory assumes that, for both parties, the costs of conflict exceed the gains. Thus, a party must boost its credibility, somehow, to prevent resistance. Thomas Schelling's (1960, 1966) thoughts—in helping to popularize the theory—converge nicely, then, with the writings of theorists who see nuclear confrontations as a game of chicken, lost by the first player to "exit" the game. In one such game, two cars race at each other. To win the game, a "rational" driver can presumably render itself "mad" by conceding control of its vehicle. The driver can rig its car to speed forward while—importantly—taking *visible* measures, such as hopping into the back seat, to communicate the lack of control to the other driver. The gambit will presumably work if the other driver knows its opponent surrendered control of its vehicle. Such thinking was behind Herman Kahn's notion of a computer-activated doomsday machine that would destroy the earth should the country possessing the machine come under attack.[45]

Richard Nixon supposedly took these principles to heart. He relished the thought that his foreign opponents would consider him "mad." In early 1969, without letting the military command in on the ruse, Nixon authorized US nuclear-armed, strategic bombers to launch a faux attack on the Soviet Union. In a chilling act of mock warfare, they flew toward the Soviet Union and down its coast—again and again in an elliptical pattern—to convince Soviet leaders that Nixon was "mad" enough to do what was necessary to settle the Vietnam conflict on favorable terms. Employing similar logic, the administration planned a massive bombing campaign against enemy targets (Duck Hook) that same year—which it eventually conducted in December 1972. It intended the disproportionate "Christmas bombing" campaign, over Hanoi, to signal that Nixon would act without restraint.

In his first year in the White House, Donald Trump—a self-professed disciple of the madman theory—chose to meet, and sometimes exceed, his North Korean counterpart—Kim Jong Un—in fiery rhetoric. Social scientists might doubt that the underlying ideas constitute a legitimate social-scientific "theory," shorn, as they are, of necessary conditions. Do they assume that one, and *only* one, of the two parties has read Schelling, or that neither of the parties is actually mad? They could wonder, moreover, whether the "theory" translates into coherent policy. A president who alternated between flattering his North Korean counterpart (and inviting him to meet) and threatening to blow his country off the map could arguably appear "mad" to some observers, but would he appear sufficiently mad—especially with conflicting statements from other officials—for the policy to work? For that matter, would he then appear sufficiently accommodating (i.e., less of a lunatic) to convince an adversary ultimately to cooperate, necessary also for the policy to work? After all, the adversary must believe that, in conceding to a US leader's demands, the leader will not then increase its demands or execute its earlier threats.[46]

True, there is virtue in a multiplicity of roles—a good cop/bad cop routine, casting Trump as the heavy—if that was in fact the strategy. Yet there was also little in Trump's decisional and consultative history to suggest that he would opt for deep strategizing over visceral emoting. The benefits of good cop/bad cop diminish considerably when it is not an intentional strategy or when the "bad cop" controls the outcomes when the *good cop's* intervention is needed. As Schelling (1966: 41) concludes, "while it is hard for a government, particularly a responsible government, to appear irrational whenever such an appearance is expedient, it is equally hard for a government, even a responsible one, to *guarantee* its own moderation in every circumstance."

Of course, the receiving party might still dismiss the "madman's" actions, perhaps as nonsensical or a bluff. The Soviets, in 1969, remained unpersuaded by Nixon's horrific nuclear signaling attempt. They could hardly take it seriously, for they had no sense of its rationale. Likewise, the Christmas bombing failed to make the terms of the 1973 Paris peace agreement significantly more favorable to the United States: Hanoi presumably knew that the Nixon administration had had enough of the war and was heading for the door. That does not mean, however, that such tactics are purely benign—that is, that they do no harm. Alexander George (1971) persuasively argues just the opposite. In his classic treatise on the *limits* of coercive diplomacy, George warns against viewing coercion as an effective tool, devoid of restraint, proportionality, realistic goals, and coordination with political and diplomatic action. Feigning madness—for instance—suffers the deficiencies, then, of all ploys in a game of chicken, which "encompasses only the crudest form and extreme methods of intimidation," ignoring diplomacy and the useful possibility of "combining a carrot with a stick" (George 1971: 29). Accordingly, George acknowledges false lessons drawn from the Cuban Missile Crisis: "Success in such international crises was largely a matter of national guts; if the president could convey resolution firmly and clearly, the opponent would back down in the face of superior American military capabilities" (1971: xi).

Following George, we more easily recognize the liabilities of rash, unpredictable, and threatening behavior when we, ourselves, are the targets. Our sense of the 1969 episode would likely differ, then, with a role reversal. How would US leaders have reacted if the Soviets sent nuclear-armed bombers to fly along the US shoreline? For that matter, how would they react had a North Korean leader placed nuclear warheads on boats bound for the California coast? Not too well, we can assume. Indeed, our response might differ had we been dealing with a North Korean rather than a Soviet leader. We might believe the North Korean action, unlike the Soviet action, begged for a forceful US response—because Kim Jong Un is irrational. If that is our reasoning, does it not expose a fallacy behind the strategy? Would it not flounder from a catch-22, a damning logical circularity? The strategy can work if we believe the party is irrational; but if we believe the party is irrational, the strategy cannot work. Under these conditions, we might conclude that we *must* respond with force.

Indeed, against a party that sees its back against the wall, we should wonder whether the strategy can ever work. When Vladimir Putin announced a nuclear alert in response to the Western response to Russia's actions in Ukraine, foreign leaders and expert observers seriously questioned his mental state given his isolation, aloofness, and emotional outbursts while considering the possibility that Putin had manufactured the persona of a madman for bargaining gains. Absent outward appearances of a change in Russian nuclear force preparedness, President Biden chose not to respond by alerting the US nuclear force. That seemed the appropriate recourse—whether, or not, Putin was mad. Why provoke him further if he was, or risk uncontrollable escalation if he was not? Another president, in some future crisis, might act differently.

Forging Reputations

For many theorists, credibility stems largely from a reputation, "a judgment of someone's character (or disposition) that is then used to predict or explain future behavior" (Mercer 1996: 6). The link between the two concepts is compelling. After all, people tend to value their own reputations and information concerning the reputations of others. A person can capitalize on their "good" reputation ("you can do business with me") and even a "bad" reputation ("you shouldn't mess with me"). In either case, whether and how that reputation serves a person's interests might depend on the reputation of another. "Although I have a reputation for being 'tough,' they have a reputation for being 'tougher.'"

Because reputations ostensibly draw from past behavior, successive US administrations fretted endlessly over the US *reputation* for standing by its commitments. Their comments and practices suggest that they viewed US credibility as hinging on what foreign leaders thought the United States might, or might not, do based on its prior conduct. President Truman justified US intervention in Korea as necessary to keep states from pursuing aggression elsewhere in the world; US leaders employed credibility concerns as justifications for going to war in Vietnam, and then sticking to that commitment; and the Bush administration's Iraq War supporters argued, in the bloody aftermath of the 2003 US invasion, that leaving Iraq would embolden US enemies (Brutger and Kertzer 2018: 5). In these many instances, US leaders feared that diminished US credibility—stemming from past US behavior—would override actual US stakes and capabilities in the thinking of adversary leaders. They might choose, then, to test the limits of US defensive commitments.

Were they wrong to project thinking about reputations to relations among states? Maybe so. True, considerable evidence backs the view that states care about their reputations—when held to account through public monitoring and international pressure—on humanitarian issues such as human trafficking (Kelley 2017). Indeed, the actions taken by some international bodies can have material consequences for targeted states in other bodies (Lebovic and Voeten 2009). But applied to matters of war and peace—military coercion and response—the concept provokes critical questions. The deficiencies in application become apparent when examining how reputations form and whether, or when and how, they matter in international politics.

How Reputations Form

Mercer observes that reputations reflect dispositional, not situational, attributions. "He lied because he's a liar," not because "they would kill his family" if he told the truth. Unlike reputational explanations, situational attributions, of the latter sort, place explanatory emphasis on conditions beyond the person's control. Indeed, they imply that *anyone* in that situation would behave the same way—and that the person would behave differently in another situation.

The distinction between reputational and situational attributions can get murky when reputational behavior appears only in certain "situations." For instance, "he lies to impress important people" suggests that he is sometimes—maybe generally—truthful but certain situations bring out the "worst" in him. The distinction gets murky, too, when a party seeks to sell the dispositional as situational, or vice versa. Indeed, establishing a "reputation for resolve" might amount to convincing audiences, through repetition, that behavior otherwise attributed to a situation instead reflects the disposition of the party.

In allowing for a dispositional *or* a situational explanation (or some combination of the two), interpreting behavior gets dicey. I might think you are cheap because, when we dine together, you never pick up a check. Someone else might attribute the same behavior to our difference in wealth ("he knows you make more money than him") or the nature of our relationship ("you're soliciting his business"). The point here is that nothing intrinsic to behavioral patterns endows them with "reputational" content. If so, we might not know whether, or how much, we can rely on the "reputation" of a party to predict another's behavior: "When the situation is very different, it is not clear whether a judgment of the state's overall resolve has much impact on others' predictions of its behavior" (Jervis 1982/83: 10).

Multiple interpretations—dispositional and situational—abound, then, even when leaders fear that others will draw the "wrong" lessons from their behavior. Although US leaders fret constantly about looking weak, a*ny* behavior is attributable to a situation, a disposition, or both. Did the United States display its weakness in leaving Vietnam short of victory? Or was the US exit situational, pressed on the United States by its poor decision to get involved in the war in the first place? For that matter, does the US "reputation"—drawn from the Vietnam experience—tell us anything about the US willingness to fight *from a distance*, instead, perhaps employing air power as a substitute for US ground troops, in other parts of the world?

Taken further, did the US reputation under the Nixon administration carry into the Ford administration, let alone the Carter or Reagan administrations? As Robert Jervis (1982/83: 9) asks, "if one president acts boldly, will other states' leaders draw inferences only about him or will they expect his successors to display similar resolve?" That is, will the audience attribute given actions to the character of a country's leaders, a state, a form of government (be it democratic or authoritarian), or the failings or strengths of a particular culture or people? "Americans are soft," "democracies lack staying power," and so forth. Policymakers (and analysts) differed, for instance, in their attributions of North Korea's behavior under Kim Jong Un. Where some read Kim's overtures as signs that the young North Korean leader sought outreach to the world to facilitate his country's development, others saw Kim as an extroverted variant of the ruthless leader that the North Korean system was bound to produce.

Whether a reputation forms—and to whom or what it is credited—is tied, then, to an understanding of the situation. A person might acquire a reputation for bravery. But supposed acts of "bravery" can result because the "hero" lacked other viable options. Is wrestling a gun from a perpetrator heroic when the alternative is certain death? Conversely, we sometimes anoint our "heroes" for acts committed when the personal risks were small or the payoff (public adulation) from acting heroically was great. How much explanatory weight we place on the actor's character (disposition) versus the situation depends on our values, general beliefs, and related personal experiences. Yet even those who had "similar" experiences could reasonably differ in their perception of events—and thus the heroicness of the behavior. What caliber gun was it? Or how capable was the perpetrator in using it? That our valuation of the situation matters is certainly revealed should our assessment of the "heroics" change upon learning the gun was not real, the "life-threatening" situation was a prank, or the supposed heroics caused the gunman to fire wildly, killing an innocent hostage. Then, we might attribute the resolution to the heroics of the other hostages or the police—or, instead, to the "cowardice" of the gunman.

Adding to the explanatory difficulties, we might recognize that reputations take various forms. If a state can get a reputation for resolve or weakness, it can certainly acquire reputations in other respects. The reputational vocabulary is greater than deterrence theorists typically acknowledge. Those who study international cooperation recognize that reputations form for trustworthiness (or reliability) in a willingness, for example, to stand by agreements. Deterrence theorists imply that reputations fall along a single continuum, with resoluteness at one end and indecisiveness or weakness at the other. But states can presumably acquire reputations for belligerence, perfidiousness, self-interestedness, altruism, rule-abidingness, or indifference. If so, that creates a more challenging interpretative problem—more challenging, still, when reputations seemingly conflict. A pennywise person might contribute time and money to a worthy cause and acquire a reputation for cheapness *and* philanthropy. Under these conditions, we *might* defer to situational considerations to reconcile the tension between the characteristics. We could offer a dispositional interpretation: the person is unwilling to spend on themselves but will gladly help others in need. Or we could offer a partial situational interpretation: the person *is* cheap but is generous when giving comes with positive recognition.

Thus, we might discount the reputational significance of actions undertaken for reputational gains. That a leader or state cares so much about its reputation that it acts for reputational gains can arguably feed a reputation for weakness (Jervis 1982/83: 12). We palpably display our weakness, for instance, when we pick our demonstrative battles transparently for a win—as do bullies, who cower from an equal fight (Morgan 1985: 125–152). Take Putin's announcement of an ICBM test, which he claimed signaled strength to Russia's enemies abroad.[47] Could global audiences not have read the test and rhetoric, instead, as a flexing of muscles to distract from Russian defeats in Ukraine? Perhaps, then, the Biden administration's earlier decision to postpone a planned ICBM test, to avoid provoking Russia, sent a more convincing signal of strength. No signal speaks more clearly to strength than one that is not meant as a signal.

To be sure, US policymakers of the Cold War period focused obsessively on the outcomes of their small global conflicts. They feared that a US failure to show strength and perseverance in these conflicts would compromise the US position, for instance, in a nuclear confrontation. We could reasonably ask whether a reputation established in these lesser battles would travel to the big stage—and, if so, how it might travel. The United States could reasonably have acquired primarily a situational reputation: it will push hard *because the risks are low*. Or else, it could have acquired a dysfunctional dispositional reputation. It pushes hard in small encounters *because it is weak.*[48]

That individuals fluidly move between dispositional and situational attribution does not sit well, however, with the works of one prominent reputational scholar. Mercer (1996) argues that individuals attribute the behavior of enemies (and friends) to their dispositions. Thus, for both enemies and friends, we attribute behavior that goes against type to situational determinants. Adversaries, even with a weak showing, might not acquire reputations for an absence of resolve.

Mercer is correct that our attributions can survive seemingly conflicting evidence. Movie heroes remain heroes, in our book, even when they show discretion in how they confront a superior (evil) force. We accept that they would act differently, but for the situation, even if their reticence to act brings pain and suffering to a sympathetic victim. We can thus protect our assumptions about a basic disposition by dismissing discordant behavior (which goes against type) as situationally determined. This principle is well supported in cognitive theory. It stresses the dominant impact of core beliefs. If you believe someone is rude, you are unlikely to take a friendly gesture at its face. "He acted nicely because others were present" or "because he knew it would upset me." Similarly, leaders who believe their foreign adversaries are aggressive remain unpersuaded by evidence to the contrary, for instance, that an adversary sought to avoid a confrontation. In a classic study, Ole Holsti (1962) established that Secretary of State John Foster Dulles was far more inclined to vary his thinking about Soviet strength than his views about Soviet malevolence. Thus, he attributed Soviet "cooperativeness" (a dearth in confrontational behavior) to Soviet weakness. In other words, he grounded his situational explanation in his dispositional verdict. Indeed, as cognitive theorists understand, his situational attributions shielded his key assumptions from evidence that would otherwise conflict with his core beliefs.[49]

We must ask, nevertheless, whether dispositional attitudes are more fluid than Mercer acknowledges. Cognitive theorists recognize, for instance, that beliefs can change in the face of overwhelming evidence like a "big event." When we deem someone a hero, we expect them to act heroically. We can excuse their unheroic actions—situationally—only to a point. That we judge them "heroes" might raise the bar in evaluating their conduct. We might expect them to risk bodily harm—life and limb—where the unheroic might put their own interests first.

When evaluating international behavior specifically, we must also ask, who gets "credit" for given behavior? States? Governments? Or their (replaceable) leaders? Put differently, did foreign leaders think the same way about the *US* reputation under the Trump administration as they did during the Obama administration? States, governments, and individual leaders can arguably acquire independent reputations. Consequently, observers might reach judgments about which (whose) reputation prevails (and, then, perhaps in what *situation*). Thus, a foreign leader might have evaluated US foreign relations and concluded that the tilt toward nationalism and isolation under the Trump administration reflected a *US* disposition rooted in cultural, social, economic, and political forces. Or they might have dismissed the Trump administration as an aberration (with its *own* reputation) that would only temporarily suppress or disguise strong, US liberal-internationalist tendencies. Whatever the administration, they might conclude, instead, that behavior that some attribute to the disposition of a leader reflects situational constraints on the government. For instance, they might conclude that a US president acts differently in a first than in a second (lame-duck) term, when the opposite party controls the House or Senate, and so forth.

In sum, international reputations are not unidimensional. Then, dispositional and situational attributions can combine in all sorts of ways, referencing different units, contrary to simplistic notions about how they form and when they might change. In fact, if reputations do form, we must recognize that they might not take the form that the acting party seeks. Although the Trump administration initially hoped that its tough-minded stance toward Iran would push North Korea to concede its nuclear holdings, the administration's actions might have served only to give the administration—*and the United States*—a reputation for reneging on agreements.

Do Reputations Actually Matter?

Reputations arguably form when a party predicts the behavior of another from its past behavior. For instance, banks determine the creditworthiness of potential borrowers based, in part, on their payment history. A credit rating expresses the borrower's reputation.

Like a borrower's creditworthiness, reputations in international relations might extend beyond the physical assets that a party brings to the table. States are incapable of delivering on all their threats and promises, much less simultaneously. Hence, a reputation for acting could serve as a capability "inflator." A party might get the "benefit of the doubt" should adversaries consider testing the party's capability or will to respond.

For that reason, the party might place great value on its reputation—even protecting or building that reputation by engaging in risky or costly actions. But are such reputational concerns warranted? Maybe not. There is little doubt that leaders try to craft a reputation for resolution, for instance—a phenomenon that Shiping Tang (2005: 40) attributes to the "cult of reputation." As he observes, however, the evidence is less convincing that other states respond to these "reputations." Others share that assessment. The path-breaking studies by Jonathan Mercer (1996) and Daryl Press (2005) challenged the prevailing wisdom that reputations, drawn from a state's behavior, influence the behavior of other states. If they are correct—and others disagree (Crescenzi 2018)—how states behave in the past has no bearing on how others think those states *will* behave. Indeed, given Mercer's conclusion that observers effectively impose reputations on the evidence, "reputations" cannot usefully explain international behavior.

Recent rejoinders to Mercer and Press emerge in scholarship that focuses explicitly or implicitly on "reputations for resolve." Resolve, as we have seen, is a slippery concept. It provides a crude sense that a party will persevere in conflict but only by begging the question of whether the party is acting, instead, on its interests and capabilities, or from the influence of still other variables. In search of reputational effects, however, some studies (Grieco 2001; Huth 1988; Huth and Russett 1988) establish that non-forceful or conciliatory actions by one state to another state provoke reciprocal challenges; other studies extend these findings to actions toward third parties. For instance, Mark Crescenzi (2007: 394) concludes that "states observe extra-dyadic behavior and incorporate this information when dealing with intra-dyadic relations." Alex Weisiger and Keren Yarhi-Milo (2015: 492) conclude that "countries that have backed down are substantially more likely to face subsequent challenges." These studies do *not*, however, establish that states respond to another's "reputation" per se rather than use information about the behavior of those states to assess, for instance, their interests and capabilities. Weisiger and Yarhi-Milo (2015) impugn the reputational content of past behavior, then, by including a "reputation" variable as the sole dynamic influence in their models. Given that stricture, *any* responsiveness revealed in one state's behavior to another's behavior must owe, by *definition*, to a "reputational" effect. Such an effect requires, however, that a party respond to another's supposed *predisposition* (they are a "rogue state" or "aggressor"). A party expects others to behave, by their "type," because that is who or what they are. Quantitative models that just disclose interdependent behaviors cannot provide the evidence necessary to make that theoretical call.

Stronger empirical evidence of reputation effects emerges from simulation exercises. In an experimental setting, Dustin Tingley and Barbara Walter (2011) establish that participants invest significantly in reputation building and—importantly—act on the "reputations" acquired by other participants from their prior plays of the game. They nonetheless perform their experiment in a vacuum of sorts, with the capabilities, interests, and other characteristics of the participants held constant. Thus, when the experimental subjects respond to discerned consistency in a player's behavior, they are effectively "typing" those players: that is the only way players can improve on random guess, or a one-size-fits-all, strategy. Put differently, their study possesses internal validity: it reveals what the experimenters seek to study. It suffers, however, in external validity because it rests on an unlikely scenario: decisional units possess no knowledge of opponents apart from their prior patterns of behavior.

In another set of experiments, Ryan Brutger and Joshua Kertzer (2018) find that hawks differ from doves when they ask subjects to judge reputational effects in various scenarios. Still, they do not establish that their experimental subjects respond to events, as they affect reputations per se. Is it all that surprising that hawks think that states help their reputation when resorting to threats, while doves believe the opposite? After all, hawks and doves disagree fundamentally over whether threats are counterproductive. Is it not possible that people care about reputations only when researchers ask them to frame their views on the appropriate response—to belligerence, for instance—in terms of reputation effects?

We should not conclude from these counterarguments that reputations are irrelevant.[50] Even if leader-specific reputations form (Lupton 2020), we must still ask: (a) whether, or when, these reputations matter more than general, country-specific (or country-type) reputations or the leader-specific reputations of others (say, a national security advisor) within the same government; (b) whether these reputational assessments are more observer-specific than leader-specific and thus exist principally in the "eye of the beholder"; (c) what types of reputations matter more, or less, in coercive bargaining; and (d) how much reputations count in such bargaining relative to beliefs about another party's risk propensities, cost aversions, issue commitments, interests, or capabilities. Indeed, to the extent that we acknowledge that capabilities and interests, psychological forces, or contextual factors influence how, or whether, reputations form, we must also recognize that these variables can influence outcomes both *directly* and indirectly. We should not assume that the direct effects are small when compared to those indirect ones that register allegedly through reputations.

At the very least, we must conclude, then, that researchers have yet to provide evidence to justify the strong reputational concerns that US leaders routinely express or the beliefs of those strategists who insist that states can readily build their reputations, and then ply them for coercive gains.[51] Researchers, here, carry a significant burden of proof. They must establish that reputations explain variance in behavior patterns when other variables, including state interests and capabilities, cannot. They can build a stronger case, then, by establishing that "reputations" travel across time or space. Instead, Weisiger and Yarhi-Milo (2015: 492) conclude that their findings weaken substantially for third-party challenges and "when the subsequent interaction less closely resembles the dispute in which the country in question earned its reputation." Although issue- or target-specific reputations can form, in principle, such specificity should have us questioning whether reputations matter in that context. Once we start asking why reputations form here and not there, or on this rather than that, we recognize a host of factors that compete with reputations as potential influences—including, most definitely, the prior beliefs and biases of the observer. By contrast, aggressive moves, by weak and strong states toward another state—one that responded ineffectually to threats from still others—would constitute more convincing evidence that reputations form, and hold, in international politics.

Much-cited *anecdotal* evidence admittedly supports arguments that states acquire unfavorable (counterproductive) reputations in the use of military force. Evidence indicates, for instance, that Saddam Hussein and Osama bin Laden read the retreats from Lebanon and Somalia—in the early 1980s and 1990s, respectively—as revealing a strong US aversion to taking troop casualties. Yet even evidence that US adversaries read US behavior thusly falls short of establishing a causal relationship. Bin Laden was primed, by virtue of his ideology, to see ever-stronger Islamist forces triumphing over corrupt and morally weak Western governments. Should we be surprised that he interpreted US behavior accordingly?[52] True, Saddam Hussein voiced contempt for the US unwillingness to fight and to persevere. But did he learn from history? Or did he instead look to history to bolster his contemptuous view of the United States? If he was impressed by prior "events," why was he no more impressed with US war threats, in 2002–2003, than he had been in 1990–1991? Presumably, the subsequent Desert Storm operation, the vanquishing of the Taliban government of Afghanistan in 2001–2002, and anti–Iraq War planning and rhetoric in Washington presented powerful behavioral evidence to boost the US reputation for using force. Why did Saddam Hussein not hedge his bets? At the very least, he could have done more to deflect US accusations that he had stockpiled illicit weapons. The answer most surely lies in persisting *beliefs* about US intentions and likely war outcomes should the United States intervene.[53]

These beliefs determine how the evidence is viewed—indeed, whether it is considered evidence. Thus, negative reputational effects of the US performance in the Vietnam War were arguably overblown. After all, the United States revealed a willingness to fight for eight long, punishing years—then, for a losing cause. Does that not count for something? Moreover, by leaving Vietnam, the United States "repositioned" itself to better respond to other global military challenges. When viewed as a "strategic" retreat, it carries far less onerous weight.

Despite fears that US credibility would suffer enormously with a US withdrawal from Vietnam, a US "reputation" for irresoluteness did not prevail when the Saigon government went down in defeat (Hopf 1995). Neither the Soviet Union nor China sought to test US global commitments in the aftermath of the war. The Soviets conceded to an increased US role in the Middle East with the 1973 Middle East War; negotiated a major strategic arms control treaty with the United States (1979 SALT II agreement) that failed to seal in a permanent Soviet strategic-nuclear advantage; and, thereafter, confined their military objectives largely to Afghanistan along the southern Soviet border. Indeed, NATO expanded greatly—at Russia's expense—twenty years after the US exodus. Presumably, the countries of Eastern Europe did not perceive the United States as an untrustworthy alliance partner. Ironically, with the United States out of Vietnam, other countries might well have perceived it as *more* willing and able to attend to important US interests.

Conclusions

US policymakers have consistently stood among the believers despite unconvincing evidence that states can obtain a coercive advantage by establishing resolve or forging reputations for action. The actions and preoccupations of US policymakers made resolve and reputation driving concerns of US national security policy. Policymakers feared that US allies and adversaries might doubt the US willingness to fight if US trepidations, or the US track record, conveyed a tendency to wilt under fire. But their faith in relevant tactics was unjustified given various perils and pitfalls.

6.1: *The evidence that US policymakers seek to show resolve or establish reputations for acting is stronger than the evidence that US adversaries respond, in international conflicts, to ostensible shows of resolve or reputations for acting.*

Behavior attributed to a party's resolve or reputation could stem instead from perceptions of the party's intent, based (at least in part) in the party's apparent interests and capabilities. Or else, behavior attributed specifically to resolve could stem from the impact of other variables that international politics scholars maintain affect (individual, governmental, or societal) behavior in international conflict.

6.2: *The alleged resolve or reputation of a party might simply express its interest and capability (for accomplishing specified goals).*

6.3: *The alleged resolve of a party could stem, instead, from the influence of other variables, including a leader's risk acceptance, government preparedness, and public support.*

To be sure, US officials *intend* to signal resolve and burnish the US reputation for acting. But US actions leave much to the target's interpretation. Signals that impress the sender might little affect the target. Or else, the target might read the "wrong" message into signals. After all, leaders focus on matters of reputation and resolve under trying military and political conditions—precisely when opponents have reason to doubt these leaders can sustain their efforts. Thus, feigning resolve or acting to promote a reputation might prove counterproductive: they might convey to opponents that leaders lack the will or capability to persevere. Why put on a show if you can succeed or if the immediate stakes at issue alone justify staying the course? At the very least, leaders might concede a bargaining advantage by suggesting that they *think the opponent believes* it has the upper hand.

6.4: *Establishing resolve or burnishing a reputation invite the communication challenges found in signaling clear commitments.*

6.5: *Where leaders engage in actions to build reputations—that is, an apparent "disposition" toward certain behavior—targets might dismiss those actions as products of "situational" proclivities; where leaders view their own actions as due to "situational" determinants, audiences might view the same actions as revealing the disposition of the leader (or their country).*

6.6: *Whether reputations form, what reputations form, which party reputations involve, and how much reputations matter varies with the beliefs and expectations of the audience.*

Whether or not communicating resolve or reputation building function as the tactician expects, these tactics can still do harm. They require work, invest resources, and bring potentially high costs—even enduring burdens. In that sense, the madman theory is but an outgrowth of thinking about resoluteness.

> ***6.7:*** *Actions meant to communicate resolve or bolster a reputation can raise the stakes and risks of a conflict with deleterious short- and long-term effects.*
>
> ***6.8:*** *Faux madness might not convince targets to back down and could provoke them.*

PART III

CASE STUDIES

In international politics, beliefs concerning matters of security often rest in (hawkish or dovish) political ideology and thus in assumptions about how deterrence works or fails. These beliefs inform choices, in policymaking, among material referents. These referents—whether asymmetries, gaps, openings, or options—grab attention, frame debate, and ultimately distract from critical assessments of adversary intent. A perverse consequence of the focus is that the alleged advantages of US nuclear superiority feature little in policy discussion or debate. Two seemingly dissimilar cases—US decision-making in the 1962 Cuban Missile Crisis and controversies attending the 2015 Iran nuclear deal—demonstrate.

7

When Tactics Consume Strategy

Decision Making in the Cuban Missile Crisis

The Cuban Missile Crisis (CMC) seems a useful starting point for any discussion of crisis behavior. For many, the Kennedy administration's judicious deliberations and careful crisis management approximates the rational ideal: a president, and his key advisors, contained their impulses, adroitly employed coercive tools, but left the door open to the eventual deal that settled the conflict (see, e.g., Janis 1972). That view receives considerable backing from the writings and comments of key participants. Yet depictions of a deliberative process inaccurately portray the critical discussions of the period, as we now know from recordings of critical meetings.

We can learn much about the handling of this international crisis, and others, from the vast literature on crisis decision-making (Herek et al. 1987; Sagan 1985)—and CMC decision-making, in particular (Bernstein 2000; Gibson 2011; Lebow 1983; Scott and Hughes 2015; Welch 1989; Winter 2003).[1] These writings highlight departures from rational practices and outcomes that ensue when stress, high stakes, and time constraints expose human decisional frailties. In so doing, they undermine the depiction of a rational policymaker who uses information appropriately in devising means to serve policy goals. They also show, then, that policymakers, much like their academic counterparts, dwell in the concrete world of capability and tactics sacrificing broader concerns about policy costs and consequences.

These decision-making shortfalls reflect a key cognitive deficiency: fundamental beliefs, drawn from two prominent conflict models—the "spiral" and "deterrence" models—inform views in a crisis.[2] They direct policymakers to *salient* options that then constrain thinking about the policy problem. These options dominate discussions, frame arguments, and dictate the terms of debate. Indeed, they impair the mean-ends analyses that could tie the options to deterrence or spiral principles. In the CMC, the deliberations thus produced a paradoxical effect: the blockade option competed at an advantage because the air-strike option *consumed* deliberations. Participants settled on a blockade, then, without duly considering (a) the consequences of "doing nothing," (b) US and Soviet objectives and

The False Promise of Superiority. James H. Lebovic, Oxford University Press. © Oxford University Press 2023.
DOI: 10.1093/oso/9780197680865.003.0007

attending means-ends relationships, and (c) the implications of employing coercion, over force, in pursuit of US goals.

Whereas analysts often extoll the benefits of nuclear advantages, for coercive effect, the CMC offers powerful evidence that officials in the throes of confrontation might lack the foresight, deliberativeness, and control to produce desired results. They might defer, then, to salient (perhaps prepackaged) options with little insight, or effort to gain insight, into the adversary's goals or the conditions that could cause a conflict to spiral beyond control. Officials latched onto conventional options—in particular, air strikes—with little thought to how they might spark a nuclear conflagration. For that matter, despite alleged US nuclear superiority in the period, US policymakers gave no attention to whether, or how, a nuclear advantage would permit a favorable resolution of the crisis.

Choosing among Options in a Nuclear Crisis

Much of scholarly thinking about crisis behavior centers on the validity of the competing "spiral" and "deterrence" models (Jervis 1976). Both rest on simple and compelling logic.

Per the spiral model, each of two parties reads offensive intent into an opponent's defensive behavior. The resulting actions—and reactions—move the parties toward war. Blight and Lang (1995: 232) surmise that, during the CMC, "(e)ach side, responding in what was felt at the time to be a completely justified, defensive reaction to actions of one or both of the other two, had its intentions misjudged. This pattern reiterated until it was almost too late to reverse the perverse momentum of misperception." The "deterrence model" predicts the opposite. Where the spiral model posits that conflict begets conflict, the deterrence model assumes that reciprocating conflict—in raising the costs to the opponent of persevering—convinces the opponent that aggression is fruitless or self-defeating. A firm response, from a position of strength, thus causes the opponent to back down. As we shall see, key members of President Kennedy's National Security Council (NSC) believed an emphatic reaction to the Soviet emplacement of missiles in Cuba was required to get the Soviets to reverse course and to prevent further Soviet transgressions.[3]

I contend that at least one of the two models informs decision makers in grave national-security crises. I argue that they draw from these models, despite (or maybe due to) an absence of factual support. They rely, then, on simplistic assumptions about the nature of the strategic challenge, and the appropriate response. I argue, further, that policymakers hold to these assumptions without seeking necessary informational backing, probing the logical interrelationship of central precepts, searching for available policy alternatives, or critically assessing the viability of these options. The reason, I maintain, is that *tactical preferences*—linked to one, or both, of the two crisis models—frame and ultimately consume the consideration of US and adversary objectives.

These myopic tendencies will likely prevail in a crisis—especially a nuclear crisis—when decision makers must act quickly under extreme pressure from time constraints, high stakes, great risk, and a lack of definitive information. With salient options serving as critical referents, "how we should address the problem" will loom larger in deliberations than "why, or whether, the situation presents a problem" and "whether the proposed means are best suited to address the problem." Thus, decisional participants will not fully consider the parties' capabilities, alternatives, and objectives. Nor will they duly assess the consistency in arguments; the compatibility of positions and viewpoints; the benefits, costs, and trade-offs of given policy options; or the nature and importance of goals and their relationship to proposed options.

If these arguments hold, various non-rational patterns will afflict decision-making. First, the mechanics of implementing one or more salient options will dominate deliberations. Second, participants will reluctantly acknowledge the costs of these options—and, even then, the ins-and-outs of implementing them will dominate deliberations. Third, preferred alternatives will detract from the (full) consideration of other options. Fourth, deliberations will center on immediate benefits and costs of an option over its longer-term consequences or overarching policy goals.

Empirical Evidence: Analyzing the Transcripts

The discovery of Soviet "offensive" missiles in Cuba shocked the Kennedy administration. The administration had resisted incriminating evidence of pending Soviet missile deployments despite warnings from Senate Republicans and even CIA Director John McCone. Soviet leaders had repeatedly assured the administration—albeit somewhat obtusely—that they would not place nuclear-armed ballistic missiles into Cuba. The Soviet ambassador to the United States had even relayed a promise that the Soviets would not "make trouble" for Kennedy before an election (Lebow 1983: 433). The charade was exposed, however, once U-2 spy planes revealed the telltale fingerprints of Soviet ballistic missile sites.

The Kennedy administration had deflected Republican criticism (before midterm elections, no less) by denying evidence the Soviets had moved missiles to Cuba. They had also warned the Soviets, through various channels, against placing offensive weapons there. With indisputable evidence now that the Soviets had done just that, the president was on the spot. He assembled his team of principal advisors (and others) over thirteen days to make key decisions and manage the crisis.

Unbeknown to early researchers of the crisis, President Kennedy was a prolific recorder of conversations. During the CMC, he taped most of the NSC sessions along with his meetings with individual advisors, former president Eisenhower,

Congressional leaders, and the Joint Chiefs of Staff. The tapes were made available in the 1990s and presented as transcripts in 1997, by Ernest R. May and Philip D. Zelikow, eds., in *The Kennedy Tapes: Inside the White House during the Cuban Missile Crisis*.[4] The (revised) transcripts in the 2002 edition of that book are subject to analysis here.

The first five NSC meetings, held on October 16 (in a morning and evening session), 18, 20, and 22—where alternative US options were discussed—are the basis of this study's coding effort.[5] Apparent in Table 7.1, the coding of "comments" centers on participant references to: [Items] (1) Soviet goals in the crisis, (2) US goals in the crisis, (3) specific options, (4) advocacy of those options, (5) the potential costs of not responding to the Soviet missiles, (6) regard for the costs of not acting (whether they are accepted, discounted, et cetera), (7) the costs of the specific option referenced, (8) the regard for those costs, (9) the benefits of the specific options referenced, and (10) past or immediate challenges.

The coding of "comments"—discrete thoughts and opinions of each participant, in order of presentation—offers an intermediate approach between a single summary of a participant's (multi-sentence/paragraph) statements and the coding of individual sentences (or sentence fragments). The former precludes the coding of specific arguments; the latter sacrifices the forest to the trees (since the actual meaning of a given statement frequently emerges only after multiple sentences). The latter also risks redundancy (inflating the comment totals) when a speaker (as a matter of style, perhaps) restates a point. Individual codes are repeated—as a new comment, however—when a speaker repeats their point while adding or subtracting important content or when a comment spans multiple categories (per item) in the coding scheme. Thus, for example, every option mentioned by a speaker is included as a separate comment.[6] When a speaker repeats their prior point, after another speaker has spoken, or simply states agreement with a prior speaker's comment, the remarks are introduced into the data as a new comment.[7] In that sense, reiteration or agreement are judged differently from mere repetition.

In general, the coding rules are conservative. Codes require *explicit* reference, for example, to US or Soviet goals or specific costs or benefits, *explicit* advocacy for an option (not just implied support), and so forth.[8]

I present the findings in graphical form. The graphs pertain to: (a) to the options discussed, (b) their assorted costs and benefits, and (c) the comments and advocacy of key participants.

The Discussion of Options

In means-driven deliberations, salient means—here, military options—dominate discussions. Figures 7.1 and 7.2 provide supportive evidence in that regard.

Figure 7.1 reveals the relative attention NSC members devoted, in their remarks, to military options (Items 3b–3m, 3u, or 3v). Although the figure establishes that

Table 7.1 **Coding Transcript Comments: Goals, Options, Costs, Benefits, and Challenges**

Item 1. Soviet crisis goal/interest *explicitly* referenced
a. Change nuclear balance
b. Coercive/political gains
c. Bargaining chips
d. Ally support/deterrence
e. Offensive advantage
f. Pretext
g. Confusion/questioning/querying evidence
h. Global aggression/expansion
i. Probing action
Item 2. US goal/interest in crisis *explicitly* referenced
a. Remove or freeze missiles/deny operational capability
b. Regime change (Cuba)/addressing Cuba "problem"
c. Preserving US credibility/alliance maintenance
d. Denying the Soviets a coercive advantage
e. Preserving the nuclear or military "balance"
f. Preparedness for Soviet countermove elsewhere
g. War avoidance/crisis control
h. Preventing an additional buildup in Cuba
i. Preventing spread of Communist influence
Item 3. Options (Tactics)
a. No options/lack of options
b. Blockade: primary
c. Blockade: primary/plus
d. Blockade: secondary
e. Blockade: secondary/plus
f. Invasion: primary
g. Invasion: primary/plus
h. Invasion: secondary
i. Invasion: secondary/plus
j. Air strike: primary
k. Air strike: primary/plus
l. Air strike: secondary

(*continued*)

Table 7.1 **Continued**

m. Air strike: secondary/plus
n. Diplomacy, general
o. Diplomacy, direct (with Soviet Union)
p. Diplomacy, public statement/declaration
q. Diplomacy, indirect (3rd party influence)
r. Diplomacy multilateral
s. Specific concessions
t. Specific demands and/or threats of retaliation
u. Military preparations/mobilization/action
v. Guerilla action/undermine Castro
w. Surveillance
x. Focused military action. Immediate threats
y. All options/none specifically cited
Item 4. Advocacy
a. Oppose
b. Conditionally oppose
c. Conditionally support
d. Support
e. Informational. Queries, facts, and speculation
f. Operational. Directing/limiting/revealing specific action
g. Decisional management
Item 5. Potential costs of NOT acting referenced
a. Shift in the military/nuclear balance
b. Loss in US credibility/political position
c. Soviet coercive advantage
d. Additional Soviet assertiveness
e. Domestic political
f. General war
g. Emboldened Cuba/strengthened Castro
h. Missile use
i. Unspecific negative consequences
j. Undermine alliances

Table 7.1 **Continued**

Item 6. Regard for costs of NOT acting
a. Discounted/rejected
b. Recognized contingently/probabilistically
c. Accepted
Item 7. Potential costs of the action referenced
a. Local Soviet or Cuban conventional response
b. Inadvertent/accidental/unauthorized nuclear launch
c. Intentional Soviet nuclear-weapons launch
d. Nuclear-weapons launch
e. General war/all-out US–Soviet conflict
f. Global Soviet political or military reaction
g. 3rd-party (in)action/*n*th-order or spillover effects
h. Potentially ineffective/open-ended/counterproductive
i. Domestic political reaction
j. Loss in control (Soviet lives)/major general effects
k. International reaction/loss in political support
l. US mistakes/erroneous assessments
m. Forces US concessions
n. Crisis escalation
o. Forces US escalation
p. Stalemate
Item 8. Regard for Item 7 costs
a. Discounted/rejected
b. Recognized contingently/probabilistically
c. Accepted as calculated risk
d. Accepted given benefits and necessity of acting
Item 9. Potential benefits of the action referenced
a. Open to diplomacy
b. Open to follow-on/ancillary military action
c. Builds political support or legitimacy
d. Allows for information acquisition/communication
e. Allows for US capability improvement

(*continued*)

Table 7.1 **Continued**

f. Permits flexibility
g. Buys time, in general; postpones choice or action
h. Increases pressure
i. Limits provocativeness/maintains control
j. Creates fait accompli
k. Contributes to/assures success
l. Avoids general war/limits risk of conflict expansion
m. Increased transparency
n. Address negative consequences (in general)
o. Information control
p. Deterrence
q. Potential de-escalation of conflict
Item 10. Past/immediate challenge
a. Changing Soviet views/attitudes
b. Effective or appropriate application of force/attack response
c. Maintaining control of crisis
d. Preserving US credibility
e. Buying time
f. Acquiring information/interpreting evidence
g. Maintaining surveillance
h. Protecting US forces/vessels/aircraft
i. Increasing US force levels/capabilities
j. Building/retaining political support
k. Domestic politics
l. Increasing pressure
m. Limiting provocativeness/de-escalating conflict
n. Addressing Soviet escalation
o. Addressing Soviet de-escalation
p. Preventing information leaks/controlling information
q. Considering consequences of action

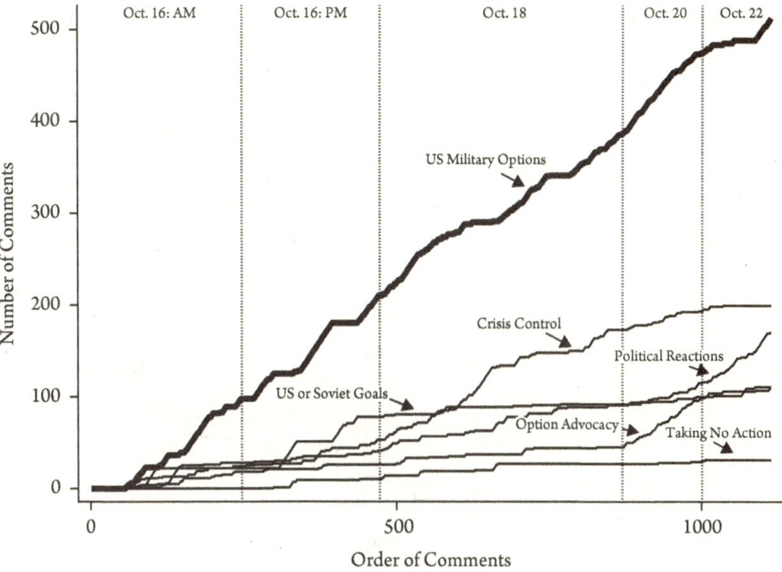

Figure 7.1 Changing decisional focus in the first five NSC sessions.

military options dominated deliberations, that finding hardly impugns the rigorousness of the discussion and debate. We should expect military options—indeed, options in general—to remain front and center in a crisis. An issue for rationality, however, is whether members considered options at the expense of broader purposes and the consequences of action.

Indeed, the figure provides evidence that "how" and "what" the United States should do was far more central to thinking than "whether" and "why" the United States should do it. US military options—albeit mostly without explicit "advocacy" (Items 4a–4d)—dominated discussions from the start, and then rose in profile throughout the sessions.

By contrast, references to US or Soviet crisis goals (Item 1 or 2) rose in the second session but then plateaued thereafter. Admittedly, that is not the entire story. The figure somewhat understates attention to Soviet goals toward the end of the fifth session when, to build support and blunt opposition, the discussion turned to a *public justification* for the blockade decision.[9] Even then, Soviet goals were not meaningfully discussed. In general, the discussion of Soviet goals reduced to rebukes of the Soviets for their brazen, flagrant, and duplicitous actions in Cuba, ready-made generalizations, or expressions of puzzlement about Soviet behavior.

NSC members also failed to probe, at any length, the option of "taking no action" (Item 5). Instead, the participants remarked—here and there—that doing nothing, under the circumstances, was unacceptable: it would either inflame domestic critics, compromise US global credibility, give the Soviets a coercive edge,

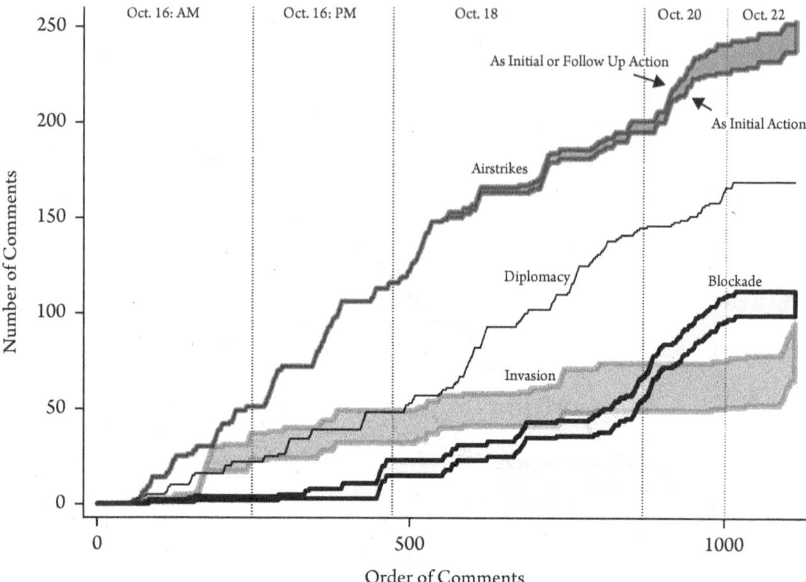

Figure 7.2 Changing attention to specific military options in the first five NSC sessions.

or reward—and thereby guarantee, future—bad Soviet behavior. Members did not consider the likelihood of any one negative consequence more than any other. Nor did they consider how they might mitigate certain negative consequences, or tolerate others; they implied, instead, that each negative effect would reinforce each and every other. They thereby created a "collective negative" that blocked serious consideration of the "doing nothing" alternative.

Although Kennedy and his advisors recognized domestic political constraints, these constraints did not drive the discussions. True, Kennedy's initial reaction to the missiles' discovery showed that domestic politics was on his mind (Dobbs 2008b: 6); NSC members groused about certain members of Congress and unplanned leaks of information; and Kennedy expressed displeasure with his own prior decision to draw a line in the sand. He wondered out loud whether the situation might have been different had he not publicly warned the Soviets against placing missiles in Cuba. More often, silence on the topic spoke backhandedly to the impact of politics. We do not need to discuss what everyone understands, and no one wants to acknowledge. After all, the members believed—or wanted others to believe—they were concerned about the *national* interest.[10]

Whereas outward expressions might understate the importance of *domestic* politics in the discussions,[11] politics is defined generously, here, to include domestic and international reactions (Items 5e, 7i, 7k, 9c, 10j, 10k, and 10p). Many germane comments pertained to building the legitimacy of the eventual US response, muting international condemnation, acquiring international support, and more generally obtaining support for any US action. Even then, politics cannot explain the relative attention that members devoted to military options or the lesser attention that members devoted to crisis goals and crisis control—that is, the potential escalation of the conflict perhaps to an all-out nuclear war (Items 7a–7f, 7j, 7n, 7o, 9i, 9l, 9q, 10c, and 10m–10o). Remarks concerning crisis control built substantially over the course of the crisis, though never faster than remarks about military options, and peaked toward the end of the third session. Why should political considerations have suppressed concerns about crisis escalation? After all, the tenor of the discussions, and the eventual decision to hold off air strikes *for now*, do not suggest that the participants believed that politically their "hands were tied." They struggled hard—rationally or not—to make a judicious decision.

Figure 7.2 shows that much of the attention to military options centered, from the beginning, on air strikes (Items 3j–3m). They seemed the most appropriate response—quick and complete—to the grave threat created supposedly by the surprise Soviet move. The assessments gave clear voice to deterrence principles: without exception, the participants agreed initially that air strikes would signal, with necessary power, that the United States would not tolerate such transgressions. Although attention to the blockade option (Items 3b–3e) built rapidly by the fourth NSC session, air strikes continued to command attention, through the last session. Even then, the participants devoted remarks to a blockade as a follow-on to air strikes, or air strikes as a follow-on to a blockade, should the Soviets hold firm in Cuba or subsequently reintroduce missiles.

The participants also considered a potential invasion of the island (Items 3f–3i). Such an invasion would presumably solve the missile problem permanently, ending Soviet hopes of establishing a nuclear position in the vicinity. Yet even an invasion became a "second-wave" option to follow air strikes on the Soviet missiles.[12] Consequently, neither a blockade nor an invasion received the prominence of air strikes in the deliberations. Indeed, the blockade option was addressed belatedly and competed for attention with air strikes even in the last two sessions.

The Discussion of Option Costs and Benefits

When means drive decisions, participants reluctantly acknowledge the costs of a preferred option. Indeed, even "costly" options can control debate by framing, or impeding, the discussion of available options.

Figure 7.3, in revealing these purported costs, offers additional perspective on the primacy of the air-strike option in NSC deliberations. The top line of each shaded area in the graph represents the running total of references to costs (Item 5),[13] while the bottom line of each shaded area subtracts the running total of references to benefits (Item 9), of the cited options.[14] The top-shaded area on the graph thus reveals the costs and net costs referenced for the full set of options; the similarly

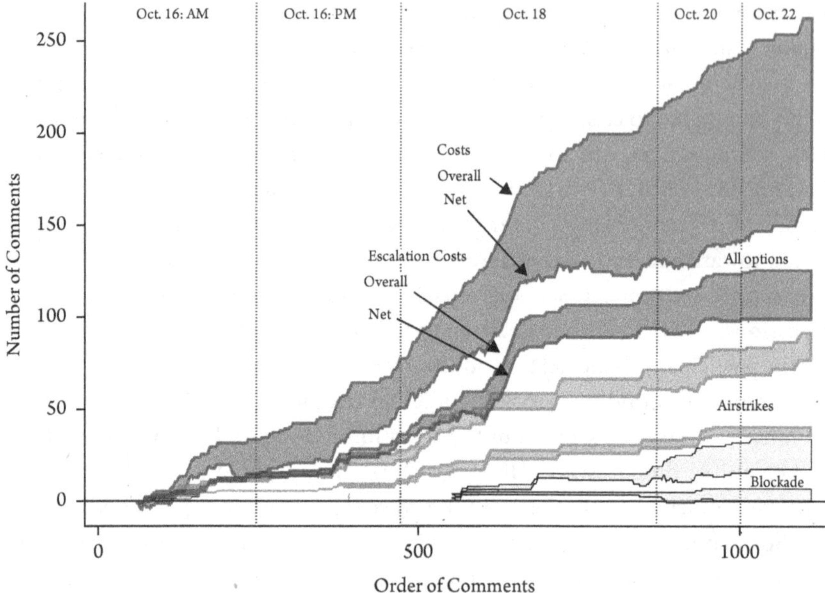

Figure 7.3 Changing attention to costs and benefits of military options in the first five NSC sessions.

shaded area just beneath it reveals the subset of these costs and net costs that pertain specifically to escalation, provocation, nuclear war, or general war (per Figure 7.1). The same set of costs and net costs are expressed below these two areas for air strikes and, then—below it, again—for the blockade option.[15] The figure permits numerous conclusions.

First, the NSC participants—slowly at first, but then more rapidly—came to acknowledge constraints on their options. Initially, the participants committed to using military force to end the threat posed by Soviet missiles in Cuba. The attributed benefits almost offset the costs of various options in the first NSC session.[16] By the beginning of the third session, however, appreciation of costs took hold and rose thereafter through the final session. Although participants, by the middle of the third session, increasingly acknowledged option benefits, subsequently reducing net costs relative to overall costs,[17] these references to benefits still did not offset the increasing number of cost references through the final session. The conclusion that cost concerns predominated is strengthened further when recognizing that these "benefits," in the figure, sometimes amounted to discounting costs predicted by other participants.

Second, these cost concerns centered often on the *escalatory* potential of the various options.[18] Roughly half of the references to costs concern their provocativeness, escalatory, or general-war potential. Indeed, the line for *net* escalatory cost closely tracks in size and direction the line for overall net costs. The pattern thus points to the ascendance in deliberations of "spiral-based," relative to "deterrence-based," assumptions.

Third, the trend lines for the costs of the air strike option appear modest when compared to the clear, and growing, predominance of that option in deliberations (as revealed previously in Figure 7.2). Whereas air strikes were referenced by the participants hundreds of times by the end of the fourth session, less than a hundred references are recorded for their costs, and only a couple of dozen references are made to their escalatory character, by the end of the final session.[19] Even then, the costs of the air-strike option constitute less than half of the references to air strikes overall (as revealed, again, through comparisons with Figure 7.2); do not rise as precipitously as the costs referenced for the full set of options; and represent only a fraction of the costs referenced for all options (including diplomacy). The collective evidence thus strongly suggests (a) that (at least some of) the participants (drawing from the spiral model) were increasingly sensitive to costs overall, including those of escalation, not just to costs attributable to air strikes, and (b) that air strikes dominated the discussion (as a source of queries and conjecture) apart from mounting concerns about the escalatory costs of that option.[20]

Finally, the blockade option, which Figure 7.2 shows rose in deliberative prominence, received little attention for either its costs *or benefits*. Despite the roughly equal number of references to costs and benefits, more striking is the limited extent—in absolute and relative terms—to which the blockade option was subject to evaluation. It is hard not to conclude from this and the prior figure, then, that the blockade option gained adherents due, in part, to *the limited consideration of that alternative*.

Augmenting these findings, Figure 7.4 displays the changing fortunes of diplomatic alternatives over the course of the deliberations. It provides further evidence, then, of the growing awareness of the costs of military options. The top shaded area in the figure juxtaposes references to direct US–Soviet diplomacy with references to other forms of diplomacy. The change in shading in the third session shows that, at that point, references to direct diplomacy with the Soviets overtook references to other forms of diplomacy. These forms include public statements, multilateral diplomacy, third-party diplomacy (including approaches to the Cuban government or some other third country), threats and demands, and considerations of specific concessions.

A similar change in shading records the improving fortunes overall of diplomacy in the discussions. By the second day of deliberations, references to the positive consequences of diplomacy exceeded references to its costs. Even the relative prominence in discussions of demands and threats (Item 3t) relative to concessions (Item 3s) changed dramatically. Although the emphasis on concessions never quite approached the emphasis on coercive tools, the gap closed between the two by the final session. The participants at that point seemed willing to bargain, reconciled to trading US missiles, in Turkey, for the Soviet missiles in Cuba.[21]

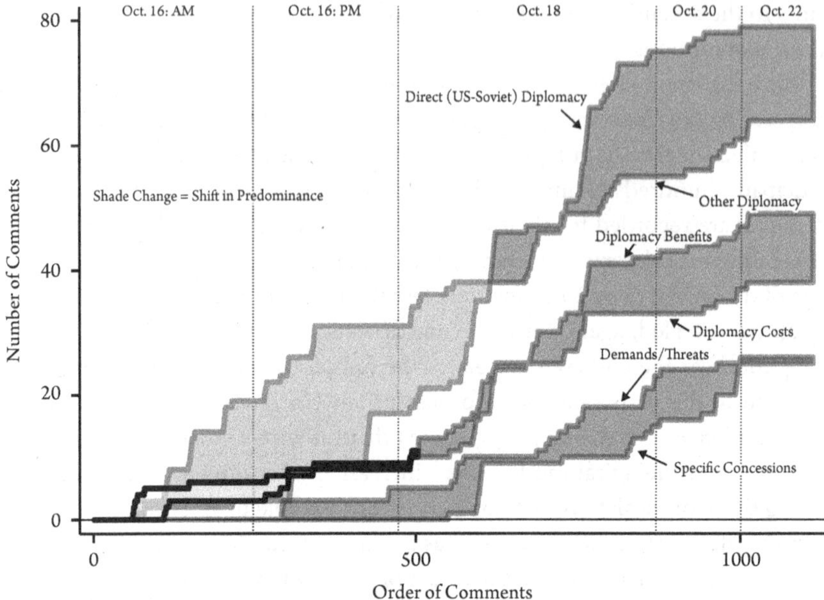

Figure 7.4 Changing assessment of the costs and benefits of various diplomatic actions.

NSC members were initially reluctant to approach the Soviets given likely Soviet resistance to non-negotiable US demands and a desire not to foreclose military options by tipping the US hand. But they came to view direct diplomacy more favorably, and the concessions that would likely accompany it, lacking better ways to resolve the dispute, or even evidence that the Soviets would agree to a compromise. In that sense, they again revealed a growing appreciation that military options invited high costs which included their escalatory potential.

The Advocacy and Concerns of Key Participants

The air strike option might have prevailed in another decisional group. We should not understate the influence of key participants on the decision to forgo air strikes, or on the crisis outcome. The participants diverged significantly in their thoughts about escalatory risks (drawing from one of the two dominant models), and thus gravitated eventually toward different options.

Figure 7.5 reveals the concerns expressed by prominent NSC members about overall option costs (the black lines), the escalatory costs of all options (the top of each light gray area), and the escalatory costs attributed specifically to air strikes (the top of each dark gray area). The figure *stacks* the results across participants to avoid many intersecting lines and to highlight the changing difference in concerns among the participants. The JFK line—as the top line—thus represents the cumulative

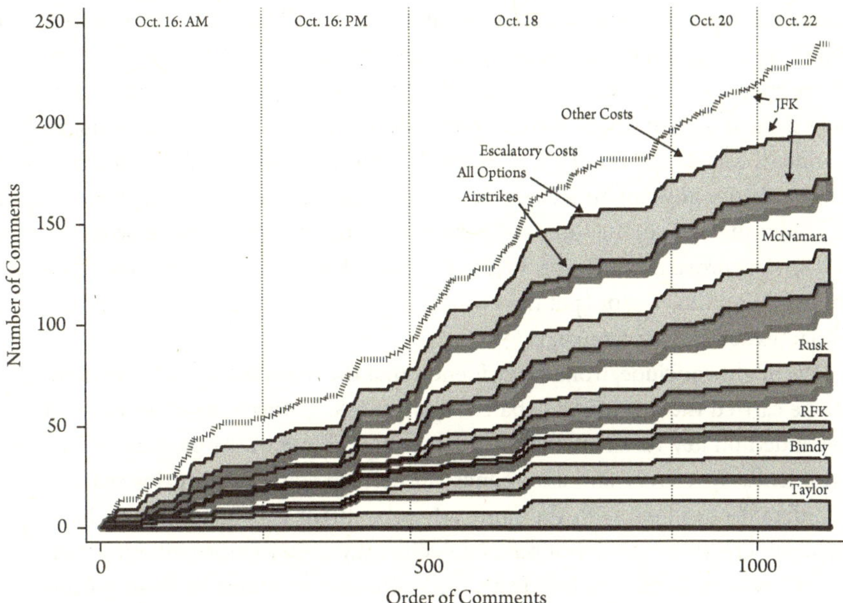

Figure 7.5 Concerns about nuclear war and escalation expressed by various NSC members.

concerns of all six participants; the difference between the JFK and McNamara lines represents the changing level of JFK's concerns (with the McNamara line serving as the "0" baseline). Thus, the level (over time) of each participant's cost concerns is revealed by the difference between adjacent lines over the course of the deliberations. For each participant, the light gray area (between the lines) represents the contribution of escalatory costs to the overall costs attributed to all options; the dark gray area represents the portion of these escalatory costs attributed specifically to air strikes.

A clear difference in opinion among the NSC participants is apparent in the figure. McNamara emerged as the champion of escalatory concerns by the second day of deliberations. By then, he was joined by President Kennedy, who repeatedly pushed back against options that might expand or intensify the conflict. They were hardly alone in these concerns. At least half of the concerns expressed by the participants about option costs concerned escalatory potential: for the other participants, the light gray area stands about halfway between the dark lines separating each participant.

The concerns expressed about the escalatory costs of the various options should not obscure two key points: First, concerns about escalatory costs constituted only around half of the cost concerns expressed; second, concerns about the escalatory costs of air strikes constituted less than half of the escalatory concerns expressed about the various nonmilitary and military options. Even JFK devoted less than a quarter of his concerns about the escalatory potential of the available options to air strikes per se; consequently, *escalatory concerns about air strikes constituted only a small percent of his overall option-cost concerns.* Three participants, Robert Kennedy, McGeorge Bundy, and Maxwell Taylor, stated few if any concerns about the escalatory costs of air strikes. Although opinion eventually converged around the blockade option, the three deliberants continued to promote air strikes should the Soviets prove intransigent. While President Kennedy kept air strikes alive, as a follow-on option, Taylor (per the position of the JCS) lent consistent support to an initial (extensive) air campaign.

We might conclude, from the evidence, that the president's position is key. After all, he carried the day in the debate. We must ask, however, whether the outcome depended on key *variables*, that is, the ideological mix of the participants and the president's own savvy and assuredness in defying some of his principal advisors. A different president, with the same advisors, might have taken policy in a more hawkish direction. That is not pure conjecture. Lyndon Johnson inherited the same set of advisors, on Kennedy's death, and—buying into their hawkish positions—brought the United States fully into the Vietnam War. That the president's position conflicted with that of his brother—the president's closest advisor—certainly testifies to JFK's independence. We must ask, as well, whether presidential experience mattered (Saunders 2017). Would the president have shown the same reserve had he not been chastened, in his first year in office, by the ill-fated Bay of Pigs invasion of Cuba? At the very least, the disastrous (CIA-backed) effort soured him on the invasion option.

Yet we must also recognize that assumptions—tied here to the deterrence and spiral models—were key. The participants drew very different conclusions from the same set of facts and they diverged, moreover, in their overall escalatory concerns almost from the start of deliberations.[22]

Figure 7.6 provides additional perspective on the decision process. It compares the tenor and pace of the NSC sessions with that of the single JCS meeting[23]—adjusting for the divergent length (in quantity of remarks) of the sessions.[24] (Kennedy's remarks are omitted from the two figures to highlight the advice and comments provided to the president.) Clear from the figure is that the two sessions varied markedly on three dimensions: the number of comments concerning US or Soviet goals in the crisis; the consequences of taking no action; and military advocacy, that is, the explicit approval or opposition (Items 4a–4d) stated about specific military options.

The last point is the most critical. NSC participants engaged initially in limited advocacy for, or against, specific military options. They alluded to option costs and benefits but without explicitly rejecting air strikes, a blockade, or an invasion. Only in the last session, when the participants were pressed by time for a decision, did option advocacy take hold—at which point it overtook, in number, comments

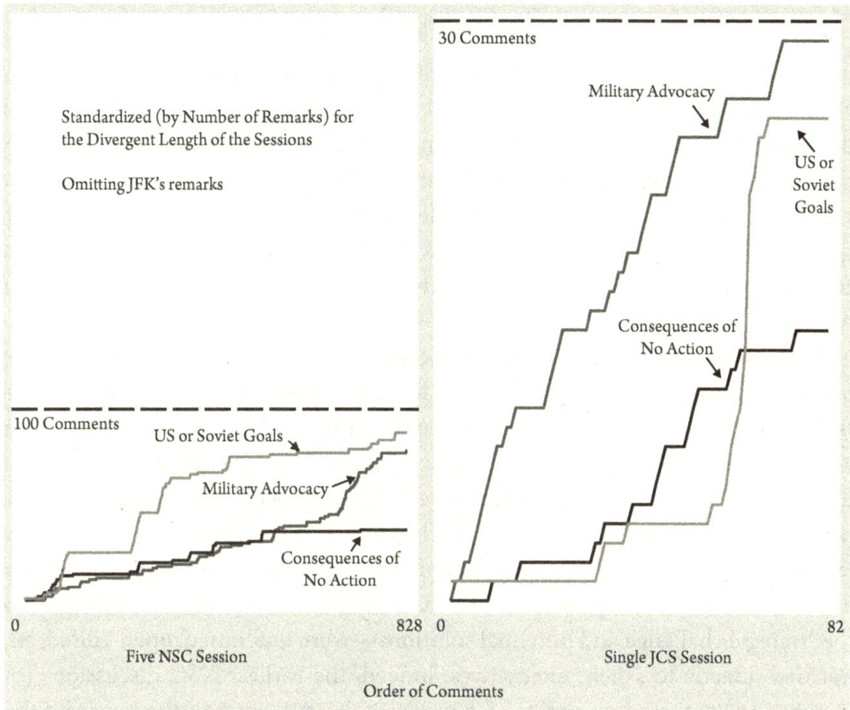

Figure 7.6 Relative attention to military options in the NSC and JCS sessions.

concerning the consequences of taking no action. Even by the end of the last session, however, comments involving explicit advocacy of a military option did not exceed, in number, comments devoted to US or Soviet crisis goals.

By contrast, military advocacy dwarfed other comments in the JCS session: indeed, the military pushed for air strikes, a blockade, and invasion as a single strategy. Only toward the end of the session did attention turn to US or Soviet goals—and then to sell the advocacy rather than to draw out opinions or contribute to productive dialogue. The JCS bought fully into deterrence assumptions: in its view, the Soviets only understood strength and preyed on weakness. Comments concerning the consequences of taking no action served much the same argumentative purpose. These comments built quickly around the middle of the meeting, when they were overtaken by comments related to crisis goals.

Throughout much of the early NSC sessions, then, the participants posed hard questions and expressed doubts without stating clear support or opposition for one or more options. Perhaps they were somewhat disingenuous given their preexisting philosophical commitments. Still, they did not slam the door shut on any of the options (except "doing nothing") and thus conveyed a level of uncertainty, and deference, not found in the JCS session.[25]

The Findings, In Sum

In the CMC, assumptions about conflict dynamics—how the conflict might abate, intensify, or spread—were critical to the decisional outcome. A change in the makeup of the decision group—under different leadership, and under intense military pressure to violently suppress a threat—could well have produced a different outcome. Yet reference to ideology—adherence to deterrence versus spiral assumptions—only partly explains the opting for a blockade. The limits of attributing the result solely to a duel between sets of assumptions are apparent in two critical respects.

First, ideological adherence does not explain why the dialogue centered largely on tactics. Reducing disagreement to an ideological dispute fails to explain the resilience, in discussions, of the air-strike option, the relatively limited attention devoted to alternatives despite a growing recognition of the costs of these strikes, and the failure to tie the air-strike option more deliberatively to policy goals or even the risks of escalation that concerned the participants. Participants focused on solutions rather than the nature of the strategic challenge. Discussions of underlying US and Soviet goals—which, in principle, could have provoked broad-based thinking about the strategic challenge and potential solutions—were unfocused, open-ended, and not tied directly to salient alternatives. Indeed, the earliest NSC discussions (on October 16) centered on options for *removing* the missiles, not the purposes that those missiles served or their overall strategic impact.

Second, ideological adherence does not explain the limited attention devoted to the blockade option. As the air-strike option absorbed valuable deliberation time, the blockade option emerged from deliberations without competition, or due scrutiny. It effectively profited as the "non–air strike" option, that is, the anointed counter to the air-strike option. The blockade seems to have become the option of choice because officials thought they had *no good options*. Important to note is that NSC members voiced concerns about the escalatory potential *of all conceivable alternatives*. What the blockade offered, then, was breathing space. It would buy time and possibly create opportunities for the administration to "feel its way" through the crisis.

Decisional Consequences: The Neglected Elements of a Means-Driven Process

The Kennedy administration deserves credit for its cautious and deliberative approach in crisis decision-making. It sought evidence, openly assessed the policy challenge and available alternatives, and considered the consequences of US actions and Soviet countermoves given potential Soviet goals and strategy. Indeed, the option selected by the administration—a blockade of Cuba—seems ideal in multiple respects. It permitted the administration to take a forceful stance while avoiding unnecessary provocation. With a naval blockade, it could control the flow of military resources to the island and send an implicit, yet strong, coercive message—a US willingness to capitalize on US regional conventional superiority—while shifting the burden of violence to the Soviet Union. Soviet ships would have to run the blockade—initiating conflict—to regain the initiative.

Still, we must avoid the temptation to link consequences to intentions by attributing a "good" outcome to good decision-making. Indeed, the decisional participants failed to show due diligence in multiple critical respects. Specifically, the participants neglected (a) the consequences of "doing nothing," (b) the US and Soviet goals at issue, and (c) the liabilities of relying on coercion in pursuit of US goals. That shortfall might have produced a disastrous outcome under less favorable conditions.

The Consequences of "Doing Nothing"

The participants failed to consider the advantages of doing nothing—that is, of simply accepting the "reality" of Soviet missiles in Cuba. They focused, instead, on ridding Cuba of the Soviet missiles.

But why were these missiles a problem? President Kennedy, among others, downplayed the possibility that the Soviets would rationally launch missiles from Cuba, even under US attack, given the inevitable consequence—an all-out nuclear war. Indeed, Kennedy and his principal advisors, without exception, believed that deterrence (i.e., the prospect of unacceptable cost) inhibited the Soviets from attacking the United States, and near unanimously (Taylor excepted) maintained that the introduction of Soviet missiles into Cuba did not fundamentally change the US–Soviet nuclear balance.[26] Why should US officials have worried then about *these* missiles? Indeed, these missiles seem far less a strategic challenge to the United States than pending increases in the size of the Soviet ICBM force or the Soviet acquisition of a significant submarine force, also capable of firing missiles close to US shores.[27] Although the actions of Cuba's government complicated matters, participants thought it quite unlikely that Russia would cede control of these weapons.

True, the missiles had coercive value: NSC members seemed to agree that the missiles presented a political challenge to the United States. If left unanswered, participants thought countries would doubt the administration's willingness to back US interests, and tough talk, with action. Leaders abroad would conclude that the United States would retreat, not resist, in the face of adversity. Thus, the Soviets would obtain a "coercive" edge in dealings with the United States and, regardless, would continue to probe and push. They saw a need, then, to stand firm.

Yet a coercive advantage requires its recognition as such. Inasmuch as the administration had stood firm in Europe and was moving to strengthen the US position in

NATO, held a significant advantage in nuclear hardware and delivery capability, and was actively backing allies around the world (including in Asia), why should Soviet nuclear missiles in Cuba have created a US coercive disadvantage? These missiles could have had the opposite effect if they had reinforced the perceived Soviet threat and thereby increased pressure on Kennedy to resist further Soviet transgressions.

The administration retained options to deflect Soviet coercion. It could obtain leverage in future confrontations, following the Soviet lead in Cuba, by engaging in high-risk behavior. For that matter, it could have answered the Soviet missiles in Cuba with deployments in kind. Instead, it rejected the option of adding ballistic missiles along the Soviet perimeter. Although such a buildup seemed provocative, and it ran counter to administration hopes to end the crisis with a missile trade, it nonetheless offered a symmetric (arguably, "proportionate") response to the allegedly unfavorable shift in the "coercive" balance, seemed a far less provocative response than did military action, and fit Kennedy's desire to trade missiles for missiles. After all, what went up could come down—especially if the additional missiles pressed the Soviets to deal.

Apart from concerns about a coercive disadvantage, could the administration rightly have feared that rewarding Soviet behavior with US inaction would encourage further Soviet malfeasance? After all, the parties could stumble into war if the Soviets pushed into areas of US vital interest. But that scenario assumed a world of continuous Cuban-type crises that seemingly came out of nowhere, with no space between them for US signals—words and deeds—that warned against Soviet encroachments. It also assumed that US vital interests, and military capabilities, would not "speak for themselves" (on this, see Betts 1987) when, in actuality, any coercive edge the Soviets *might* obtain might still not outweigh the "credibility" of US foreign commitments. At some level, Kennedy administration officials conceded as much. They gave no thought in the CMC sessions, for example, to whether US actions, or inactions, in Cuba would provoke a Soviet attack on Western Europe.

Of course, the Soviets might have "guessed right," if continuing to probe, establishing positions in parts of the world that the United States was unwilling to "defend." Would that not put the United States in a bind? But that question is both asked and answered. We can doubt the loss to the United States should the Soviets have gained position in countries that the United States had deemed of limited strategic value. We can even question the cost-effectiveness of such Soviet "gains" given the aid and political concessions required to sustain them.

What about the (domestic) political fallout of not standing up to Soviet "aggression," especially after Kennedy had publicly warned the Soviets not to place offensive weapons in Cuba? He was arguably caught in a "commitment trap" (Sagan 2000) which left him with two bad options. He faced a loss in credibility at home, and abroad, if he failed to act decisively, but he also incurred unnecessary risks if acting only to preserve US credibility. Why did he not try, then, to finesse his public pledge—by seeking some political way out? Yes, the missiles were politically problematic: by merit of their proximity to the United States, the missiles drove home the "reality" of what was otherwise an abstract, existential Soviet threat. But the administration was otherwise finessing the issue. It focused attention on the threat from Soviet ballistic missiles, not Cuba-based, nuclear-capable, Soviet bombers that could hit collocated US population centers and military bases in a surprise attack.[28]

Maxwell Taylor, as JCS chair, made the case for air strikes by asserting that "the risk of these missiles being used against us was less than if we permitted the missiles to remain there" (Gibson 2011: 398). But what was the basis of his calculations? Whether or not the missiles in Cuba gave the Soviets a military or coercive advantage, NSC members still needed to weigh the risks of removal against the risks of doing nothing.

The Neglect of Underlying Purposes

Assumptions about Soviet goals did matter in the NSC assessments. NSC members frequently wondered why the Soviets acted as they did, or how the Soviets might respond to US military and diplomatic moves. They nevertheless answered largely through bold assertions that were rarely subject to scrutiny. In consequence, policy rested on specious and, at times, inconsistent assumptions about how the Soviets might react and whether deterrence was secure.

Soviet Reactions

The air strike option suffered, at times, in deliberations from fears that the Soviets would use such an attack as a pretext for acting elsewhere in the world. A Soviet attack on Berlin, or a move against the US missiles in Turkey, loomed large in the discussions.

But such fears hardly justified the US blockade decision. If the Soviets had placed missiles in Cuba to bait a US attack, why would the US decision to blockade Cuba have made a difference? Would not a US blockade of Cuba give the Soviets an even better pretext for acting in Berlin as a "proportionate" response to the US action? The 1948 Soviet *blockade* of Berlin, which did not produce a US military response, certainly set a precedent for such a Soviet move.

These fears also lacked a firm logical foundation. The pretext argument rested on a mismatch between Soviets means and purposes. If the Soviets were looking for a pretext to attack Berlin, a gambit in Europe—or a provocation more directly linked to Berlin—would have better served that purpose. The pretext argument assumed, moreover, that the Soviets would take high risks for limited gains. The idea that these missiles were pawns to be sacrificed for Berlin or, perhaps, the US-nuclear armed (Jupiter) missiles in Turkey did not suit the large Soviet investment in, or the high risks of, the Cuban venture (Lebow 1983: 438). The Soviet investment was too significant, and the risks to the Soviets simply too great, to have been designed only to bait the United States. The small number of Jupiter missiles on the Soviet perimeter posed little threat to the Soviets compared to the many hundreds of ICBMs and SLBMs in the US arsenal—a threat that would only grow into the 1960s. For that matter, the size of the Soviet arsenal would also increase to dwarf the threat from the Turkish-based missiles.[29]

Finally, the pretext argument supplanted more reasonable explanations for Soviet behavior. NSC members routinely discounted Soviet claims that the missiles were meant to provide political and military support to the Cuban government. That explanation, though still incomplete, better fits the evidence than does the pretext argument. Indeed, not known to the United States at the time, the Soviets had sent troops armed with tactical nuclear weapons to Cuba. These weapons, though useful for Cuba's defense, had no influence on the global nuclear or coercive balance and were unlikely US-attack targets, given their numbers and low profile, if the Soviets were seeking a pretext for action elsewhere. The Soviets, in accepting considerable risk that they would use these weapons, in the event of a US invasion, surely indicated that the Soviets valued their Cuban commitment.

The Stability of Deterrence

NSC members seemed confident that the Soviets would not simply launch these missiles when they became operational. That is, they did not believe that the Soviets would attack the United States *when they could*. Yet they suggested, at times, that deterrence was fragile, at best.

The participants acknowledged that the Soviet deployments best served a deterrence function given the limited size and capabilities of the force. It might increase the costs that the Soviets could inflict in war but not limit the damage that the United States could inflict in a retaliatory strike. They acknowledged—and believed the Soviets knew—that the United States would read any missile attack from Cuba as a *Soviet* nuclear attack and would respond accordingly.

Still, participants waffled implicitly in their deterrence commitment. Some NSC members expressed concerns early that, if alerted to the US detection of the missiles, the Soviets might take the nuclear offensive. Partly for that reason, participants initially rejected diplomatic initiatives, including prior announcement of the discovery of the missiles, and even low-level reconnaissance flights over the missile sites that could provide invaluable intelligence. Although the participants soon put those concerns to rest, they remained troubled that the Soviets had armed these missiles. Why the arming of these missiles presented a graver threat to the United States, *if deterrence was secure*, was left unaddressed. A similar logical problem afflicted concerns that a US air strike would spare some of the Soviet missiles. Whereas General Taylor made clear that a US strike would likely fall short of perfection, NSC members seemed convinced that the Soviets would still not risk an all-out US nuclear response by launching the surviving missiles. Indeed, Taylor, and the JCS, pushed for a broad-based air campaign while expressing no outward concern about a potential Soviet missile launch. Again, the question is, "Why would *armed*-Soviet missiles directed at the United States present a problem if deterrence held?"

An answer, of course, is that the Soviets could launch these missiles accidentally or inadvertently—or even transfer their control to the (less rational) Cubans. The participants focused their attention instead on what Soviet leaders might *intentionally* do and explicitly discounted the likelihood of a missile transfer.[30] So, again, why was an incomplete strike, like an operational Soviet missile force, so troubling for US officials?

The Liabilities of a Coercive Option

The deliberations eventually left air strikes in loose contention with the blockade option. Whereas air strikes were viewed as a *forceful* solution to the missile problem, a blockade was viewed, instead, as a *coercive* solution that might create space (and time) for a non-forceful resolution of the dispute. Kennedy and his advisors hoped that, with careful crisis management and the right set of signals, the Soviets might back down.[31] A blockade could signal US resolve while, paradoxically, permitting flexibility in implementation. Indeed, Kennedy initially let Soviet ships pass and ordered the Navy quarantine line moved closer to Cuba to give the Soviets time to reflect. Still, the deliberants did not fully consider the trade-offs required in adopting one option—air strikes or a blockade—over the other.

The participants most definitely acknowledged the limits of air strikes as a *forceful* solution. Yet none of them suggested that less-than-perfect air strikes might still send the convincing *coercive* message that the surviving missiles were a severe Soviet liability. The Soviets might have withdrawn them, then, to avoid an escalation of the conflict in and around Cuba, where the United States enjoyed a significant military advantage. Administration officials could reasonably expect such successful messaging if they were true to their assumption that deterrence was secure.

Of course, administration officials had good reason to doubt that the Soviets would interpret the message as intended. But the participants nonetheless showed little recognition that the nature and severity of the US response in Cuba might (a) increase Soviet insecurity, (b) push the Soviets to respond to recoup their credibility, or (c) boost Soviet concerns about a strategic imbalance, spurring rapid growth in Soviet conventional or nuclear capabilities. For all their attention on Soviet gamesmanship—the Soviet missiles in Cuba as bait for a trap—they virtually ignored the escalatory potential of air strikes per se. Although they occasionally fretted that air strikes would inflict Russian or Cuban casualties, the deliberants did not ask whether the US response might trigger the Soviet countermoves that so troubled them.

The implications of a blockade also remained unaddressed. Once a blockade became the less risky alternative to air strikes, it provoked no deep probing or second-guessing. Left unexplored was whether a *coercive* action—meant to retain control and avoid the use of force—might result in a loss of control, and a forceful Soviet response. A half-century after the momentous events, ongoing disclosures—as discussed in Chapter 6—reveal in frightening respects how *little control* Kennedy and Khrushchev effectively exercised over events.

Conclusions

In sum, assumptions drawn from the deterrence and spiral models help explain the administration's answer to the Soviet challenge in Cuba. These assumptions survived though lacking necessary scrutiny and critical factual support.

These assumptions nevertheless offer but a partial explanation for the decisional outcome. Attributing outcomes to one set of assumptions or the other fails to account for the preoccupation with air strikes in the initial NSC sessions, the reduction of these two models in practice to a binary choice between the air-strike and blockade options, or the under-examining of conditions under which an air strike or a blockade would succeed or fail in the crisis. True, the ascendance of deterrence principles soon made air strikes the option of choice, and the ascendance of spiral concerns eventually made the blockade the "winning" option. President Kennedy's more consistent reticence toward escalating the crisis certainly contributed to its peaceful denouement. Yet the deliberations, throughout, were consumed by operational issues—whether and when air strikes could destroy Soviet missiles, how various audiences would respond to air strikes, and so forth—at the expense of broader issues related to the utility of force. Whether the United States had more to gain, or less to lose, by accepting the regional presence of Soviet missiles was rejected out of hand. What the Soviets sought to achieve by placing missiles in Cuba was addressed mainly by assumption, treated as a subject of puzzlement, or reduced to a "frame" for selling the preferred military option. Whether and when coercion might carry the risks of a forceful resolution was left unexplored. In the void, a blockade became the option of choice, as the salient alternative to air strikes.

Evidence of US decisional failings in the Cuban Missile Crisis requires recognition, then, in the form of apparent perils and pitfalls.

7.1: *In a crisis, officials might lack the foresight, deliberativeness, and control to produce desired effects.*

7.2: *In a crisis, officials might give insufficient thought to whether, or how, their preference among conventional options might spark a nuclear conflagration.*

> **7.3:** *In a crisis, officials might defer to salient (perhaps prepackaged) options with little insight, or effort to gain insight, into the adversary's goals or the conditions that could cause a conflict to spiral beyond control.*
>
> **7.4:** *In a crisis, officials might give little thought to whether and how US nuclear advantages might yield a coercive edge.*

That leaders and leadership mattered in the most dangerous nuclear crisis in human history provides reason for optimism, but also pessimism, that nuclear-armed parties can avoid open warfare. The specter of nuclear destruction might prove an insufficient deterrent when confidence, stemming from optimism, invites considerable risk—or when available military options—conventional, nuclear, or both—become *the* options. If so, they will dictate the direction of discussion, limit the latitude for choice, and draw attention from the consideration of consequences and goals. Under these conditions, the *illusion of choice* will pose the biggest threat to control in the crisis.

8

When Red Lines Consume Debate

Thwarting Iran's Nuclear Ambitions

US efforts to constrain Iran's nuclear ambitions, culminating in the 2015 Joint Comprehensive Plan of Action (JCPOA) agreement in the Obama administration, tell us much about the practical implications of US nuclear superiority. Controversy centered largely on the material dimensions of the challenge—that is, establishing thresholds and allowances for Iran's nuclear program. Yet underexamined assumptions about Iranian *intentions*—based on prior ideological commitments—determined whether members of the large US policy community accepted or rejected the deal. Thus, the material focus—far from highlighting enormous US ratios of advantage—had just the opposite effect. Ironically, US policy hawks who had long stressed the value of relative nuclear advantages were the ones who most stressed the unsettling impact should Iran acquire *any* nuclear weapons. With the focus on what Iran *could* do, lost in discussion and debate was what Iran *would* do given the incentives and disincentives ostensibly created by the fact of US nuclear superiority.

Backdrop to an Agreement

In the lead-up to the signing of the JCPOA, in July 2015, Iran's unwillingness to offer meaningful concessions fueled controversy over their pace and substance. Indeed, Iran largely controlled the negotiations through drawn-out bargaining with the EU-3 (France, Germany, and the United Kingdom), the P5 + 1 (China, France, Russia, the United Kingdom, and the United States; plus Germany), the International Atomic Energy Agency (IAEA), and assorted other countries, including Turkey and Brazil. The Western powers strove in the mid-2000s for a comprehensive settlement that would constrain Iranian nuclear options, seeking a deal that would end Iranian enrichment and commit Iran to tight safeguards. Iran tried to keep its options open, however, by eschewing specifics, narrowing commitments to certain facilities and points in time, and tying "concessions" to nonnuclear issues.[1] With the resumption of the P5 + 1 talks in February 2013, Iran proved unwilling to respond

The False Promise of Superiority. James H. Lebovic, Oxford University Press. © Oxford University Press 2023.
DOI: 10.1093/oso/9780197680865.003.0008

in any detail to Western proposals or to schedule a follow-up meeting when the talks ended without agreement.

Iran's outward cooperativeness increased considerably when, in mid-2013, Hassan Rouhani assumed the Iranian presidency. By year's end, his outreach to the West, eleventh-hour compromises, and hard bargaining produced an interim agreement (the Joint Plan of Action)—the first respite in the Iranian program since negotiations began a dozen years earlier, a period in which Iran's holdings increased from a couple of hundred to almost twenty thousand centrifuges.[2] As a step toward a comprehensive agreement, the six-month deal froze and rolled back critical portions of the Iranian nuclear program. Under the terms of the deal, Iran had to halt the installation of new centrifuges, cap low-grade (5-percent) enriched-uranium production, cease work on a heavy-water reactor, deplete stocks of 20-percent enriched uranium, and accede to daily inspections of its nuclear facilities. In exchange, Iran received only modest financial concessions: limited reduction of some sanctions and access to some frozen funds.

In pronouncing their country's right to enrich uranium, Iran's negotiators still edged closer to the demands of the country's hard-liners than to the positions of P5 + 1 negotiators; the latter insisted that Iran significantly reduce its enrichment capabilities, shut down its enrichment facility at Fordow and heavy-water reactor, account for its full range of prior nuclear work, and accede to far-reaching inspections. So, the actual significance of Iran's concessions in the negotiations would remain unclear. As Iran's defenders could note, the Fordow complex was a logical place for an enrichment facility because it was hardened to a preventative attack; an expansive enrichment program would allow Iran to meet "future" nuclear-energy needs, the increased transparency from nuclear inspections should reduce the need for constraints on Iranian enrichment, Iran should not have to compromise its nuclear programs without actual sanctions relief, and so forth (ICG 2014: 17–19). For that matter, Iran could create doubts about its sincerity in these talks by complying with some, but not all, of the terms of the interim agreement. It required that Iran address the IAEA's concerns over the country's prior nuclear activities, which Tehran had long resisted.[3]

After weeks of arduous bargaining in which Iranian negotiators allegedly withdrew prior concessions and increased their demands, a breakthrough of sorts occurred in early April 2015 with the signing of a general framework agreement, intended as a step toward a more detailed agreement.[4] The framework's strenuousness exceeded the expectations of many skeptics in requiring that Iran:

1. reduce its number of centrifuges from around 19,000 to 6,000 and then limit enrichment activities, for ten years, to roughly 5,000 older and less-efficient (IR-1) centrifuges operating in a single (the Natanz) facility;
2. reduce its stockpiles of low-enriched uranium from 10,000 kilograms to 300 kilograms;
3. forgo uranium enriched beyond the 3.67 percent levels required to fuel a nuclear power plant, for a fifteen-year period;

4. restrict the hardened Fordow complex to research, involving no fissile material for fifteen years;
5. convert the Arak nuclear reactor, to reduce its plutonium production, and forgo plutonium reprocessing;
6. accept far-ranging inspections under the Additional Protocol; and
7. acknowledge the contingency of sanctions relief on Iran's compliance with an agreement.

As always, however, the devil was in the details, and these were largely unsettled. The parties had agreed on a short, joint text for public release but that each side could separately publicize the agreement's specifics as "fact sheets" without the prior approval of the other. Although some residual ambiguity is typically necessary to overcome differences to forge international agreements (especially involving sensitive, domestic issues), the extent of the discrepancies between the US and Iranian specifics—or, at least, the unwillingness of one or both parties to own up to their concessions—led many critics to wonder, justifiably, whether the agreement would truly curtail Iranian options.[5] Even major issues remained unresolved. Iran had not agreed to export its uranium stockpiles or inalterably convert them to prevent their reuse in a bomb program, destroy its unused centrifuges, ban advanced centrifuges (for "research") from the Qom facility, or allow full and permanent access of inspectors to all suspect (including "military") facilities. Iran also insisted on immediate sanctions relief with the signing of a final agreement and the end to all controls with the expiration of the agreement.

So, the question stood, did Iran's obstructionism amount to inflexibility or, instead, to good (hard) bargaining?[6] More generally, the question for those negotiating with Iran remained, "Will Iran foreclose its nuclear options?" Answering both questions left the negotiators tying ambiguous evidence to their own assessments of Iranian intent.

Setting Red Lines

The policy debate surrounding Iran's nuclear program appears to rest on concrete criteria for determining dangerous levels of nuclear progress. Such progress fuels a controversy among nonproliferation experts and concerned policymakers: at what point should a country be considered a significant proliferation threat and, therefore, where should states place red lines that, when crossed, signal a clear and present danger, perhaps requiring a forceful military response?

As was true in US–Soviet arms control, however, policymakers draw proscriptions (and prescriptions) to halt nuclear proliferation implicitly from the intentions of the suspect country. Although Iran's intentions inform all debate, even experts obscure the central issues by structuring these red lines around key metrics. As we shall see, these metrics cannot provide conclusive standards, for they all leave considerable room for disagreement and debate. The issue of line setting is perplexing, in part, because a country can adhere to the 1968 Nuclear Nonproliferation Treaty (NPT), maintain robust enrichment capabilities, and position itself to acquire nuclear weapons once renouncing its NPT obligations. Discussion and debate center on three basic standards.[7]

Nuclear testing emerges as one prominent line in debate. In this regard, Jacques Hymans argues that the NPT generally embodies the best standard—the performance of a nuclear test—for judging whether a country has crossed a critical threshold toward becoming a nuclear-weapons state (Hymans 2010). The testing standard has the advantage of requiring that countries demonstrate a nuclear-weapons capability given the regularity with which states have announced their nuclear-weapons programs with tests; the potential for test failures, such as those in North Korea; the reality that countries, including Japan and possibly South Korea, acquire fissile-material stockpiles without intending or deciding to go nuclear; the useful warning that a test by a country provides before it stockpiles bombs and makes them deliverable; the uncertainties of judging progress in earlier (pretesting) stages of a nuclear program; and the incentive that earlier thresholds give states to acquire nuclear weapons—since they are presumed "guilty" when crossing those thresholds. Thus, as a consistent feature of nuclear-weapons development and a shiny bright signal, with an undeniable meaning and impact, the explosion of a nuclear device overcomes challenges of perception and uncertainty for parties that must monitor a country's nuclear progress from a distance.

Many of these arguments hold up to criticism. Although critics argue, for instance, that a state can acquire a nuclear-weapons stockpile, as Israel did, without ever having tested a weapon, the Israeli case might well be unique. As Jacques Hymans and Matthew Gratias conclude, testing is virtually inevitable in a nuclear program: current nuclear aspirants lack the will and capability to duplicate Israel's "bomb in the basement" strategy of secretly deploying nuclear weapons without ever testing them (Hymans and Gratias 2013). Iran, for one, would likely test a device to ensure it works and to advertise the country's nuclear prowess for deterrence benefits. It might do so recognizing that its fragmented government would undercut the broad consensus that makes the strategy work. Critics also maintain that a state can hide the true purposes of a test by claiming that it had peaceful purposes. Backing these claims, the global reaction to India's 1974 peaceful test was notably tame in comparison to the reaction to India's 1998 military test.[8] Assertions that a nuclear test is peaceful are likely to remain unpersuasive, however, when made by countries like Iran that have invested heavily in delivery systems and heretofore denied seeking nuclear weapons. Critics maintain, moreover, that a state can move rapidly from a successful test to weapons that might then be hidden or used. Again, the strategy might produce little net gain. After any such test, a country might confront considerable developmental challenges—in a lengthy process of trial and error—before acquiring a deliverable weapon.

The most compelling retort to the testing standard is that aspirants could gain an edge by acquiring and hiding large amounts of enriched material before a test explosion. Iran could position itself, then, to build a multiple-warhead nuclear arsenal—following the North Korean model—by hiding, shielding, and dispersing its enriched material and bomb-making and delivery capabilities from any military retaliation that a nuclear test would invite. Indeed, Iran could conceivably stockpile uranium, construct numerous less-efficient nuclear devices, and test one to ensure it works. Having dispersed its nuclear materials or devices and acquired a weapons reserve to guard against retaliation, it could proceed then to develop more-efficient warheads. Iran could benefit after a bomb test, from the large array of targets an attacker would have to hit in a preventative strike to set back the country's nuclear weapons program—as compared to the smaller number of perhaps more vulnerable targets (plutonium-based reactors, uranium enrichment facilities, and so forth) that could have been hit in the earlier enrichment phase. For that matter, Iran might benefit from a post-test, global hesitancy to attack Iran given residual uncertainty about the actual extent of its nuclear program, its vulnerability to attack, and the strategic implications of targeting nuclear weapons.[9]

Given these limitations, some critics have explicitly and implicitly proposed an alternative threshold: the possession of a significant quantity of fissile material. Israeli Prime Minister Benjamin Netanyahu, in a 2012 United Nations General Assembly speech, set the red line for Iran at the accumulation of medium-enriched uranium sufficient for one bomb. With a significant quantity of material, presumably most of the hard work has been done; by comparison, the transition from a significant material quantity to nuclear-weapons status is relatively short, unproblematic, and unobtrusive. The fissile material can be hidden somewhere, for as long as necessary, until it becomes part of a deliverable weapon. Still, critics rightfully ask whether a significant quantity of material is the real issue. After all, some non-nuclear-weapons states possess sizable material stockpiles or could acquire them quickly with the necessary infrastructure in place. Although global attention has focused, for example, on Iranian stocks of 20-percent enriched uranium that could, with further enrichment, supply material for a bomb, enlarging these stocks is no more a proliferation threat than is expanding centrifuge capacity for producing low-enriched uranium. The latter could eventually fuel a large nuclear arsenal.[10]

With the risks and limits of the more technical standards, hawkish critics of US policy have insisted that countries like Iran cross the critical line early through actions that impugn their stated peaceful intent, such as reneging on NPT obligations.[11] When North Korea withdrew from the NPT in 2003 and Iran suspended its observance of the Additional Protocols (though not legally bound by them) in 2006, the international community was thereby placed on notice that these countries had "bad intent" and would pursue their nuclear options. These critics are inclined, then, to set lines somewhere before the hardening, dispersal, or development of a suspect nuclear program renders it impervious to destruction.

This approach to line drawing fueled the very public US–Israel dispute over the wisdom of attacking Iran—sooner rather than later—to destroy its nuclear infrastructure. Israel set red lines for using force earlier than is warranted from the US perspective. The divergent reasoning of the United States and Israel reflected their relative exposure to an Iranian bomb and the greater vulnerability of the Iranian nuclear infrastructure to a US attack as compared to an Israeli one.[12] Israel's red lines would keep the Iranian program ostensibly within reach of Israel's destructive capabilities, as Israel lacks the logistical and deep-penetration capacities of the US Air Force—for example, refueling and bunker-busting abilities. Israel's fear, shared by US policy hawks, has been that Iran is playing for time—to take the Iranian program beyond some point of no return—by making false promises and feigning compromise. Although, under Israeli and domestic pressure, the Obama administration responded by pledging that the United States would not tolerate a nuclear Iran, the administration left itself some wiggle room, and Israel ultimately chose to placate its more powerful ally.[13] The latter conceded—by default—that an attack on Iran would occur on the US timetable, as dictated by US capabilities and threat assumptions.

Setting the red line around the limits of preventative-strike capability assumes, however, that outside parties can judge the location and vulnerability of key sites when nuclear-weapons programs are hidden from scrutiny. These programs involve activities that "take place in secret on computers, in small shops and labs, and in bunkers and underground, and they may not be revealed until long after the program has been terminated."[14] It could also push these parties to act despite being highly uncertain about the suspect country's intent given the ambiguity of available information. In the Iranian case, the evidence was sufficient to convince the US intelligence community, as evinced in its 2007 national intelligence estimate, that Iran ceased work in 2003 on its nuclear-weapons program. Indeed, Iran had subsequently allowed the international monitoring of its uranium enrichment facilities and kept enriched uranium amounts below a threshold—even before it agreed to extend the limits and increase transparency in late 2013 under an interim agreement. But observers also had grounds for more dire conclusions. Iran only admitted to constructing enrichment facilities at Natanz and Qom after these sites became known, continued to expand its uranium-enrichment beyond the country's energy needs, and maintained an active program to develop ICBMs (Kroenig 2014).[15]

Danger exists in over-reading the signals in noncooperation (Jervis 1976, 1982/83). Moving against non-cooperating states has a significant downside if requiring that the United States and its allies shun rule violators when engaging them instead could reveal options, generate useful information, and overcome misunderstandings.[16] The chances for compromise are hurt when parties view outcomes in zero-sum terms, lock into their positions, and fail to see the conflict from an alternative perspective.[17] A lack of informational access can cause outsiders to exaggerate a threat. That was certainly the case with the now infamous October

2002 national intelligence estimate, "Iraq's Continuing Programs for Weapons of Mass Destruction."[18] Used to justify the 2003 war in Iraq, the report expressed the general view within the US intelligence community that Iraq had substantial holdings of weapons of mass destruction (WMD) and was reconstituting its nuclear program.[19] Bush administration officials, who ardently believed that Iraq had WMD, reinforced this view. Accordingly, they trumpeted impugning evidence, readily accepted the intelligence agencies' judgments, and implicitly established a standard of proof that inhibited professionals from challenging the administration's conclusions (Rovner 2011: 137–184; Pillar 2014). Post-mortem assessments established, however, that US intelligence was a captive of the belief that Iraq had not destroyed its illicit weaponry and production capabilities.[20]

Given differing and ambiguous threshold positions, and the limitations of all of them, whether (not just under what conditions) the United States might strike Iran remains an open question. The Obama administration's stated red line—not allowing Iran to acquire a "nuclear weapon"—left doubt about exactly when the United States might act militarily to disrupt a suspected nuclear-weapons program.[21] The administration certainly had good reasons to avoid specificity. Risks exist to the line drawer when much remains unknown about the target's intentions and capabilities and the full effects of acting on a threat. The equivocations of the administration in setting a clear red line for the Iranian nuclear program were thus an understandable response to the difficult challenges of deterring and compelling adversaries in international politics. But they also stemmed from its struggles to respond to a difficult question: "What kind of Iranian nuclear program could the administration accept, and under what conditions?"[22] The answer rested on assumptions about Iranian intent.

Certainly, relative capabilities inform the red-line debate. While proponents of a precipitous US military strike against Iran's nuclear assets accentuated the dangers of delaying an attack, opponents emphasized the confounding implications of an attack and the incompleteness of the military solution.[23] After an attack, Iran might have an even greater incentive, and public backing, to reconstitute its program (an attack will set back a program, not end it), seek a nuclear weapon, engage in terrorism, and act aggressively to undermine the attacking countries' regional positions. For that matter, Iran would have even less incentive, after an attack, to open the country to inspections, which would, from its perspective, assist the future targeting of Iran's nuclear and military infrastructure. But capability considerations are only part of threat assessment, and not typically the biggest part given the range over which presumed intentions can vary.

Thus, the essential disagreement among policymakers, and states, was not over the disutility of force or the precise criteria for determining nuclear-threshold status—however critical these criteria might appear. More important to policymakers were the nature and urgency of the threat—whether, how, when, and against whom a country might use a nuclear weapon. For them, the underlying issue was whether decisive preventative action was required, and sooner rather than later. The specifics of progress fueled debate but remained at most a secondary concern.

Policymakers who doubted that nuclear weapons serve Iran's strategic ambitions (except under dire circumstances, such as deterring an attack) preferred vague, faint, or distant lines based on a belief that Iran has little reason to pursue nuclear weapons. They argued accordingly that Iran had expanded and maintained its regional influence effectively through nonnuclear means, including its support for Hezbollah and other regional militant groups, and had shown little desire for a direct military confrontation with Israel, the region's only nuclear power. They argued also that the principal threat to Iran's leadership is internal, not external. Thus, in opting to acquire a bomb, Iran's leaders would have to accept continuing sanctions that could weaken the leadership's grip on power. Furthermore, Iran would have to pay a prohibitive price should it target or threaten its powerful adversaries with nuclear weapons. The United States and Israel were unlikely to back down and would certainly retaliate—perhaps with annihilative force—if attacked.

Policymakers who argued that nuclear weapons serve Iran's objectives instead preferred proximate red lines, though these policymakers disagreed implicitly over exact line placement. Pushing the line back, perhaps far back, were those who believed that nuclear weapons serve more traditional purposes—that is, that nuclear weapons would allow Iran to acquire status by joining the exclusive global club of nuclear-armed countries and to deflect major security threats that include Western-imposed regime change. Moving the line forward, perhaps considerably so, were those who maintained that a nuclear Iran would use its weapon(s) to harm the country's adversaries (regardless of the retaliatory consequences) or, at least, to coerce other states and pursue regional aggression with impunity. Unsurprisingly, Israel showed zero tolerance for any nuclear program in a hostile Middle East country, as demonstrated by its precipitous attacks on Iraq's Osirak reactor in 1981 and Syria's al-Kibar nuclear facility in 2007 and its hardline position toward the Iranian program. Hard to ignore, from Israel's perspective, was that Iran's leaders had called repeatedly for Israel's destruction and that Iran had strongly supported militants in Lebanon and Gaza and a Syrian regime that had targeted Israel directly.

The point is that important indicators of nuclear progress fuel debate but do not determine the essential positions of policy advocates. Why else has Iran attracted global attention when Japan and South Korea have more developed nuclear infrastructures and, by various metrics, present the greater proliferation threat? For that matter, why were India and Pakistan, despite their alleged nuclear aspirations, allowed to stand outside the proliferation regime, and why, after the Indian nuclear test, did the George W. Bush administration sign a civil-nuclear agreement with India? The answers, obviously, are that the United States and its allies consider motives when determining which countries deserve exceptional scrutiny and the timing and form of any retaliatory measures. The metrics,[24] in shifting attention from critical assumptions about these motives, can well serve as a distraction.

Giving Green Lights to Nuclear Activities

The Cold War ended, but its pattern of reasoning remains. Then as now, policymakers defended their agreements by arguing that they have everything to do with restrictions and verification and nothing to do with trust. But they have *everything* to do with trust when understood to mean that another, from a reading of its intent, will not act as it is capable (Hoffman 2006). Even those who believed that the agreement controversy was an unnecessary distraction—that deterrence would ultimately stop a nuclear-armed Iran from achieving aggressive goals—*trusted* that Iran will not willingly accept the costs of aggression.

Of course, intentions provide a deficient basis for national-security policymaking. Intentions are opaque and variable, as many realists are quick to note. Realists are wrong, however, when they insist that the viable alternative to considering intentions is to ignore them and to rely, instead, on the worst-case assumption that others act as they are capable. Agreement is impossible under these conditions—for no agreement is ironclad or exempt from interpretation. The critical issue is whether laxities *or* safeguards matter given a party's incentives to exploit or adhere to the terms of the agreement.

The basic differences in perspectives and interests proved challenging to overcome. In the ensuing months, old issues resurfaced, and new issues emerged. Each side accused the other of backtracking, and deadlines for an agreement came and went. In July 2015, after a week of dashed hopes that a deal was "imminent," the negotiators delivered a detailed agreement that largely built on the April framework.[25] Among the provisions affecting Iran, the agreement

1. retained the framework's limit on centrifuge numbers over a ten-year period (now, with a staggered [eight-and-a-half- to fifteen-year] schedule for introducing advanced centrifuges at Natanz, the only permissible enrichment site for the fifteen-year period);
2. limited low-enriched uranium stocks to 300 kilograms, severely curtailed plutonium generation, and prohibited plutonium reprocessing capacities for the same fifteen-year period;
3. permitted inspectors access to all suspect sites, with a dispute-arbitration process under the effective control of a Western voting majority;
4. delayed the loosening of sanctions until Iran's initial compliance was confirmed by the IAEA; and
5. outlined a process permitting sanctions to "snap back" into place with evidence or suspicions of Iran's noncompliance.

In return, Iran could challenge inspections of suspect sites and delay access for a matter of weeks; would receive an estimated $100 billion in frozen oil-sale assets; and would have all nuclear-related, multilateral sanctions on the country lifted (likely within a matter of months), along with the embargo on conventional arms within five years and restrictions on Iranian missile-technology acquisition within eight years.

In critical respects, the agreement drew from the advice of nuclear experts who argued that various restrictions could work in tandem to foreclose Iranian options.[26] The negotiators thereby sought the monitoring of Iran's full fuel cycle—mining, uranium conversion, and centrifuge production, operation, and storage—to boost the probability of detecting illicit Iranian activities. Their goal was to lengthen the time required for Iran to accumulate the materials to construct a nuclear weapon. Thus, the P5 + 1 crafted the JCPOA framework and the July 2015 agreement that followed to give countries a *full-year's warning* before Iran could obtain a nuclear weapon. Presumably, a year gave the P5 + 1 time to bring Iran into compliance with the agreement through assorted threats and sanctions or to disable or destroy its nuclear infrastructure by force, should Iran race for a bomb. Secretary of State John Kerry testified before the US Senate that increasing US warning time by six to twelve months was "significantly more" than the current window.[27] Whether Kerry was right or wrong obviously depends on whether these controls gave the United States and its allies additional warning time; breakout time is only "a useful proxy for the obstacles a deal might create for an Iranian sprint to the bomb" (Robb and Wald 2014: 30). But it also depended on whether any additional time improved the US position *significantly* to counter Iranian transgressions. Accordingly, answers to two basic questions informed all readings of the agreement.

First, would Iran simply wait out the agreement, expecting that it could acquire a nuclear arsenal in short order once the agreement has expired? A reasoned response required that analysts assess both Iran's current and future commitment to obtaining a nuclear weapon; and, given an affirmative commitment, Iran's willingness to postpone acquisition to some point in the future. In making the required judgments, analysts had to consider Iran's openness to the beliefs of hard-liners versus reformers, domestic and strategic conditions that press for and against acquisition in the near and long term, willingness to concede the country's nuclear ambitions to obtain resources to pursue other military or subversive political goals, and acceptance of the risks of conducting research and constructing facilities in secret. Definitive judgments in these regards were elusive, of course, which left policymakers and skilled analysts alike to rely on rather general assumptions about Iran's objectives.

Proponents of the agreement maintained, then, that a fifteen-year sunset provision provides considerable room for Western cooperation with Iran to grow and that the risks to Iran from endangering the agreement override any temptation to cheat. In this view, Iran had made the costly commitment of conceding the country's nuclear prerogatives by agreeing to very stringent terms that would essentially cut off all pathways to a bomb for a full decade and a half. During that period, Iran might reform under pressure from a growing middle class (strengthened by economic growth), acquire good cooperative habits, and receive ever-greater economic and political incentives, through ongoing relationships, to build bridges to the West.

In turn, the agreement's critics feared that Iran made short-term concessions to realize the country's long-term goal of acquiring a nuclear weapon. That is, Iran might prepare, through ongoing research, development, and accumulation of wealth, to rush for a bomb as the agreement expires. After fifteen years, Iran would be freer to increase and expand its nuclear enrichment capabilities without restriction. Under the deal, Iran's program "will be treated in the same manner as that of any other non-nuclear-weapon state party to the NPT," as stated in the Agreement's Preamble and General Provisions. Critics asked why a stronger Iran (now, a "nuclear threshold state") would presumably be a more compliant Iran.

Second, would Iran violate the terms of the deal? In other words, would Iran incur the costs of a breakout from the agreement with a transparent push for a bomb, or seek, alternatively, to minimize the risk of premature exposure by conducting necessary research, developing relevant technologies, and enriching uranium in secret facilities? A reasoned response required analysts to judge Iran's risk propensities under the agreement, again by considering Iran's goals.

Proponents concluded, accordingly, that Iran is unlikely to test the will of Western countries by engaging in prohibited nuclear activities when the chances of detection are high. Iran carries the burden of providing access and information to allay Western suspicions, and any one party to the agreement can take its concerns to the UN Security Council where a consensus is required to *block* the automatic re-imposition of sanctions within a matter of weeks. Knowledgeable proponents argued further that the possession of a significant quantity of fissile material is but a single step toward a survivable nuclear arsenal. Thus, by violating the nuclear deal, Iran invites potentially high political and economic costs without compensatory gains in security. Proponents maintained, then, that a cautious Iran would concede its nuclear prerogatives to come out from under the threat of sanctions or military attack.

In contrast, US policy hawks opposed any agreement that provided less-than-complete transparency and allowed Iran latitude to pursue its nuclear ambitions. If Iran's technological knowledge and capabilities could improve over time, *increasing* vigilance was also necessary, backed by a credible threat to impose costs on Iran for any lack of transparency. Critics worried, in fact, that Iran would repeatedly block inspections by insisting that "credible evidence" of violations is lacking, or they might delay access to suspect sites for a number of weeks (in the name of "managed access") to hide incriminating evidence.[28] Through obstruction and deceit, Iran might position itself to pursue a bomb before the agreement had expired. The opportunity to do so actually increased at the mid- to far end of the agreement's life span, as the time that Iran needs to acquire the nuclear material to build a bomb reduces under the terms of the deal.

Iran might bet, then, that it could eventually violate the agreement without cost due to favorable political conditions. Critics worried also, then, that foreign leaders would remember the arduous negotiations that led to the July 2015 agreement and seek not to reopen old debates, fearing that Iran would renounce all constraints on its nuclear program (the *nuclear* "snapback" option). Experts would disagree over whether the incriminating evidence is convincing, Iran's actions reflect "legitimate" alternative interpretations of the agreement, or the potential developments bring Iran meaningfully closer to a bomb (Hymans 2013). Military and intelligence officials would maintain that a US attack on the Iranian nuclear infrastructure can only damage known facilities and set back—not stop—an Iranian nuclear program. US allies might argue that a significant quantity of nuclear material is different from a weapon in hand. Regional experts might urge caution, warning that attacking Iranian facilities would provoke a regional (maybe global) conflict and would weaken the position of Iranian moderates who could impede Iran's march toward a bomb. Finally, commentators throughout the world could insist that countries that acquire nuclear weapons can still be deterred and have strong reasons to act responsibly.

Iran could benefit further if it had planned a breakout from an agreement to catch foreign opponents flatfooted, that is, when sanctions have ended, the counterproliferation coalition has splintered or eroded, and the military option has lost viability with the hiding, hardening, or dispersion of Iranian nuclear assets. The risks to Iran at that point are potentially small. Iran might sprint toward the finish line, expecting countries to accept one more nuclear-armed state, as they had a nuclear-armed North Korea. In time, the United States and its allies might well accommodate the "new reality" rather than sacrifice trade and investment opportunities or accept the risks of forcefully resolving the dispute. Iran had reason to expect a favorable resolution. By pursuing a one-year window to respond to Iran's violations, the United States implicitly conveyed its own uncertainty about its willingness to act and ability to build a supportive international coalition. After all, the United States does not require a full year to pre-position US forces in the region to attack known Iranian nuclear facilities and requires considerably more time for new sanctions to work.

Supporters of the July 2015 agreement insisted, however, that intentions are beside the point. They were arguably correct if any agreement with Iran is the *best* that the P5 + 1 could achieve under the circumstances and better for the P5 + 1 than no agreement. Therefore, they maintained that, with an agreement, controls and checks on the Iranian nuclear program will increase. Indeed, the US capability to damage the Iranian nuclear infrastructure will only improve under the agreement with the information that is obtained from monitoring critical sites, the reduced size of the Iranian program, and the program's concentration in a smaller number of facilities.[29] They further maintained that, without an agreement, the sanctioning regime will fracture, the transparency of the Iranian nuclear program will dramatically decline, and the military option would remain as the sole—bad—alternative. These very conditions, according to President Obama, left the US Congress with no viable reasons to oppose the agreement.

Supporters and opponents undoubtedly said what they must to sell or to kill a deal. One prominent opponent, former ambassador Eric Edelman, noted accurately that the Obama administration once deflected criticism with the mantra, "a bad deal was worse than no deal," yet defended the final agreement by suggesting that "this deal, whatever its flaws, is better than no deal and the only alternative is war."[30] Others argued that, should Iran violate the deal, UN sanctions will fully snap back into place, while insisting nonetheless that states will ignore these same sanctions should the United States reject the agreement.[31] In turn, critics, who once insisted that "sanctions would not work," now championed the retention of sanctions to get a "better deal." They also challenged the agreement by implying that the alternative was a better deal, not—perhaps—*no* deal, which could leave the world without a window on the Iranian program or control over its direction.

Salesmanship aside, even reasoned judgments about whether the agreement is the "best that we can do" derive in no small part from assessments of Iranian intent. For a reasoned judgment, supporters had to consider what Iran will ultimately concede to get a deal, whether Iran will abide by the terms of the agreement or violate it brazenly or artfully to thwart the re-imposition of sanctions or a preventative military strike, how Iran will respond to the renunciation of the agreement or a military strike, and whether Iran will build the infrastructure to rush for a bomb from a stronger strategic position at the far end of the agreement. In fact, US policymakers had grounds to reject the JCPOA if concluding that Iran will effectively violate the deal at some moment of strategic advantage and that the agreement could breed complacency, an overriding commitment to making the deal "work," or a desire to avoid confrontation at all costs by those who are charged with holding Iran accountable. To avoid that trap, the United States could renounce the agreement, press for further concessions, exert economic pressure on Iran, and try—through various means—to impede its nuclear progress. Should the United States stand alone, its disruptive influence and potential might give US allies and the business community pause in their dealings with Iran and provide Iran reason to placate the foreign opposition by holding, at some level, to the terms of the agreement.

The implications of these various arguments are simple—and perhaps disconcerting. Like it or not, the agreement came with risk, and the risk grew or receded with assumptions about Iranian goals. Obviously, stringent constraints on Iranian nuclear prerogatives were preferable to lax constraints. Tighter constraints could only increase the risks to Iran should it try to violate the terms of the agreement. But support for a nuclear deal within US policy circles was far more sensitive to assumptions about the intentions of Iran than to its opportunities to reap gains, illicit or otherwise, from the agreement. Assumptions about these goals, as shaped and charged for political effect, will determine whether an agreement's presumed benefits are worth the costs.

Critics certainly tried to scuttle the agreement by focusing on its laxities. They suggested, for example, that Iran will exploit any openings to its advantage, that these openings constitute prima facie evidence of Iran's bad faith in the negotiations, and that Iran's prior compliance with agreements surely indicates that negotiations work to Iran's favor. Focused thusly, critics made two incompatible assumptions about Iranian objectives. When challenging the agreement's safeguards, critics assumed that Iran will pursue nuclear weapons with urgency; it will secretly or blatantly cheat on the agreement because these weapons serve the country's coercive or destructive goals. Conversely, when excoriating the agreement's effective expiration date, critics suggested that Iran will postpone nuclear-weapon acquisition to some point in the future. By then, the sanctions regime will have eroded, Iran's economy will have improved, Iran's nuclear infrastructure will have matured (as it introduces new centrifuge models and reaps benefits from permissible research and development), and the onerous constraints of the agreement will have loosened. Taken together, these assumptions present a logical conundrum.[32] An Iran that is plotting to acquire nuclear weapons in secret will act with haste and take high risks and presumably seek one or more nuclear weapons for their inherent game-changing potential. An Iran that is plotting a long-term nuclear challenge to Western interests is presumably postponing—maybe, compromising—its nuclear aspirations in deference to cost. At the very least, such an Iran seems unlikely to exploit all potential avenues to acquire a bomb, let alone use it to harm the United States, Israel, or any other country simply *because it can*. Rather than refining their positions, however, critics resorted to grand assumptions. For instance, an open letter to Congress from two hundred retired US general and admirals recounts the litany of short- and long-term failures of the nuclear deal and concludes, with insufficient support, that the "agreement will enable Iran to become far more dangerous, render the Mideast still more unstable and introduce new threats to American interests as well as our allies."[33]

In making their case, supporters of the agreement constructed a wobbly edifice of their own. In emphasizing the challenges confronting Iran should it secretly pursue a bomb, they focused on near-term treaty safeguards that permit a one-year warning period. Supporters thereby answered critics who argued that Iran will relentlessly pursue its nuclear objectives through all available means. They did so, however, only by de-emphasizing long-term risk. Supporters noted correctly that *some* safeguards will continue for two decades and beyond and that Iran committed to additional long-term monitoring of its nuclear program by agreeing to *seek* ratification of the NPT Additional Protocol under the agreement (see, e.g., Davenport 2015). Still, negotiators would most definitely have rejected these more limited long-term restrictions had they been proposed as sole, near-term constraints on the Iranian program. What will have changed during the duration of the agreement to justify relaxing the restrictions? If the unprecedented short-term constraints are required because Iran might accept great risks and costs to acquire a bomb, does not that preclude weakening these constraints at the far end of the agreement?

Supporters offered answers that begged for further development. Some advocates inside and outside of the Obama administration pinned their hopes for the coming years on Iran's willingness to reform and opt for cooperation with the West.[34] One nuclear-proliferation expert concluded, for example, that "the JCPOA provides a solid formula for blocking Iran's ability to build nuclear weapons for at least 15 years, and the time necessary to pursue and implement complementary initiatives to head off the possibility that Iran will try to pursue an expansion of its nuclear program over the long-term."[35] But why should Iran's leaders moderate their goals as they become increasingly realizable? If they "have been on a superhighway, for the last 10 years, to create a nuclear weapon or a nuclear weapons program, with no speed limit," as former Secretary of State Colin Powell put it in praising the "remarkable" short-term restrictions of the agreement, why would they not just hit the gas when these restrictions are lifting?[36]

Elsewise, supporters focused on the safeguards entirely in suggesting that Iran's goals are irrelevant. Indeed, three dozen former admirals and generals, who supported the Iran deal, signed an open letter to Congress that highlighted the deal's ability to block "the potential pathways to a nuclear bomb" and strictures for "intrusive verification" yet simply rejected insinuations, also without sufficient backing, that the agreement was "based on trust."[37] As a result, supporters downplayed two plausible scenarios. Iran might seek to weaken US resolve and capability to confront Iranian transgressions, at home in its nuclear program and abroad by playing to widespread desires to preserve the nuclear arrangement; or, instead, Iran might simply wait out the agreement and push for a bomb once the deal has expired.

Thus, opponents and supporters heatedly dueled over laxities and safeguards in the agreement. Despite the tenor and substance of the debate, both sides relied on their unexamined assumptions about what Iran is likely to do in the near and long-term future.

Postscript: Leaving the Agreement

As we see, red lines can acquire formal status in agreements that also explicitly green-light some nuclear activities. Yet the red lines, no less than the green lights, will remain controversial: Whether they are duly restrictive or overly permissive depends on assumptions about a potential proliferator's intent.

The Trump administration's rejection of the Iran agreement was thus perhaps preordained. In April 2018, Donald Trump fulfilled a campaign promise by walking away from what he deemed "a horrible one-sided deal that should never, ever have been made."[38] The administration reimposed sanctions on Iran's economy—which had been lifted under the agreement—while adding 1,500 new ones. Its justifications were as follows.

First, the administration accused Iran of numerous violations. Most notably, these included impairing IAEA access to nuclear facilities and exceeding limits on centrifuge numbers and heavy water stocks. For the agreement's defenders, however, the alleged violations were but technical, materially insignificant, or remedial infractions—if infractions at all, given valid interpretations of provisions.[39] In their view, controversies concerning a party's adherence to an agreement inevitably arise in implementation. The JCPOA (much like strategic nuclear arms agreements) include negotiating structures to address these kinds of issues.

Second, the administration insisted that Iran's past behavior showed that Iran could not be trusted. The administration jumped on the incriminating evidence when, in 2018, Israel released a treasure trove of Iranian documents that disclosed Iran's past nuclear-weapons activities. Although US intelligence had concluded (in 2007) that Iran had ended its nuclear-weapons program in 2003, the program was more advanced than previously credited and continued beyond that year (Albright et al. 2018). The agreement's proponents could argue, however, that the evidence still did not reveal how long the program continued—and thus whether Iran's *ongoing* activities violated the terms of the JCPOA.

Third, the administration pointed to the years down the road when (given "sunset clauses") some constraints on the Iranian program would expire. Then, in Trump's words, the United States and its allies could not stop Iran "under the decaying and rotten structure of the current agreement."[40] Still, the agreement's defenders noted correctly that, in those outlying years, Iran was not "free" to pursue its nuclear options. Under the JCPOA, Iran explicitly committed not to build nuclear weapons; and, as an NPT signatory, Iran accepted restrictions on its nuclear program. Thus, the international community could hold Iran accountable should it move to acquire nuclear weapons. Regardless, they asked, what constraints would impair Iran in the absence of an agreement? Iran provided an answer. By 2021, with the US exodus from the JCPOA, Iran was enriching uranium at 60 percent enrichment and employing advanced centrifuges at the Fordow facility, among other breaches of the agreement.

Fourth, the administration charged that Iran had continued its ballistic missile programs in violation of UN Security Council resolutions, and that the agreement, itself, failed to impose restrictions on Iran's capability to develop missile-delivery capabilities. Likewise, it charged that the agreement failed to constrain Iran's military and subversive activities throughout the Middle East, which the United States deemed hostile to its interests. Although these deficiencies arguably spoke to a need for a broader agreement, the administration most certainly knew that such an agreement was "unlikely" with the existing Iranian regime. Even so, the administration was asking for broad restrictions which Cold War–era administrations rejected in strategic arms control negotiations. These administrations sought, instead, to constrain the strategic-nuclear competition, not to end strategic *competition* with Russia nor even to curtail the deployment of non-strategic nuclear weapons.

Notwithstanding its public posturing, the administration did not actually seek a better agreement. Instead, it sought to disrupt Iran's struggling economy, through the reimposition of sanctions, hoping to spur a change in regime.[41] Although the administration focused its ire on the JCPOA's supposedly lax constraints, and their violation nonetheless by Iran, its evidence was neither clear-cut nor convincing. It suggested, then, that Iran's intentions, not the agreement's restrictions, were in fact the issue. Put simply, the administration assumed that Iran sought the agreement to cloak the country's rush for the bomb, from a more advantageous economic, political, and technological position, at some point in the future. Thus, the evidence—its selection and interpretation—could only validate the administration's prior assumptions, not change them.

Conclusions

Nuclear-proliferation experts recognize that restrictions can work in tandem to foreclose the options of potential proliferators, even those that remain determined to maintain a nuclear infrastructure. The solution resides in a diverse range of measures that include limiting uranium stocks and imports of critical technologies; restricting the numbers, sophistication, and configurations of centrifuges and the production and reprocessing of plutonium; continuous monitoring of known nuclear facilities and intrusive inspections of suspect sites; and exchanging relevant information among national intelligence agencies and IAEA inspectors.

For their part, arms-control experts recognize, importantly, that a verification system can work despite its imperfections. Negotiators need not close every loophole nor strive for a fully verifiable agreement. Even a small probability of detection is adequate for enforcing an agreement if the monitored party is risk-averse or highly values the benefits of the agreement. Thus, monitoring a portion of the fuel cycle well, or multiple portions less well, can strengthen an agreement by *increasing the chances* of detecting a violation. The odds of detecting noncompliance only improve when interdependencies exist between a permissible and an illicit program that could expose irregularities or diversions of labor, material, and supplies or when any discovered violation can trigger more rigorous or exhaustive inspections or impugn the monitored party's adherence to jeopardize the agreement.

Although the JCPOA broke new ground in its exhaustive restrictions, assessments of the negotiating progress nonetheless reflected *implicit* readings of Iran's current and potential goals. Whether the agreement's terms were seen as duly restrictive or overly permissive depended on presumed assumptions about Iran's intentions that remained largely buried in the debate. Policy hawks could push, then, for the most severe restrictions on the Iranian program with little regard for ostensible US nuclear advantages. In other words, they paid little heed to the inconsistency between their own: (a) fears that Iran *might* capitalize on the agreement's laxities to acquire *but a small nuclear arsenal* and (b) beliefs that a quantitatively and qualitatively superior US nuclear force conferred practical political and military advantages. The deficiencies in such thinking deserve recognition in the final set of perils and pitfalls.

> **8.1:** *Given a capability focus, US national-security policy debates center inordinately on what adversaries* could *do rather than what they* would *do.*
> **8.2:** *Despite the attention that strategists devote to nuclear superiority, US nuclear advantages might little affect how US policymakers identify and address perceived threats.*

Indeed, US nuclear superiority had no apparent influence on Iran's willingness to bargain or on the terms of the deal that Iran ultimately accepted. If anything, Iran's desire to free the country from the oppressive effects of *even some* of the Western-imposed economic sanctions explain Iran's past willingness to deal and current hints of a willingness to return to serious negotiations (under still-unspecified conditions). Of course, Iran's decision to forgo a bomb could reflect the positive effects of nuclear deterrence: the recognition by Iran that it was ill prepared to compete in a nuclear arms race. Yet, if that were Iran's logic, Israel—a nuclear superior *to Iran*—was also well positioned for that (deterrence) purpose. That is, we could not attribute Iran's choice of strategy to *US* nuclear superiority.

> **8.3:** *Despite the attention that strategists devote to nuclear superiority, US nuclear advantages might little affect the decisions of non-nuclear armed adversaries to conflict or cooperate with the United States.*

9

The Case for Nuclear Superiority

Assessing What We Know (and Do Not Know) about Nuclear Deterrence

The Cold War stands as a golden age for thinking about nuclear deterrence and nuclear-war strategies. Academics and policy analysts cite deterrence theory, then, as a rare success story of bridge building between the academic and policy worlds. The arcane debate between hawks, doves, and those who sought some middle ground preoccupied the policy community and dominated academic studies of international politics. Scholars mastered the technological complexities of weapons development, deployment, targeting, and performance and moved freely into think tanks and government positions to help devise and sell US nuclear strategies. Yet the strategies suffered under scrutiny.

The dubious assumptions of the past now undergird claims of US nuclear supremacy. We are led to believe that, for the United States at least, deterrence is secure. The United States can negate or (at least) significantly limit the disastrous effects of a nuclear conflict with the right US technologies, in the right numbers, in the right places. The United States could thereby dissuade an adversary from attacking, and perhaps even disarm them if they foolishly planned to attack. At worst, the United States could play off its nuclear strength with coercive tactics that lend credibility to US threats to employ nuclear force.

With such claims, contemporary strategies echo the logic of their Cold War–era predecessors. As before, the claims rest on compelling but indeterminate math—inconclusive ratios or differences of advantage—and ungrounded assumptions about the relative willingness of the combatants to absorb cost. No less problematically, they exaggerate US control over nuclear-crisis outcomes by downplaying the confounding influence of political, social, psychological, and organizational forces *on all parties to the conflict.*

The False Promise of Superiority. James H. Lebovic, Oxford University Press. © Oxford University Press 2023.
DOI: 10.1093/oso/9780197680865.003.0009

The Flawed Case for US Nuclear Superiority: The Biases of Strategic Reasoning

Nuclear capabilities and compensatory tactics continue to drive our thinking about deterrence requisites and challenges. US analysts inflate the benefits of seemingly overwhelming nuclear advantages and, lacking a capability to impose the US will, the compensatory benefits of coercive tactics.

Table 9.1 underscores these deficiencies by grouping the logical perils and pitfalls found in the prior chapters. The first set draws from Chapters 2, 3, 7, and 8 in exposing the exaggerated influence, in the Cold War and post–Cold War years, of capability-centered reasoning. It reduced critical deterrence issues to matters of weapon options, numbers, types, and locations. The second set of deficiencies draws exclusively from Chapter 3. It highlights the aversions of nuclear states to using nuclear force, which limit the coercive influence of nuclear weapons. The third set of deficiencies draws from Chapters 4 and 6. It references the exaggerated effectiveness of coercive tactics in Cold War and post–Cold War strategic thinking, while the fourth set draws, in turn, from the same chapters to underscore the ineffectiveness and counterproductivity of these very tactics. The final set of perils and pitfalls draws from Chapters 5 and 7 to recognize the understated challenges of maintaining control in a crisis.

To be sure, social science provides reasons for crafting theory around nuclear capabilities and compensatory bargaining tactics. First, we must always "bracket" some aspects of problems, assuming away their effect, to keep intellectual problems manageable. Even a talented juggler can only keep a finite number of balls in the air at one time. Second, understanding progresses through advocacy, not through equivocation. Other factors can always confound conclusions, but theorists must nonetheless make their best case—pushing their preferred arguments to their limits. Third, we advance knowledge by favoring general and parsimonious explanations—then, for broadly defined phenomena—to achieve the benefits of efficiency. After all, we could argue that all factors have at least *some* explanatory impact under some circumstances. Fourth, we build knowledge through coherent—that is, unambiguous and logically consistent—explanations. No theory can, or should, explain all evidence or incorporate the assumptions and arguments of rival theories. How can you logically reconcile rational deterrence theory (that allows for "uncertainty") with a theory that presumes that existing beliefs color, filter, and thereby override evidence?

We lose much, however, when we doggedly adhere to explanations to the exclusion of available evidence and explanatory alternatives. Efforts to preserve the integrity of a theory could reflect a profound bias. Indeed, an "instrumental" bias and "linear" (as opposed to "dialectical") thinking (Luttwak 1987) have long colored US theorizing about nuclear advantages.

Table 9.1 **Perils and Pitfalls in Assessing Nuclear Advantages and Coercive Tactics**

Overstating the utility of nuclear capability by focusing on:
2.1: The salient capabilities of protagonists at the expense of their less-salient objectives.
2.2: Weapons in numbers, types, and locations to address asymmetries, without regard for their actual significance or the range of available options.
2.3: Conflict escalation narrowly as a product of weapon attributes, deployment, and employment.
2.4: Flexibility and matching in force employment to avoid hard questions pertaining to when, how, and where to use nuclear force.
3.1: Simple math, though insufficient for assessing relative nuclear advantages given the complexities of determining outcomes, with different force sizes, usages, and capabilities, over the course of a nuclear conflict.
3.4: Selective history, suggesting that the United States obtained coercive leverage from nuclear advantages over US nuclear- or non-nuclear-armed adversaries.
7.4: US nuclear advantages, when US officials might give little thought, in a crisis, to whether and how their nuclear advantages yield a coercive edge.
8.1: What US adversaries *could* do rather than what they *would* do.
8.2: US nuclear advantages, which might little affect how US policymakers identify and address perceived threats.
8.3: US nuclear advantages, which might little affect whether a non-nuclear-armed adversary conflicts or cooperates with the United States.
Understating the coercive limits of nuclear capability by ignoring or downplaying (the effects on intentions of):
3.2: The relative cost acceptance and shared aversions of the conflicting parties.
3.3: The impact of deterrence on ostensibly "superior" parties that do not believe they can fully disarm their opponent with assuredness.
3.5: The effects of a nuclear nonuse "tradition," which constrains behavior through fear of the cascading effects if bucking the tradition.
3.6: The deterrence benefits available to states with relatively small nuclear arsenals, given even the small chance of landing a warhead on the adversary's territory.
3.7: The implication that, by settling for "inferiority," US rivals impugn the practical impact of nuclear "superiority."
3.8: The risks of using nuclear weapons against a US ally, which weigh decisively in an adversary's cost-benefit analyses.
3.9: The behavior and incentives of "rogue-state" leaders, which suggest they reject the potentially devastating costs of attacking the United States or its allies with nuclear weapons.

Table 9.1 **Continued**

Overstating the utility of coercive tactics by relying on:
4.1: The signaling effectiveness of words or deeds—or one over the other—given the strengths and deficiencies of both.
4.4: The actual opaqueness of ostensibly ambiguous commitments.
4.5: The value of ambiguous commitments, which might offer a "worst of both worlds" solution: a commitment that is too soft to impress potential challengers yet sufficiently strong to pull the defender into action to confront an actual challenge.
6.1: The strong evidence that policymakers seek to show resolve or establish reputations for acting at the expense of weaker evidence that US adversaries respond, in international conflicts, to shows of resolve or reputations for acting.
6.2: The reputed effects of a party's resolve or reputation in crises though the effects are attributable to the perceived interests and/or capability of the party.
6.3: The reputed effects of a party's resolve though the effects are attributable to a leader's risk acceptance, government preparedness, and public support.
6.8: The utility of faux madness, which might not convince targets to back down and could provoke them.
Understating the limits of coercive tactics by ignoring or downplaying (the effects of intentions on):
4.2: The ineffectiveness of a defender's commitments when: a) a defender cannot accurately gauge the scope and nature of future challenges. b) a challenger believes that the operant conditions do not hold. c) a defender speaks without clarity and precision. d) a defender's messages are lost to contradictory signals.
4.3: The counterproductive nature of commitments when: a) viewed by challengers as provocations. b) foreclosing acceptable exits. c) trapping the defender by limiting its future options. d) implying concessions through exclusion. e) viewed as a "challenge" or demand by a challenger which "compels" it to act.
4.6: The potential for ambiguous commitments to (a) lead incrementally to firm commitments that "trap" the committing party or (b) reflect or enable indecisiveness which weakens commitments (in appearance and execution) when challenged.

(*continued*)

Table 9.1 **Continued**

6.4: Ineffective signaling when efforts to establish resolve or burnish a reputation invite the challenges of communicating clear commitments.
6.5: Ineffective signaling, in reputation building, when a target attributes to a disposition what an actor attributes to the situation, and vice versa.
6.6: Confounding results when the beliefs and expectations of the audience determine whether reputations form, what reputations form, which party reputations involve, and how much reputations matter.
6.7: Counterproductive results when behavior meant to communicate resolve or bolster a reputation raises the (short- and long-term) stakes and risks of a conflict.
Understating the challenges to crisis control by ignoring or downplaying:
5.1: The fact or likelihood that policymakers will accept risks and tradeoffs of precipitous action without duly considering arguments, information, and available alternatives that weigh toward a cautious response.
5.2: The fact or likelihood that the rules and practices of military organizations will remain largely unknown to government leaders and confound policy implementation.
5.3: The constraints on the gradual or discriminate employment of nuclear force given military efforts to maximize force capabilities in conflict.
5.4: The interdependent actions of competing military organizations, which could fuel escalation toward war, and even an all-out nuclear conflagration.
5.5: A critical trade-off: Constraining the nuclear-launch authority of a US president compromises the US capability to respond as quickly, and lethally, to a nuclear attack, but conceding complete launch authority to the president could empower a poor, venal, or malign decision maker.
5.6: A fading opportunity for control: At some point, maybe early in a nuclear confrontation, the parties might stop manipulating risk and start preparing for the worst to hedge against the inherent risks of the confrontation.
7.1: The limits of decisional capability: In a crisis, officials might lack the foresight, deliberativeness, and control to produce desired effects.
7.2: The limits of foresight: In a crisis, officials might give insufficient thought to whether, or how, their preference among conventional options might spark a nuclear conflagration.
7.3: Deference to salient options: In a crisis, officials might defer to conspicuous (perhaps, prepackaged) options with little insight, or effort to gain insight, into the adversary's goals or the conditions that could cause a conflict to spiral beyond control.

Instrumental Bias

With "instrumental" bias, individuals attend to policy means more than policy goals or policy consequences; indeed, they think about policy means as if they were policy objectives. These tendencies are perhaps understandable in thinking about nuclear conflicts. After all, the goals in such conflicts are elusive. They will depend on where, and for what, the war is fought, the presumed cost acceptance of both parties, and assumptions about the likely course and collateral effects of war.

By contrast, nuclear capabilities command attention—and then, for various reasons. First, military capabilities are *tangible*. Their physical form makes them observable. Their quantities and qualities give them salience when states prepare for war, vie in combat, or flaunt their power with military parades, flyovers, missile tests, and naval shows of force. Second, military capabilities are *measurable*. Analysts can predict changes in capabilities by assessing their growth rate, trajectory, or likely value based on changes in other critical variables. Third, military capabilities are relatively *stable*. They typically do not change rapidly, tied as they are to a country's economic scale and resources, technological prowess, and population base. Fourth, military capabilities are *manipulatable*. States can increase their capabilities to counter the capabilities of other states. Conversely, they can decrease their capabilities, signaling peaceful intent, to defuse conflicts and reduce ambient levels of threat. Finally, military capabilities are *transparent*. Realists argue, for instance, that states must defend against the capabilities of other states, not their intentions, because intentions are opaque and subject to change (see Waltz 1979).

Yet a focus on military capability can bias the assessment of policy problems when analysts read more into that capability than is warranted. That effect is apparent when analysts conflate intentions with capability and assume that the adversary will act as it is capable. Thinking that the adversary *will do* what it *can do* has stoked fears of the consequences should certain states acquire a single nuclear weapon. Conversely, the same mode of thinking instills an excessive faith that US nuclear capabilities can overcome basic deterrence challenges. Efforts to impose solutions through force, or to coerce through some bold use of tactics, might not work, and could prove counterproductive.

The strength of means-centered bias is apparent in its pervasiveness. Organizations routinely pursue surrogate goals—crafted to reflect organizational means—without due concern for broader purposes (March and Simon 1993). Wartime officials thereby read success into deceptive military indicators such as the body count, sorties flown, weapons captured, and the intensity and pace of combat. Academics, too, fall prey to the same basic tendency. We saw this, with the Cold War's end, when international politics theorists made much of the ostensible transition from a bipolar to a unipolar global system.

We could benefit from dwelling a bit on that example. Theorists made a seemingly compelling case for a now "unipolar" global system by highlighting relative US military capabilities—as measured, most prominently, by the US share of global military expenditures (Wohlforth 1999). But the case for "unipolarity"—like that for "bipolarity," before it (Wagner 1993)—weaken when we assess the standards in use. For instance, by one definition, the system is unipolar when a state is sufficiently strong "that no other single state is powerful enough to balance against it" (Pape 2005: 11–12). But why a mere "plurality of power"? Could some grouping of states, even if not allied in opposition, successfully oppose it? After all, the United States was tested severely when it fought two wars simultaneously in Iraq and Afghanistan. Raising the standard—to say, a "majority" of capability (Thompson 2006: 12)—does not solve the problem. How can we assess the utility of US capability—US preeminence aside—without evaluating the goals of the parties, their relative cost acceptance, and where and how any war is fought?

As with nuclear superiority, a simple statistical case for US conventional-military preeminence was easier to provide than logical support for the capability share central actors must possess to establish preponderance (Mansfield 1993). In a word, the logic used to pronounce systemic unipolarity was "indeterminate." We cannot discern polarity logically from aggregate capability distributions, even should they appear related to implicating behaviors. Indeed, what some scholars took as post–Cold War "balancing" against the unipolar state, others read justifiably as normal interstate competition and cooperation. (On polarity, and unipolarity, see Lebovic [2017].)

We lose touch, then, with the *limits* of capability when we work with numbers, blind to their practical significance and implications. The United States enjoys an unquestionable qualitative edge against all nuclear opponents in command and control, communications, intelligence, and force performance. It also presents an enormous quantitative challenge to all its nuclear adversaries. But we must predict all conflict outcomes with large margins of uncertainty. The US wars in Vietnam, Iraq, and Afghanistan testify to the hazards of forecasting based on prewar military balances. The consequences of nuclear wars are hardly more predictable. Indeed, they invite phenomenal costs, regardless of scale, that are lost in thinking about relative nuclear advantages.

Linear Thinking

Strategic theorizing is afflicted further by "linear thinking." In such assessments, researchers ask, for instance, how a party can strengthen the credibility of its threats—perhaps as it progresses through some sequence of programmatic steps to achieve a desired outcome. If verbal threats do not work, a show of force is deemed necessary, with intensities that grow along some established ladder of escalation.

In dialectical assessments, by contrast, researchers confront the complexities and challenges that weaken *both parties'* control over conflict outcomes. What one party does will depend on what the other party does—so that neither party fully controls the outcomes.

With a dialectical view—acknowledging both parties and their interdependent beliefs and behavior—researchers can more fully explore when and how parties misread each other's actions. They can also assess how the dynamics of conflict, as it changes form and severity over time, promotes misperception and stress, feeds on diverging beliefs, alters the stakes, reduces time horizons, constrains choices, and creates "self-fulfilling prophesies." A US adversary could easily see US "deterrence" demands as "compellence" threats—even take them as blackmail. The identity of the "defender" and "challenger" reduces inevitably to the biased viewpoints of the participating parties. Neither party will need to dig deep into its own history to establish why it, not its opponent, is the "victim." Besides, all parties to a conflict understand their own considerations—the political pressures and material constraints—far better than they understand the adversary's situation. Each party is inclined to attribute its own actions to internal constraints, and pressures, and the adversary's prior provocations and current actions to nefarious purposes. The "cure" offered by coercive tactics is worse than the disease, then, if it reduces US bargaining space and increases the target's resistance.

A dialectical view can also increase sensitivity to the narrowing of decisional groups, and the shifting of command authority, that could occur with grave conflict. We can appreciate, then, that options will remain significantly constrained, on both sides of the conflict, by the planning criteria and operating procedures of military organizations. Prepackaged options will likely become *the* options when stress, low decision-time, uncertainty, and miscommunication take their toll. Leaders themselves might stand outside the decisional loop should they predelegate launch authority over nuclear weapons to subordinates who will possess only a grounds-eye view of the conflict and a rule-driven sense of available options. We worry about troubling encounters that "scream" crisis—the Cuban Missile Crisis, for example. We must also worry about the silent threats—in preparatory, preprogrammed, or defensive actions—that can drag parties unwittingly into war.

Lessons from (Almost) Three-Quarters of a Century of Deterrence Theorizing

Robert Jervis' seminal distinction between the "deterrence" and the "spiral" model (as discussed in Chapter 7) provides a useful starting point in assessing what we know, and do not know, about nuclear deterrence. The deterrence model holds when states choose not to conflict, given its costs. The model draws from the realist view of a world characterized by profound threats and hardheaded cost-benefit analyses. States will challenge others—even if it means war—when opportunities for gain present themselves, but states will back down, in the face of threats, when the balance of forces is unfavorable. By contrast, the spiral model holds when states choose unwisely to answer threats with threats, and conflict with conflict, regardless of the underlying payoff. The outcome—even if fueled by *defensive* concerns—worsens accordingly, for all parties. Conditions spiral beyond control as threats by one party reinforce the opponent's perception that the party is untrustworthy and dangerous (a "self-fulfilling prophesy"). Conflicts spiral further when threats activate detrimental domestic forces. Leaders will act forcefully to deflect charges that they are "soft" on defense. They might answer with scripted behavior—mobilizing military forces, for example—which the opponent takes as an act of aggression, even war.

The irony, of course, is that deterrence threats induce spirals because the defender mistakenly believes that the deterrence model holds. The defender does not understand that, by acting forcefully in response to threat, it is exacerbating conditions, making its situation potentially worse. To the contrary, it believes that, by standing firm, it can avoid conflict.

For academics and practitioners, however, the key question is not "Which model is best?" Instead, it is "Under what conditions (level of conflict, asymmetry of forces, and so forth) does one of the two models better predict outcomes?" Despite decades of research, and theorizing, then, we remain ill equipped to predict whether and how coercive tactics will work, and when they will yield the consequences that policymakers hope to avoid. So, where does that leave us? With these severe limitations in mind, I offer policy prescriptions that acknowledge the limits of our understanding, and the dangers that accrue from bold leaps into the unknown. These (albeit modest) prescriptions are meant—first, and foremost—to reduce the risk of unwitting escalation and inadvertent war.

First, stay clear of the brink. Because crises are defined by surprise, high threat, and limited decision time, parties are least able to process information at the very point when it is most needed. At that point, they might lack critical information, receive it in overwhelming quantities, and flounder trying to separate the informative from the noise. The risk, then, is that they will simply react out of fear, only adding to a cacophony of signals—all the while believing that they, or their opponents, can control crisis outcomes.

To avoid these conditions, we should let reality—here, the destructiveness of any nuclear war—"speak for itself." We risk losing the message—maybe killing all the messengers—when we manufacture stakes, manipulate risks, show false bravado, and make commitments that ultimately invite future cost that we accept only by downplaying risk. As crises create stakes for the parties that inhibit retreat, conflicts become less about specific issues of contention and more about preserving or correcting the balance in a relationship, and the credibility and reputation of the parties. The parties themselves can lose sight, then, of the actual stakes, the costs of war, and the attending risks.

Parties should intervene, as necessary, to acquire or preserve options or remove doubts about the consequences of a potential challenge. But they should avoid precipitous responses and exaggerated threats which can fuel deleterious conflict dynamics. They should seek to constrain adversary behavior, rather than provoke it, and to reinforce the visceral tendency to exert caution on matters pertaining to nuclear war. Although analysts might assume that the United States can manipulate risk to its advantage, the opposite might hold true. Perversely, the uncertainty that US leaders might manipulate for gain can intensify ambient threat. The weaker party might act rashly, fearing the consequences of acting second.

Second, nuclear weapons serve best as instruments of deterrence, not compellence. Although parties are more reticent to concede what they possess than what they do not (yet) possess, policymakers can reduce a compellence challenge by transforming compellence problems into deterrence problems. A nuclear aspirant or nuclear-armed state, with its options or ambitions constrained (deterred), might eventually concede to (internal and external) forces that reduce the incentive to arm. Or better, policymakers can act with foresight before issues become compellence problems. They should not shy, then, from identifying their critical interests (assuming they are critical interests) to avoid major conflicts. To be sure, demarking interests *might* tempt adversaries to strike outside these designated areas of interest. But "line drawing"—delineating priorities through candid remarks and routine behavior—is an inevitable feature of international interaction. Although occasions arise when parties must make their interests explicit, they should not inflate the value of previously excluded zones when—that is, *because*—they are under assault.

Finally, we should not exaggerate the political or material benefits of relative nuclear advantages. Certainly, the logic of damage limitation is seductive. Why not target adversary nuclear capabilities to reduce the damage an adversary can inflict in return? Why not build missile defenses that can limit damage to US allies or the US homeland? But dangers exist in any competition between nuclear offenses, or nuclear offenses and defenses. Counterforce targeting, like missile defenses, can foster an illusion that perfection is possible or that damage in nuclear war is reducible to acceptable levels. Ironically, damage limitation, whether by offense or defense, can spur compensatory actions. Imperfect US offenses and defenses might prove ill equipped to handle the now graver threat that those systems inspired.

Put differently, we must avoid the temptations of denial given the potentials of punishment. "Perfect" solutions likely rest on an *imperfect* understanding of operant conditions. Even if these systems work as advertised, the costs are non-negligible. Nuclear wars will not reduce to surgical counterforce strikes, with limited human effects. A high human toll is a likely inevitable consequence of such wars given the colocation of targets, miscalculation and accidents, the relaxing of prohibitions and inhibitions, a heightened state of preparedness, and the ever-present danger that the "weaker" of two parties will lash out when anticipating an attack, whatever the apparent payoff.

In short, policymakers must acknowledge that nuclear weapons are exceptional in the damage they can inflict and the unknowns and uncertainties of use. We must fervently hope that we will never reduce these unknowns and uncertainties through practical experience.

NOTES

Chapter 1

1. Jonathan Swan and Margaret Talev, "Scoop: Trump Suggested Nuking Hurricanes to Stop Them from Hitting U.S.," *Axios* (August 25, 2019), https://www.axios.com/trump-nuclear-bombs-hurricanes-97231f38-2394-4120-a3fa-8c9cf0e3f51c.html.
2. Alex Ward, "Trump Says He Could Wipe Afghanistan Off Face of the Earth in 10 Days," *Vox* (July 22, 2019), https://www.vox.com/world/2019/7/22/20704248/trump-afghanistan-10-days-war.
3. Carrie Dann, "Donald Trump on Nukes: 'Let It Be an Arms Race,'" NBC News (December 23, 2016), https://www.nbcnews.com/politics/politics-news/trump-nukes-let-it-be-arms-race-n699526.
4. Lauren Gambino, "Donald Trump Boasts that His Nuclear Button Is Bigger than Kim Jong-un's," *The Guardian* (January 3, 2018), https://www.theguardian.com/us-news/2018/jan/03/donald-trump-boasts-nuclear-button-bigger-kim-jong-un.
5. Ryan Browne and Barbara Starr, "Trump Touts New 'Super Duper' Missile but Pentagon Won't Confirm Details," CNN (May 16, 2020), https://www.cnn.com/2020/05/16/politics/trump-super-duper-missile/index.html.
6. Infamously, Trump's outburst provoked his Secretary of State, Rex Tillerson, to refer to the president as a "[expletive] moron" in a post-meeting outburst (a comment that the secretary never denied or retracted). Courtney Kube, Kristen Welker, Carol E. Lee, and Savannah Guthrie, "Trump Wanted Tenfold Increase in Nuclear Arsenal, Surprising Military," NBC News (October 11, 2017), https://www.nbcnews.com/news/all/trump-wanted-dramatic-increase-nuclear-arsenal-meeting-military-leaders-n809701.
7. Nuclear Posture Reviews are the Pentagon's quadrennial reviews of US nuclear forces.
8. Biden's NPR currently remains classified, apart from a one-page unclassified summary, so its full content is not publicly known. Its public unveiling has been delayed reportedly because of the challenge posed by the Russian invasion of Ukraine, China's nuclear buildup, and political pushback from one major NPR recommendation—the cancellation of the submarine-launched cruise missile program (which currently remains in political limbo). International Institute for Strategic Studies (IISS), "The US Nuclear Posture Review in Limbo," *Strategic Comments* (June 2022), https://www.iiss.org/publications/strategic-comments/2022/the-us-nuclear-posture-review-in-limbo. For an early assessment of the NPR, see Daryl G. Kimball, "Biden Policy Allows First Use of Nuclear Weapons," *Arms Control Today* (April 2022), https://www.armscontrol.org/act/2022-04/news/biden-policy-allows-first-use-nuclear-weapons. See also Adam Mount, "The Biden Nuclear Posture Review: Obstacles to Reducing Reliance on Nuclear Weapons," *Arms Control Today* (January/February 2022),

https://www.armscontrol.org/act/2022-01/features/biden-nuclear-posture-review-obstacles-reducing-reliance-nuclear-weapons.

9. Mark Melamed, Lynn Rusten, and Jill Hruby, "Russia's New Nuclear Weapon Delivery Systems: An Open-Source Technical Review," NTI (November 13, 2019), https://www.nti.org/analysis/articles/russias-new-nuclear-weapon-delivery-systems-open-source-technical-review/.
10. Vladimir Isachenkov, "New Russian Policy Allows Use of Atomic Weapons against Non-Nuclear Strike," *Defense News* (June 2, 2020), https://www.defensenews.com/global/europe/2020/06/02/new-russian-policy-allows-use-of-atomic-weapons-against-non-nuclear-strike/. See also Lieber and Press (2020: 112–113).
11. Then, Belarus effectively renounced its non-nuclear status by revising its constitution to permit the stationing of Russian nuclear weapons on the borders of Lithuania and Poland.
12. Reuters, "Russia's Lavrov: Do Not Underestimate Threat of Nuclear War," Reuters (April 25, 2022), https://www.reuters.com/world/russia-says-western-weapons-ukraine-legitimate-targets-russian-military-2022-04-25/.
13. Hypersonic glide vehicles have grabbed much attention for their novel performance: they hug the terrain, travel many times the speed of sound, and defy attack-warning and defensive systems. David Martin, "Exclusive: No. 2 in U.S. Military Reveals New Details about China's Hypersonic Weapons Test," CBS News (November 16, 2021), https://www.cbsnews.com/news/china-hypersonic-weapons-test-details-united-states-military/.
14. In this book, I employ "superiority" as an umbrella term, though I distinguish the assumptions and implications of "superiority," as Kroenig understands it, from those of the damage-limitation posture that Lieber and Press associate with primacy.
15. I use the terms "analysts," "theorists," and "strategists" interchangeably. I do so for convenience but also because individuals often wore multiple hats in deterrence debates, making it difficult to describe them with a single label. Indeed, "theorists" frequently became strategists; and theorists and strategists implicitly or explicitly engaged in policy analysis.
16. Hawks often present their argument as a lesson of "appeasement." Although appeasement literally refers to actions meant to satisfy some demand, it acquired its ugly connotation from British Prime Minister Neville Chamberlain's 1938 concession to Hitler, at Munich, of the German-populated portion of Czechoslovakia. It came to mean cowardly attempts to placate an insatiable adversary. For hawks, the attending lesson is clear: attempting to avoid conflict, at all costs, only emboldens and strengthens the adversary. Indeed, it only postpones the day of reckoning. By contrast, doves reject the idea that threats require an answer with threats. They see danger in treating every conflict as a "battle with Hitler." In the dovish view, most state leaders voice finite demands and accept compromise to achieve cooperative benefits. Doves thus warn of the dangers—to all sides—should conflict dynamics take hold. When misperception and misunderstanding prevail, parties fail to recognize their common interest in avoiding conflict.
17. For a review of contemporary writings on the subject, see Early and Asal (2018).
18. Indeed, these concerns inspired the Kennedy administration's "flexible response" strategy, which would presumably give NATO's conventional forces the capability to serve as a more effective, first line of defense (with tactical nuclear weapons now held in reserve).
19. For an important exception, see Sechser and Fuhrmann (2017). The authors critically assess the theory and evidence behind assertions that nuclear states can harness their absolute or relative nuclear advantages to obtain coercive benefits from other states. Although I emphasize the illusionary benefits of US "superiority," I find little daylight between my views and the more general set of arguments and assumptions offered in that excellent volume.
20. For the seminal outlines of deterrence principles, see Schelling (1960, 1966), Schelling and Halperin (1961), and Snyder (1961).
21. On intentions in state behavior, see Edelstein (2002) and Yarhi-Milo (2013).
22. See, e.g., Knopf (2010), Lebovic (2007), and Lupovici (2010).

23. Todd S. Sechser and Matthew Fuhrmann (2017: 8–9) view these positions as representing opposing schools of thought: "absolutists" assume that the mere possession of nuclear weapons gives a country a coercive advantage; "relativists" assume, instead, that advantages stem from favorable nuclear-weapon ratios.

Chapter 2

1. Fred Kaplan, "JFK's First-Strike Plan," *The Atlantic* (October 2001), https://www.theatlantic.com/magazine/archive/2001/10/jfks-first-strike-plan/376432/
2. Although that solution initially favored the Air Force, which controlled both missiles and bombers, the Navy's submarines—with the greater invulnerability to attack—increasingly carried the destructive load. Aided by arms-control restrictions, the US nuclear force moved out to sea.
3. William Burr, "U.S. Nuclear War Plan Option Sought Destruction of China and Soviet Union as 'Viable' Societies," National Security Archive, Briefing Book #638 (August 15, 2018), https://nsarchive.gwu.edu/briefing-book/nuclear-vault/2018-08-15/us-nuclear-war-plan-option-sought-destruction-china-soviet-union-viable-societies.
4. Indeed, like Assured Destruction, Sufficiency had a budgetary rationale: it too was intended in part to control the budgetary demands of governmental interests (Burr 2005: 54).
5. William Burr, "U.S. Cold War Nuclear Target Lists Declassified for First Time," National Security Archive, Briefing Book #538 (December 22, 2015), https://nsarchive2.gwu.edu/nukevault/ebb538-Cold-War-Nuclear-Target-List-Declassified-First-Ever/.
6. William Burr, "U.S. War Plans Would Kill an Estimated 108 Million Soviets, 104 Million Chinese, and 2.6 Million Poles: More Evidence on SIOP-62 and the Origins of Overkill," Unredacted (November 8, 2011), https://unredacted.com/2011/11/08/u-s-war-plans-would-kill-an-estimated-108-million-soviets-104-million-chinese-and-2-3-million-poles-more-evidence-on-siop-62-and-the-origins-of-overkill/.
7. William Burr, "U.S. Nuclear War Plan Option."
8. For the basis of this argument, see Davis (1975).
9. After all, would US leaders have respected Soviet "restraint," in 1962, had the Soviets hit the United States only with Cuba-based missiles?
10. The principles were articulated in National Security Decision Memorandum-242, in January 1974.
11. Indeed, the prior targeting of "war-supporting industry," by the military's reckoning, would result in the killing of 30 percent of the Soviet Union's population (Burr 2005: 41).
12. By the mid-1980s, "counter-recovery" targets no longer appeared in the SIOP. See Ball and Toth (1990: 70).
13. That is, "military strategic planners generally found the concept of limited nuclear options, as propounded by civilian analysts, to be an abstraction that could not be translated into realistic targeting plans" (Terriff 1995: 112). Indeed, "guidance provided by political appointees had typically been seen by the targeteers as so out of sync with reality, or so vague, that there was little to be done but to ignore it" (Nolan 1989: 255).
14. On how the execution of US nuclear plans would have played out in the 1960s and subsequent decades, see Robert S. Hopkins, III, "How the Strategic Air Command Would Go to Nuclear War," National Security Archive, Briefing Book #663 (March 13, 2019); Bruce Blair, "How SAC Would Have Gone to Nuclear War in the 1970s," National Security Archive, Briefing Book #663 (March 13, 2019). https://nsarchive.gwu.edu/briefing-book/nuclear-vault/2019-02-25/how-strategic-air-command-would-go-nuclear-war. For a discussion of process and considerations in developing the SIOP, see US GAO (1991).
15. William Burr "Looking Back: The Limits of Limited Nuclear War," *Arms Control Today* (August 29, 2008), https://www.armscontrol.org/act/2006-01/looking-back-limits-limited-nuclear-war. See also Nolan (1989: 124).

16. During the Kennedy administration, McNamara considered such city trading—a nuclear game of chicken—as a potential response to the Soviet acquisition of some assured retaliatory capability. He shelved the idea, however, when conceding that the Soviets, as the weaker party, would have little interest in playing a game that would effectively disarm them. They would run out of weapons (for destroying US cities) before the United States had depleted its arsenal.
17. The strategy was codified in 1980 in Presidential Directive-59 (PD-59). On PD-59, see Nolan (1989: 126–139). For an excellent critique of the strategy, see Jervis (1984).
18. William Burr, "Jimmy Carter's Controversial Nuclear Targeting Directive PD-59 Declassified," National Security Archive, Briefing Book #390 (September 14, 2012), www.gwu.edu/~nsarchiv/nukevault/ebb390.
19. Slocombe served as Deputy Undersecretary of Defense for Policy Planning.
20. It replaced the single-warhead SS-4 and SS-5 missiles. They were less accurate, liquid-fueled (requiring loading before launch), and often silo-based (Anderson and Nelson 2019: 92).
21. That logic would probably make most hawkish analysts uncomfortable. Similar logic was used by hawks, however, to oppose Kennedy's flexible-response strategy. They argued that conventional options deflated the deterrent value of theater nuclear weapons.
22. US allies in NATO certainly hoped that was not the case, though they retained their suspicions.
23. Although these missiles could hit Soviet bases and command centers, whether they had sufficient range to hit Soviet *strategic* command and control centers in Moscow (for an affirmative view, see Haslam 1990)—and thus the degree of threat—was a matter of contention. Whatever the reality, Soviet fears and perceptions would count in a crisis.
24. Charles Mohr, "Pershings Put Moscow on 6-Minute Warning," *New York Times* (February 27, 1983), https://www.nytimes.com/1983/02/27/weekinreview/pershings-put-moscow-on-6-minute-warning.html.
25. With the country's insistence that other NATO countries must also accept the intermediate-range missiles, Britain, the Netherlands, Italy, and Belgium accepted the basing of the cruise missiles on their national territory. For documentation on the NATO dispute, see William Burr, "Thirtieth Anniversary of NATO's Dual-Track Decision: The Road to the Euromissiles Crisis and the End of the Cold War," National Security Archive (December 10, 2009). https://nsarchive2.gwu.edu/nukevault/ebb301/.
26. This outcome was not for lack of foresight. The United States had earlier lobbied for answering the new Soviet threat, instead, by replacing older shorter-range systems with newer versions. After all, as Paul Warnke (Carter's first director of the Arms Control and Disarmament Agency) recognized, the number of new warheads the United States planned for Europe was dwarfed by the numbers available in the US strategic-nuclear arsenal (Talbott 1985: 43). Indeed, a portion of the US SLBM force was dedicated for use in the European theater.
27. Drawing from AD principles, US policy doves took comfort in the deterrence potential of the surviving legs of the US triad—the mobility and evasiveness of the substantial US bomber and submarine-launched ballistic missile (SLBM) force upon which the United States had come to depend.
28. Both conclusions became mainstays of US National Intelligence Estimates of Soviet strategic capabilities until the Cold War's end. On the exaggeration of the Soviet threat, see Rovner (2011: 89–136).
29. Bombers and submarine-launched ballistic missiles were reputedly not up to the task throughout much of the Cold War.
30. Their dark assessment of Soviet intentions would inform future US National Intelligence Estimates (the authoritative judgments of the US intelligence community).
31. For a fuller expression of these views, see US Senate (1979b).
32. Because their criticism focused mainly on the *coercive* edge the Soviets would obtain before or after a nuclear exchange, however, we cannot conclude that critics were generally uncomfortable with Carter-administration *doctrine* per se. After all, it ostensibly addressed that concern. For that matter, it is difficult to make much of the treaty's criticism, at its face, given

the behavior of the administration to follow. Although hawks found in Reagan a kindred spirit—he campaigned for president in 1980 as a treaty opponent—his administration soon announced it would adhere to the (unratified) treaty, if the Soviets did too, and it remained generally in compliance though accusing the Soviets of treaty violations.

33. They thus defied dovish claims that the Soviets had only to provide the United States with *necessary* data for verifying Soviet treaty compliance under the terms of the SALT II Treaty and that the Soviets themselves lacked unfettered access to US telemetry, which the US military was loath to provide.
34. Suspect actions involved the construction of a phased-array radar facility that was not located on the Soviet border facing outward and the use of radar at Soviet missile test facilities simultaneous with tests of antiballistic missiles.
35. On the compliance disputes, see Schear (1985).
36. By designing strikes solely around blast damage, US war planners depreciated the overall destructiveness of US attacks on the Soviet populous (Burr 2005: 43, 74).
37. Such thinking was a product of reverse imaging: "We are good and care about our people; they are bad and care only about their own lives or ideological commitment to preserve the (oppressive) Soviet-state system."
38. Tactical warning would alert a party that adversary bombers and missiles were approaching their targets; strategic warning, by contrast, would alert a party that the adversary was preparing to attack.
39. Thus, AD advocates tended not to insist, for example, that a US interest in missile defense would feed Soviet suspicions that US policymakers were unwilling to incur the destruction of US cities, as promised in retaliation by AD doctrine.
40. Some hard warfighters shared the AD aversion to urban defense but for quite different reasons. For them, such defenses represented an unnecessary distraction and diversion of resources from the offensive requisites of warfighting. Stated more crudely, why build defenses when you can build missiles? For a supportive hard-warfighting view, however, see Van Cleave (1986).
41. Even US policy doves, who worried that US defenses might spark an arms race with Russia and China, were uncomfortable challenging defensive systems that might reduce the costs of war when the alternative—a preventative US attack—was far less palatable.
42. Indeed, that was the concern behind Russian fears that less-capable US defenses would become more-capable ones.
43. The Trump administration's aspirations were articulated in the Pentagon's much-delayed missile defense review. Paul Sonne, "A Missile Plan Akin to Reagan's 'Star Wars,'" *Washington Post* (January 17, 2019), A1.
44. On this, see Nolan (1999: 29–30).
45. The START I (Strategic Arms Reduction Treaty) was signed in 1991 and went into effect in December 1994.
46. The New START agreement was signed in 2010 and entered into force in early 2011. It was set to expire ten years later.
47. For an argument against the counterforce emphasis in contemporary nuclear targeting, see Kristensen, Norris, and Oelrich (2009).
48. Whereas the Obama administration NPR ruled out a nuclear attack on any country deemed in compliance with the Nuclear Nonproliferation Treaty, the Trump administration significantly loosened various prohibitions. Nuclear retaliation was now permissible in response to adversary use of weapons of mass destruction, a class that pointedly included chemical and biological weapons, as well as attacks on US cyber and space assets that could weaken US nuclear-response capabilities.
49. Lawrence Korb, "Why Congress Should Refuse to Fund the NPR's New Nuclear Weapons," *Bulletin of the Atomic Scientists* (February 7, 2018), https://thebulletin.org/commentary/why-congress-should-refuse-to-fund-the-nprs-new-nuclear-weapons/.

50. Stephen Collinson, "Biden Sends a Careful but Chilling New Nuclear Message to Putin in CNN Interview," *CNN* (October 12, 2022), https://www.cnn.com/2022/10/12/politics/joe-biden-nuclear-message-putin-cnntv-analysis/index.html.
51. On the discrimination challenge, see Vipin Narang, "The Discrimination Problem: Why Putting Low-Yield Nuclear Weapons on Submarines Is So Dangerous," *War on the Rocks* (February 8, 2018), https://warontherocks.com/2018/02/discrimination-problem-putting-low-yield-nuclear-weapons-submarines-dangerous/.
52. Associated Press, "Russia Slams US Arguments for Low-Yield Nukes," *Defense News* (April 29, 2020), https://www.defensenews.com/smr/nuclear-arsenal/2020/04/29/russia-slams-us-arguments-for-low-yield-nukes/.
53. Paul Sonne, "Pentagon Looks to Adjust Missile Defense Policy to Include Threats from Russia, China," *Washington Post* (March 2, 2018), https://www.washingtonpost.com/world/national-security/pentagon-looks-to-adjust-missile-defense-policy-to-include-threats-from-russia-china/2018/03/01/2358ae22-1be5-11e8-8a2c-1a6665f59e95_story.html.
54. Noteworthy, then, the review borrowed the verbiage of a Cold War–era decision memorandum (NSDM-242), without the critical terminology. In the words of that memorandum, "should conflict occur, the most critical objective is to seek early war termination, on terms acceptable to the United States and its allies, at the lowest level of conflict feasible. This objective requires planning a wide range of limited nuclear employment options which could be used in conjunction with supporting political and military measures (including conventional forces) to control escalation."
55. Paul Sonne, "Mattis Portrays Nuclear Strategy as A Check on Russia," *Washington Post* (February 7, 2018), A7. Mattis's claim seems odd given the NPR's conclusion that an aversion to nuclear war's horrendous destructiveness had long kept the peace: "Non-nuclear forces also play essential deterrence roles, but do not provide comparable deterrence effects—as is reflected by past, periodic, and catastrophic failures of conventional deterrence to prevent Great Power war before the advent of nuclear deterrence" (US Department of Defense 2018: 2).
56. Take, e.g., the Trump administration's decision, in 2019, to withdraw from the Intermediate Range Nuclear Forces (INF) Treaty ostensibly in response to Russian violations of the treaty. (Negotiated in the final Cold War years, the 1987 INF Treaty banned an entire class of weapons: land-based cruise and ballistic missiles with ranges of between 500 and 5,500 kilometers.) Seeking the missiles that Russia is pursuing would seem a suitable response, at first glance, to the alleged violations. Why should a party *unilaterally* abide by the provisions of a treaty? The relevant question, however, is "are we better off with the treaty than without it?" Although people routinely violate speed limits on US interstate highways, would conditions improve if the government lifted all limits? Thus, we must consider whether the US benefits from acquiring intermediate-range, ground-launched missiles—given available substitutes—offset the significant costs of dumping a treaty that imposes *some* constraints on a major strategic competitor.
57. As Hans Kristensen (2010) puts it, "Obama's OPLAN 8010 is the product of strategic targeting and war planning principles developed during the Cold War, the Clinton administration's expansion of nuclear doctrine, and the Bush administration's preemption and 'New Triad' vision."

Chapter 3

1. A Cold War–era observation is prescient now: "The strategic debate has focused on numbers of missiles and warheads as if they were living creatures whose survival was of value in their own right." It thereby ignores the implications of war "in terms of values about which [statesmen] do care" (Schilling 1981: 72).
2. Kroenig claims to address the failings of the more standard brinksmanship argument, which supposedly focuses on the "payoff to winning" but not also the "expected costs of disaster."

3. Remarks by the President at the United Nations Security Council Summit on Nuclear Non-Proliferation and Nuclear Disarmament, https://obamawhitehouse.archives.gov/the-press-office/remarks-president-un-security-council-summit-nuclear-non-proliferation-and-nuclear-.
4. On these points, see the H-diplo review of Kroenig's book by Charles Glaser in "Roundtable 10-25 on *The Logic of American Nuclear Strategy: Why Strategic Superiority Matters*," https://issforum.org/roundtables/10-25-nuclear.
5. Beyond payoffs, costs, and risks, the resulting calculations assume that all warheads are equal. Kroenig thereby ignores the host of variables that will determine the outcome of a nuclear exchange. Results are sensitive to the timing of the attack (whether a party struck first or second); weapon survivability, reliability, accuracy, and promptness in delivery; command, control, communications, and intelligence capabilities; the distribution and temporal spacing of explosions; the characteristics and choice of targets; and so forth.
6. For a critique of Kroenig in that regard, see Long and Green (2015: 66).
7. For a critical evaluation of the meaning and consequences of the "nuclear revolution," see Bell (2019).
8. For a contrary view suggesting conditions under which non-nuclear states consider the nuclear capabilities of opponents, see Avey (2019).
9. Efforts by the Truman administration to transfer nuclear technology (and, eventually, US nuclear weapons) to an international authority offer lessons in this regard. The United States reserved the right to increase its arsenal until it was assured that an international authority—in which the United States enjoyed a voting majority—could adequately verify and control global nuclear capabilities. The Soviet Union refused to play by those rules, ending the hopes of those who sought early to avoid any global nuclear-arms competition. Whether or not the Soviets actually sought a surreptitious nuclear advantage, their response to the United States—disarm first, and then we will talk—is hardly shocking. Indeed, for students of international politics, the outcome serves as a textbook case of prisoner's dilemma.
10. We can reasonably ask whether, rather than validating the hypotheses and measures simultaneously, we actually have validated either. A hypothesis test requires valid measures; and a deficient hypothesis test, reliant on questionable measures, cannot validate those measures.
11. Disingenuously, perhaps, some warfighting critics did have a point, then, when they challenged the morality of AD doctrine.
12. That South Korea and Taiwan have chosen to forgo these weapons, even when close adversaries (North Korea and China) possess them, and most states have remained non-nuclear countries despite potential security gains from their acquisition, provides reason to suppose that states reject the nuclear option out of hand. On this, see Solingen (2009).
13. Warfighters went to great length to discount these factors in no small part because they played to AD claims that negative effects would rebound with certainty to afflict the initiator of a nuclear conflict.
14. Truman maintained civilian control of nuclear weapons through the Atomic Energy Commission.
15. See Halperin (1966) and Charlie Savage, "Risk of Nuclear War over Taiwan in 1958 Said to Be Greater than Publicly Known," *New York Times* (May 23, 2021), https://www.nytimes.com/2021/05/22/us/politics/nuclear-war-risk-1958-us-china.html.
16. US leaders also have strengthened those constraints, albeit unintentionally, by seeking to strengthen the nuclear nonproliferation regime and thereby giving voice to the principles (in the NPT Treaty) that the nuclear haves work toward nuclear disarmament.
17. William I. Hitchcock, "Trump Threatened to Nuke North Korea. Did Ike Do the Same?" *Washington Post* (August 11, 2017), https://www.washingtonpost.com/news/made-by-history/wp/2017/08/11/trump-threatened-to-nuke-north-korea-did-ike-do-the-same/.
18. See Tannenwald (2007: 309–310). Even President Bush took pains not to suggest that a US nuclear response to a chemical weapons attack was automatic (Tannenwald 2007: 311).

19. See Scott D. Sagan and Allen S. Weiner, "Would the U.S. Really Answer Cyberattacks with Nuclear Weapons?," *Washington Post* (July 11, 2021), B1.
20. In a 1999 Gallup Poll, 66 percent of Americans rated the atomic bombing of Hiroshima "one of the most important" events of the century when specifically asked. Yet, in a Pew Research Center for the People & the Press Millennium Survey, less than .5 percent of American respondents listed the atomic bomb when asked for a word or phrase that describes their impression of the 1940s. Although that is understandable given the overriding importance of the Second World War, only around a third of the public volunteered answers directly related to that war. Likewise, a 1999 Gallup/CNN/USA Today Poll that asked respondents to list the single most important event to have occurred in that century showed only 2 percent listing the invention of nuclear weapons or the atomic bombing of Hiroshima—a small number in absolute but also relative terms. It rated behind space exploration, the Kennedy assassination, presidential scandals, the Berlin Wall coming down, and—for that matter—the "birth of my children." These data were obtained from The Roper Center for Public Opinion Research.
21. In a 2005 Associated Press Poll, 68 percent of Americans indicated that the "bombings were unavoidable," versus 29 percent who maintained that the "bombings were not necessary," to bring the war to a "quick end." In a 2009 Quinnipiac University Poll, for example, 61 percent of Americans indicated that the United States did the "right thing" in dropping the bombs on Japan. In a 2015 Pew Global Attitudes Project Poll, 56 percent of Americans said that the US use of nuclear weapons against Japan was "justified." Americans split more evenly, however, when asked if they "approve" or "disapprove" of the decision. In a 2005 Associated Press Poll, 48 percent of Americans either strongly or somewhat approved of the decisions versus 47 percent who either strongly or somewhat disapproved of the decision. A 2016 CBS News Poll produced roughly similar results, with 43 percent of Americans approving versus 44 percent disapproving of the bombing. These data were obtained from the Roper Center for Public Opinion Research.
22. In a 1991 CBS News/New York Times Poll, only 39 percent of Americans agreed (strongly or somewhat) the statement that dropping the bombs on Japan was "morally wrong"; 55 percent of respondents disagreed with that statement. These data were obtained from the Roper Center for Public Opinion Research.
23. A 1995 CBS News Poll indicated that 76 percent of Americans rejected the idea that the United States "should formally apologize to Japan and its people" for the bombing, though a Gallup/CNN/USA Today Poll indicated, the same year, that (a lesser) 59 percent of Americans approved of the US use of the bomb. These data were obtained from the Roper Center for Public Opinion Research.
24. Some legal experts argue that the stockpiling of nuclear weapons, given their destructiveness, is a violation of international law. See, e.g., Vail (2016).
25. See Hans M. Kristensen, "Falling Short of Prague: Obama's Nuclear Weapons Employment Policy," Arms Control Association, 2013, https://www.armscontrol.org/act/2013-09/features/falling-short-prague-obama%E2%80%99s-nuclear-weapons-employment-policy.
26. Stephen Mihm, "Putin's Nuclear Threat Is Terrifying Even If He's Bluffing," *Washington Post* (February 28, 2022), https://www.washingtonpost.com/business/putins-nuclear-threat-is-terrifying-even-if-hes-bluffing/2022/02/27/5ce6cbbc-9817-11ec-9987-9dceee62a3f6_story.html.
27. On the uses and misuses of "precedent," as a term, in one important US national-security debate, see Lebovic (2022).
28. Still, the "norm" of nuclear nonuse and a nuclear "taboo" are often conflated in the literature. See, e.g., Sagan and Valentino (2017). A prohibition, however, need not amount to a taboo.
29. Hitler showed—very quickly, in fact—that he was not reluctant to employ force to achieve his goals. He soon moved against all of Czechoslovakia, Poland, and indeed much of Europe.
30. National Security Archive, "First Documented Evidence that U.S. Presidents Predelegated Nuclear Weapons Release Authority to the Military," National Security Archive (March 20, 1998), https://nsarchive2.gwu.edu/news/19980319.htm.

31. Of course, decision makers are not just slaves to machines or to prior practices and choices. Both the United States and the Soviet Union have responded to false nuclear alarms with restraint. Averting a nuclear cataclysm in 1983, a year of high US–Soviet tension, a Soviet military officer responded to an alert of a US ICBM attack by attributing it to a system malfunction (given the small number of missiles involved and a history of system malfunctions).
32. On collateral damage ensuing from such attacks, see Glaser and Fetter (2005: 93–97).
33. These analyses are based on binomial probabilities—the likelihood that a given number of warheads will or will not survive an attack—where the number of warheads in the inferior party's arsenal represents the original sample size (n), the survival ratio (1-the kill ratio) represents the selection probability (p), and the number of warheads that survive represents the number of successes (x). The likelihood that a given number of warheads will survive is thus understood as the probability that a certain number of cases will be selected (x) from a population of a given size (n), given the chance that any one case in that population will be selected (p). Probabilities are calculated from: $(n!/x!(n-x)!))*(p^x(1-p)^{n-x})$.
34. On assured retaliation as a doctrine, see Narang (2015).
35. For a discussion of internal and external factors that could lead to substantial increases in the size and capabilities of the Chinese nuclear force, see Heginbotham et al. (2017).
36. Joby Warrick, "Signs that China Is Busy Building Silos for ICBMs," *Washington Post* (July 1, 2021), A1. See also Matt Korda and Hans Kristensen, "A Closer Look at China's Missile Silo Construction," Federation of American Scientists (November 2, 2021), https://fas.org/blogs/security/2021/11/a-closer-look-at-chinas-missile-silo-construction/.
37. US Department of State, Office of the Historian, Foreign Relations of the United States, 1961–63, Volume XIV, Berlin Crisis, 1961–1962, Memorandum of Conversation, Paris, May 31, 1961, https://history.state.gov/historicaldocuments/frus1961-63v14/d30.
38. On the merits of a preventative attack on Iran and North Korea, see, respectively, Kroenig (2012) and Edward Luttwak, "It's Time to Bomb North Korea," *Foreign Policy* (January 8, 2018), https://foreignpolicy.com/2018/01/08/its-time-to-bomb-north-korea/.
39. Indeed, hawks confess as much when they deride these leaders for their avarice and ruthlessness.
40. Rory McCarthy, "I Am Saddam Hussein the President of Iraq and I Am Willing to Negotiate," *The Guardian* (December 15, 2003), https://www.theguardian.com/world/2003/dec/16/iraq.rorymccarthy.
41. Gerald F. Seib, "Amid Signs of a Thaw in North Korea, Tensions Bubble Up," *Wall Street Journal* (January 9, 2018), https://www.wsj.com/articles/amid-signs-of-a-thaw-in-north-korea-tensions-bubble-up-1515427541.
42. The administration took the strategy sufficiently seriously that it withdrew the nomination of a prominent North Korean expert as US Ambassador to South Korea who, though a proponent of a coercive bargaining strategy, argued in closed session that such a policy was provocative and counterproductive.
43. On administration resistance to the tactic and claims that it misrepresented administration policy, see David Nakamura and Greg Jaffe, "The White House's 'Bloody Nose' Strategy on North Korea Sounds Trumpian. So Why Do His Aides Hate It?," *Washington Post* (February 26, 2018), https://www.washingtonpost.com/world/national-security/the-white-houses-bloody-nose-strategy-on-north-korea-sounds-trumpian-so-why-do-his-aides-hate-it/2018/02/26/9ec20744-18b5-11e8-b681-2d4d462a1921_story.html
44. Indeed, the administration might have cleared the way for that response. Its bloody-nose attack could send the message that *not* "all options" were on the table. In other words, it might communicate that the United States was reluctant—given the consequences—to use disarming force and that North Korea, too, could intensify the conflict without fearing a US nuclear response.

Chapter 4

1. Understood accordingly, the term appears synonymous with "resolve" (the subject of Chapter 6), and thereby exhibits the same conceptual deficiencies.
2. Alex Ward, "Trump Says He Could Wipe Afghanistan Off Face of the Earth in 10 Days," *Vox* (July 22, 2019), https://www.vox.com/world/2019/7/22/20704248/trump-afghanistan-10-days-war.
3. Jeffrey Heller, "Netanyahu's Iran Cartoon Bomb Times to Make Big Impact," *Reuters* (September 28, 2012), https://www.reuters.com/article/us-israel-iran-netanyahu/netanyahus-iran-cartoon-bomb-timed-to-make-big-impact-idUSBRE88R12K20120928.
4. Trump now claims to have more directly threatened to leave NATO in his meeting with other members. See Aaron Blake and Michael Birnbaum, "Trump Says He Threatened Not to Defend NATO against Russia," *Washington Post* (April 22, 2022), https://www.washingtonpost.com/world/2022/04/22/trump-says-he-threatened-not-defend-nato-russia/.
5. Secretary of Defense Donald Rumsfeld once explained the failings of the US effort in Iraq by claiming, "You go to war with the army you have, not the army you might want or wish to have at a later time." In truth, you might go to war *because* of the army that you have.
6. This rather straightforward principle resonates through Robert Putnam's (1988) theorizing about two-level negotiating games. In his analysis, negotiators play the strategy simultaneously at the international and domestic levels. They attempt to confine the negotiating space domestically to reap concessions internationally, and vice versa.
7. David L. Stern and Robyn Dixon, "Ukraine's Zelensky's Message Is Don't Panic. That's Making the West Antsy," *Washington Post* (February 7, 2022), https://www.washingtonpost.com/world/2022/01/30/ukraine-zelensky-russia-biden/.
8. See Gordon and Trainor (2006: 66); Stein (1992: 174–175); and Trager and Zagorcheva (2005: 104).
9. Perhaps he was right. However, he wrongly assumed that his next war would resemble his last (with Iran) and that he could count on political support from other Arab countries given his plans to draw Israel into the fight (Stein 1992: 176).
10. Whether or not the Syrian government discounted the credibility of Obama's "red line" is another matter. After all, it did concede to a Russian plan that removed much—but not all—of the chemical weapon stocks from the country.
11. Paul Sonne and John Hudson, "U.S. Reviews Military Drills in Europe amid Russia Tension," *Washington Post* (November 22, 2021), A3.
12. Tom Balmforth and Andrew Osborn, "Russian Drills Seen as Play for Biden Talks," *Washington Post* (November 25, 2021), A10.
13. Missy Ryan, "NATO Talks Begin amid Rising Tensions with Russia," *Washington Post* (December 1, 2021), A19.
14. Zachary Cohen, "US Destroyers, Subs, Jets Could Answer Trump's Syria Strike Call," CNN Politics (April 12, 2018), https://www.cnn.com/2018/04/11/politics/us-military-strike-options-syria-trump/index.html.
15. On this, see Karen DeYoung and Shane Harris, "Trump Instructs Military to Begin Planning for Withdrawal from Syria," *Washington Post* (April 4, 2018),https://www.washingtonpost.com/world/national-security/trump-instructs-military-to-begin-planning-for-withdrawal-from-syria/2018/04/04.
16. Veronica Stracqualursi, "Trump Declares Mission Accomplished in Syria Strike," CNN Politics (April 14, 2018), https://www.cnn.com/2018/04/14/politics/trump-syria-strike/index.html.
17. John Hudson and Louisa Loveluck, "With Airstrikes, Biden Lowers Bar for Use of Military Force," *Washington Post* (July 2, 2021), A15.
18. That prospect led the Biden administration to limit the range of rockets provided to Ukraine. John Hudson, "U.S. Prepares to Send Long-Range Rocket Systems to Kyiv," *Washington Post* (May 28, 2022), A11.

19. Emma Bowman, "Putin Calls Sanctions a Declaration of War as Zelenskyy Pleads for More Aid," NPR (March 5, 2022), https://www.npr.org/2022/03/05/1084764302/putin-calls-sanctions-a-declaration-of-war-as-zelenskyy-pleads-for-more-aid.
20. Amy Cassidy, "'All Options Are on the Table' for the West's Response If Russia Uses Chemical Weapons," CNN (April 12, 2022), https://www.cnn.com/europe/live-news/ukraine-russia-putin-news-04-12-22/h_dd41610dfdd31cf7b76827bdc14eb9f2.
21. Daniel Villarreal, "Biden Says 'It's Clear' Russia Targeting Civilians; Unsure of 'War Crimes,'" *Newsweek* (March 2, 2022), https://www.newsweek.com/biden-says-its-clear-russia-targeting-civilians-unsure-war-crimes-1684354.
22. Ashley Parker, "President Uses Term 'War Criminal' for Putin on a High-Drama Day," *Washington Post* (March 17, 2020), A1.
23. Natalia Zinets, "Russian Forces Take Chernobyl Workers' Town, Biden Calls Putin a 'Butcher,'" Reuters (March 26, 2022), https://www.reuters.com/world/europe/ukraine-forces-counter-near-kyiv-russia-scales-back-goals-2022-03-26/.
24. Ashley Parker and Tyler Pager, "Biden: Putin 'Cannot Remain in Power,'" *Washington Post* (March 27, 2022), A1.
25. Cate Cadell, Dan LaMothe, Mary Ilyushina, Karoun Demirjian, and David L. Stern, "War Will Persist, Defiant Putin Says as Inquiries Begin," *Washington Post* (April 13, 2022), A1.
26. Shane Harris and Paul Sonne, "Russia Planning Massive Military Offensive against Ukraine Involving 175,000 Troops," *Washington Post* (December 3, 2021), https://www.washingtonpost.com/national-security/russia-ukraine-invasion/2021/12/03/98a3760e-546b-11ec-8769-2f4ecdf7a2ad_story.html.
27. Paul Sonne and John Hudson, "On Ukraine, Putin Keeps Washington and Its Allies Guessing," *Washington Post* (December 11, 2021), A6.
28. Robyn Dixon and Paul Sonne, "Russia Outlines Security Demands for U.S., NATO," *Washington Post* (December 18, 2021), A1.
29. BBC, "Russia-Ukraine Crisis: Putin Says Ball in West's Court," BBC News (December 23, 2021), https://www.bbc.com/news/world-europe-59766810.
30. Robyn Dixon, Catherine Belton, and Mary Ilyushina, "Putin Drafts Up to 300,000 Reservists," *Washington Post* (September 22, 2022), A1.
31. Conor Humphries, "Putin Says Russian Culture Being 'Cancelled' Like J.K. Rowling," Reuters (March 25, 2002), https://www.reuters.com/world/europe/putin-says-west-trying-cancel-russian-culture-including-tchaikovsky-2022-03-25/.
32. Prospect theory references this as an "endowment effect" (Kahneman 2011: 292–297).
33. On these decisional failings, see Janis (1972); Jervis (1982/83: 19–30); and Yetiv (2013).
34. As Alexander George (1971: 20) noted, "the defending power thus succeeded in reversing the expected outcome of the test of capabilities without having to escalate the conflict, thereby transferring back to its opponent the onerous decision whether to engage in a risky escalation or to accept defeat."
35. Even the Eisenhower-era "massive retaliation" strategy—while never actual administration strategy, posited that the United States would respond massively (with nuclear might) to Communist transgressions in the world—assumed that the United States was retaliating for a Soviet-led attack.
36. See, e.g., Houston Keene and Evie Fordham, "Biden's 'Off-Limits' List for Russian Cyberattacks Criticized as 'Green Light' to Target Everything Else," Fox News (June 18, 2021), https://www.foxnews.com/politics/biden-putin-russian-cyberattacks-list-16-off-limits-criticism.
37. Michael Birnbaum, Karoun Demirjian, and John Hudson, "Splits Emerge in NATO about How Best to Deter Russia," *Washington Post* (March 24, 2022), A17.
38. People sign legal contracts all the time, expecting to adhere to them. But when they lose their jobs or become ill, the costs and benefits of the agreement change adversely. The signatory is now stuck with a commitment that has become a burden.

39. Kissinger was surprisingly candid, however, about his willingness to accept some eventual settlement that would preserve US credibility. As long as the Saigon government survived in the short term (known as the "decent interval"), he was content to leave the war with a pretense of US victory.
40. Robyn Dixon, Catherine Belton, and Mary Ilyushina, "Putin Drafts Up to 300,000 Reservists," *Washington Post* (September 22, 2022), A1.
41. Amy B. Wang, "Bipartisan Calls for Biden to Send in Air-Defense Systems and Jets," *Washington Post* (March 14, 2022), A1.
42. Nicola Slawson, "First Thing: Biden Flags 'Clear Sign' Russia Considering Chemical Weapons," *The Guardian* (March 22, 2022), https://www.theguardian.com/us-news/2022/mar/22/first-thing-biden-flags-clear-sign-russia-considering-chemical-weapons.
43. On this, see, e.g., Scott D. Sagan and Allen S. Weiner, "Would the U.S. Really Answer Cyberattacks with Nuclear Weapons?," *Washington Post* (July 11, 2021), B1.
44. Robyn Dixon, Natalia Abbakumova, David L. Stern, and Karoun Demirjian, "Russia Fights to Hold the Line," *Washington Post* (September 26, 2022), A1.
45. Adam Taylor, "Did Biden Misspeak on Taiwan—or Deliver Straight Talk?," *Washington Post* (October 24, 2021), A25. Biden reiterated that pledge, months later, when asked whether the United States would defend Taiwan militarily in response to a Chinese attack. Seung Min Kim, Michelle Ye Hee Lee, and Cleve R. Wootson Jr., "Biden Takes Aggressive Stance toward China," *Washington Post* (May 24, 2022), A1.
46. Dan Lamothe, "U.S. Has Few Options If China Were to Seize Islands Administered by Taiwan," *Washington Post* (October 27, 2021), A13.

Chapter 5

1. For a good discussion of the nature and sources of escalation, see Zartman and Faure (2005).
2. For an accessible discussion of related issues and research, see Kahneman (2011).
3. Together, these outcomes and probabilities amount to a "probability distribution" which "is not a collection of competing hypotheses: it is a single hypothesis about the way uncertainty is spread across multiple possibilities" (Friedman and Zeckhauser 2012: 833).
4. On this, see Friedman and Zeckhauser (2012, 2018) and Friedman, Lerner, and Zeckhauser (2017).
5. Lemay headed the Strategic Air Command from 1948–1957; he served as Air Force Chief of Staff from 1961–1965.
6. See also James M. Lindsay, "TWE Remembers: Black Saturday—Near Calamities Abound as JFK Offers Khrushchev a Deal (Cuban Missile Crisis, Day Twelve)," blog post (October 27, 2012), Council on Foreign Relations, https://www.cfr.org/blog/twe-remembers-black-saturday-near-calamities-abound-jfk-offers-khrushchev-deal-cuban-missile.
7. "Office of the Historian, US Department of State, The Cuban Missile Crisis, October 1962," https://history.state.gov/milestones/1961-1968/cuban-missile-crisis.
8. On Able Archer, see Nate Jones and J. Peter Scoblic, "The Week the World Almost Ended," *Slate* (April 13, 2017), https://slate.com/news-and-politics/2017/06/able-archer-almost-started-a-nuclear-war-with-russia-in-1983.html. For related documentation, see National Security Archive, *The Able Archer 83 Sourcebook*, https://nsarchive.gwu.edu/project/able-archer-83-sourcebook.
9. On this, see Tevi Troy, "Don't Worry, America: The 3 A.M. Phone Call Is a Myth," *Politico* (September 10, 2016), https://www.politico.com/magazine/story/2016/09/3am-phone-call-myth-trump-214208.
10. On the controversy, see Dan Lamothe, "Air Force Swears: Our Nuke Launch Code was Never '00000000,'" *Foreign Policy* (January 21, 2014), https://foreignpolicy.com/2014/01/21/air-force-swears-our-nuke-launch-code-was-never-00000000/.

11. On positive and negative control, see Feaver (1992/93: 168–170). These dueling concerns amount to the Type II and Type I errors of statistical inference—validating a claim by rejecting the null hypothesis versus forgoing a claim by accepting the null hypothesis. See Wohlstetter and Brody (1987: 186). Whereas the ethics of research are conservative (increasing the difficulty of supporting claims), the presumed stakes and military culture biased preparation toward ensuring the availability of US nuclear forces when needed. US bombers, for example, would fly to predetermined positions awaiting further instructions.
12. We can argue, of course, that the Bush administration launched its 2003 war on Iraq to prevent it from acquiring nuclear weapons. That war was initiated, however, only on the pretense of a preventative strike. The administration never settled internally on the rationale for the attack; instead, Iraqi weapons of mass destruction (WMD) proved a convenient selling point for the war. When US troops moved into Iraq, securing alleged WMD sites was not a military priority.
13. William Burr, "The 'Launch on Warning' Nuclear Strategy and Its Insider Critics," National Security Archive, Briefing Book #43 (June 11, 2019), https://nsarchive.gwu.edu/briefing-book/nuclear-vault/2019-06-11/launch-warning-nuclear-strategy-its-insider-critics.
14. Ibid.
15. For an insightful analysis of the relationship between target coverage and the precipitous launch of US nuclear weapons, see Blair (1993).
16. Indeed, in the 1980s, a US nuclear response only became more dependent on strategic, relative to tactical, warning with the effect that LOW and preemption "became almost indistinguishable in operational terms under some conditions." When on high alert, the evidentiary requirement for judging whether the United States was under attack effectively lessened as strategic indictors could now substitute for one of two (previously required) tactical indicators of attack (Blair 1993: 192).
17. "Memorandum from Seymour Weiss, State Department Policy Planning Council, to Undersecretary of State John Irwin and Deputy Secretary of State for Political Affairs U.A. Johnson, 'Luncheon Conversation October 2 with Paul Nitze on SALT,' 7 October 1970, Top Secret/Nodis/Sensitive" [Declassified], https://nsarchive2.gwu.edu//dc.html?doc=6144714-National-Security-Archive-Doc-14-Memorandum-from.
18. "Minutes, National Security Council Meeting, 'SALT (and Angola),' 22 December 1975, Top Secret" [Declassified], https://nsarchive2.gwu.edu//dc.html?doc=6144722-National-Security-Archive-Doc-22-Minutes. See also "Leon Sloss, Director, Nuclear Targeting Policy Review, to Director, Joint Staff et al., 'Nuclear Targeting Policy Review,' 13 December 1978, Top Secret" [Declassified], https://nsarchive2.gwu.edu//dc.html?doc=6144731-National-Security-Archive-Doc-29-Leon-Sloss.
19. "National Security Decision Directive 13, 'Nuclear Weapons Employment Policy,' 13 October 1981, Top Secret" [Declassified], https://nsarchive2.gwu.edu//dc.html?doc=6144759-National-Security-Archive-Doc-34-National.
20. "Memorandum from Leonard Weiss, Deputy Director for Functional Research, Bureau of Intelligence and Research, to Leon Sloss, Bureau of Politico-Military Affairs, Office of International Security Policy and Planning, 'Your Memorandum on 'Launch-on-Warning,' 29 January 1971, enclosing memorandum from Frank H. Perez, Office of Strategic and General Research, to Leonard Weiss, 'Thoughts on Launch-on-Warning,' 29 January 1971, Secret" [Declassified], https://nsarchive2.gwu.edu//dc.html?doc=6144716-National-Security-Archive-Doc-16-Memorandum-from; "C. H. Builder, D. C. Kephart, and A. Laupa, 'The U.S. ICBM Force: Current Issues and Future Options,' RAND Corporation, PR-1754-R, October 1975, Secret" [Declassified], https://nsarchive2.gwu.edu//dc.html?doc=6144721-National-Security-Archive-Doc-21-C-H-Builder-D-C.
21. William Burr, "The 'Launch on Warning' Nuclear Strategy and Its Insider Critics," National Security Archive, Briefing Book #43 (June 11, 2019), https://nsarchive.gwu.edu/briefing-book/nuclear-vault/2019-06-11/launch-warning-nuclear-strategy-its-insider-critics.

22. On the Johnson administration's directive, see William Burr, "U.S. Nuclear War Plan Option Sought Destruction of China and Soviet Union as 'Viable' Societies," National Security Archive, Briefing Book #638 (August 15, 2018), https://nsarchive.gwu.edu/briefing-book/nuclear-vault/2018-08-15/us-nuclear-war-plan-option-sought-destruction-china-soviet-union-viable-societies. Contrary to popular mythology, the president does not possess the secret codes (the "button") that will trigger the launch. See Blair (2018).
23. On this, see Terriff (1995: 139–140).
24. Argyris and Schon (1978: 29) maintain that learning most often occurs in single loops where participants "respond to error by modifying strategies and assumptions within constant organizational norms" rather than in double loops where "response to detected error takes the form of joint inquiry into organizational norms themselves."
25. See Argyris and Schon (1978: 85) and Lebovic (2005).
26. On these concerns, see Boyd (2019). For a chronical presentation of safety failures involving the US nuclear arsenal, see Schlosser (2013).
27. "Because of the time-dependent processes, invariant production sequences, and lack of slack in these [tightly coupled] systems, there is little opportunity to improvise when things go wrong, and fortuitous recovery aids rarely have time to emerge" (Sagan 1993: 34).
28. On coup-proofing a regime, see Talmadge (2015).
29. Indeed, predelegation of authority to launch nuclear weapons was essential to Cold War–era operational planning given fears that a Soviet strike could seriously impair US command and control capabilities. US military commanders had the authority to use their nuclear weapons under certain circumstances and retained "wide discretion to change the disposition of their forces in anticipation of possible conflict," even to determine the "validity of launch orders that may have arrived incomplete or garbled" (Blair 1993: 33, 48, 217).
30. Joby Warrick and Walter Pincus, "Missteps in the Bunker," *Washington Post* (September 23, 2007), https://www.washingtonpost.com/wp-dyn/content/article/2007/09/22/AR2007092201447.html.
31. Alex Horton, Karoun Demirjian, and John Wagner, "Milley Defends Calls Made to His Chinese Counterpart, Saying They Were Sanctioned and Briefed across the Administration" (September 28, 2021), https://www.washingtonpost.com/national-security/general-milley-china-calls/2021/09/28/2b319e44-2099-11ec-8200-5e3fd4c49f5e_story.html. Milley's actions were reminiscent of efforts by Richard Nixon's advisors to insert themselves into the chain of command, with the president's declining mental health at the end of Nixon's presidency.
32. On these concerns, see, e.g., the *Washington Post*, "Step Back from the Nuclear Precipice" (January 14, 2021), A24.
33. General Michael Hayden, quoted in Woolf (2021).
34. How much China's readiness has changed is controversial. On change in China's nuclear strategy, see Austin Long, "Myths or Moving Targets? Continuity and Change in China's Nuclear Forces," *War on the Rocks* (December 4, 2020), https://warontherocks.com/2020/12/myths-or-moving-targets-continuity-and-change-in-chinas-nuclear-forces/.
35. On China's attitude toward a no-first-use policy, see Zhang (2007, 2008).
36. The escalatory problem is certainly exacerbated by the vulnerability of Chinese submarines to US antisubmarine warfare. On this, see Riqiang (2013).
37. Kim Jong Un, speaking at a 2022 military parade, offered vague support for that potential use. Having referenced deterrence of war as the "fundamental purpose" of North Korea's nuclear arsenal, he nonetheless warned, "If any forces try to violate the fundamental interest of our state, our nuclear forces will have to decisively accomplish its unexpected second mission." Min Joo Kim, "N. Korea's Kim Ramps Up Nuclear Rhetoric," *Washington Post* (April 27, 2022), A22.

Chapter 6

1. Such surprising (come-from-behind) victories testify, observers argue, to the character or emotional strength of the players, their desire, or their ability to "dig deep" and tap an inner reserve.
2. Classical realists do speak, however, of an impetus akin to resoluteness when they write of the "national will," which presumably favors some nations or cultures compared to others.
3. Josh Rogin, "Pence: The United States Is Not Seeking Negotiations with North Korea," *Washington Post* (April 19, 2017), https://www.washingtonpost.com/news/josh-rogin/wp/2017/04/19/pence-the-united-states-is-not-seeking-negotiations-with-north-korea/.
4. Liz Sly, "NATO Forces Unite, Grow as Putin Tries for Opposite," *Washington Post* (February 2, 2022), A1.
5. Paul Sonne, Ellen Nakashima, and Missy Ryan, "Threat of Russian Invasion of Ukraine Tests U.S. Officials," *Washington Post* (November 30, 2021), A17; Robyn Dixon, "Six Ways Russia Views Ukraine—and Why Each Should Worry the West," *Washington Post* (December 12, 2021), A1.
6. Anthony Faiola, "In a Rambling Speech, the Typically Cagey Putin Flashes a Rare Moment of Candor," *Washington Post* (February 24, 2022), A12.
7. Perry Stein and Quentin Aries, "E.U. Struggles with How Tough Sanctions on Russia Should Be over Ukraine," *Washington Post* (December 17, 2021), A14.
8. Liz Sly, "European Division Stymie Response on Ukraine," *Washington Post* (January 24, 2022), A12.
9. Robyn Dixon and David L. Stern, "Kremlin Expects Quick Start to Talks in Wake of Biden-Putin Call on Ukraine," *Washington Post* (December 9, 2021), A12; Paul Sonne and Ashley Parker, "Biden Seeks a United Front with Ukraine, Allies in Europe," *Washington Post* (December 10, 2021), A11.
10. Paul Sonne, "President Predicts Putin Will 'Move In' to Ukraine," *Washington Post* (January 20, 2022), A1.
11. He notes further that, with resulting "confusion," "some theorists *contrast* the balance of resolve and balance of power as explanations of outcomes" (Betts 1987: 14).
12. The supposed key to the effectiveness of these statements of resolve is a "clear ability 'to follow through' on them" (McManus 2017: 2), that is, the capability to act and domestic political credentials to establish that they will act. But might these statements not amount to "cheap talk" or might leaders not become more verbally "resolute" because they anticipate or realize positive outcomes (which creates an endogeneity problem for analysis)?
13. Yarhi-Milo (2018: 7) recognizes that a reputation for resolve and credibility are distinct but confesses to using the terms synonymously.
14. An assessment of these conclusions requires that we acknowledge the limits of survey experiments. Researchers control the relative salience of competing variables (for instance, statements conveying "toughness") and assume that informing participants of a party's capabilities, for example, has the same impact as "observing" or "experiencing" those capabilities in action, whether in a nuclear confrontation or not. In this, we should acknowledge that survey experiments tend to have strong internal validity but more limited external validity (generalizability to the outside world).
15. Yarhi-Milo (2018), for instance, focuses on the motivational predispositions of a country's leaders.
16. Donald J. Trump, quoted in David Nakamura and John Wagner, "Trump and Kim to Meet in Singapore," *Washington Post* (May 11, 2018), A1.
17. David Jackson, "Trump: Killing Iran Nuclear Deal Will Send 'Right Message' to North Korea ahead of Talks," *USA Today* (April 30, 2018), https://www.usatoday.com/story/news/politics/2018/04/30/trump-killing-iran-nuclear-deal-send-right-message-north-korea-ahead-talks/565744002/.

18. Fox News Insider, "Bolton: With Iran Deal Exit, Trump Sent 'Very Important' Message to North Korea," Fox News Insider (May 9, 2018), http://insider.foxnews.com/2018/05/09/john-bolton-iran-deal-exit-sends-very-important-message-north-korea.
19. Contrast the US coercive position vis-à-vis Hanoi with that against Tokyo in World War II. With sixty-four Japanese cities already in ruin, and a US incendiary bombing campaign meant to maximize the human toll, would Japanese leaders have doubted that the United States would drop atomic bombs on Japanese cities if the United States had somehow demonstrated the power of the bomb, to those leaders, before actually using it?
20. Simon Denyer and Min Joo Kim, "Kim Pledges to Visit Seoul, but Nuclear Promises Lack Substance," *Washington Post* (September 19, 2018), A9.
21. Simon Denyer and Anne Gearan, "N. Korea Offers to Destroy Nuclear Site, If U.S. Acts First," *Washington Post* (September 20, 2018), A13.
22. John Hudson, "Trump Says There's No Need to Rush a Nuclear Agreement with North Korea," *Washington Post* (September 27, 2018), A17.
23. Jeff Mason and Khanh Vu, "Trump and Kim to Meet for Dinner at Colonial-Era Hanoi Hotel," Reuters (February 24, 2019), https://www.reuters.com/article/us-northkorea-usa/trump-and-kim-to-meet-for-dinner-at-colonial-era-hanoi-hotel-idUSKCN1QE09R.
24. Julie Hirschfeld Davis and Rick Gladstone, "Trump Cites 'Great Progress' in North Korea Nuclear Talks," *New York Times* (July 12, 2018), https://www.nytimes.com/2018/07/12/world/asia/north-korea-kim-jong-un-trump-letter-nuclear.html.
25. David Sanger and Edward Wong, "How the Trump-Kim Summit Failed: Big Threats, Big Egos, and Bad Bets," *New York Times* (March 2, 2019), https://www.nytimes.com/2019/03/02/world/asia/trump-kim-jong-un-summit.html.
26. David Sanger and Edward Wong, "How the Trump-Kim Summit Failed: Big Threats, Big Egos, and Bad Bets," *New York Times* (March 2, 2019), https://www.nytimes.com/2019/03/02/world/asia/trump-kim-jong-un-summit.html.
27. Michael Cohen, Trump's former lawyer and "fixer," was testifying publicly before Congress while Trump was in Hanoi for the summit.
28. Everett Rosenfeld, "Trump-Kim Summit Was Cut Short after North Korea Demanded an End to Sanctions," CNBC (February 28, 2019), https://www.cnbc.com/2019/02/28/white-house-trump-kim-meetings-change-of-schedule.html.
29. Associated Press, "Trump Overstated Kim's Demand on Sanctions, State Department Says," NBC News, https://www.nbcnews.com/politics/white-house/trump-overstated-kim-s-dem and-sanctions-state-department-says-n978016.
30. Everett Rosenfeld, "Trump-Kim Summit Was Cut Short after North Korea Demanded an End to Sanctions," CNBC (February 28, 2019), https://www.cnbc.com/2019/02/28/white-house-trump-kim-meetings-change-of-schedule.html.
31. David Nakamura, "Trump Signals Openness to Smaller Deal with North Korea," *Washington Post* (April 12, 2019), A4.
32. On these latter events, see Simon Denyer and David Nakamura, "Trump Again Appears to Take N. Korea's Side against His Own Military, Allies," *Washington Post* (August 11, 2019), A17.
33. All the while, North Korea built up its uranium-enrichment capabilities. After the Bush administration reneged on the agreement to provide fuel to the country—with evidence that North Korea maintained secret enrichment sites—North Korea restarted the reactor and harvested the sealed fuel rods for nuclear weapons. Glenn Kessler, "Cotton's Misguided History Lesson on the North Korean Nuclear Deal," *Washington Post* (March 13, 2015), https://www.washingtonpost.com/news/fact-checker/wp/2015/03/13/cottons-misguided-history-lesson-on-the-north-korean-nuclear-deal/.
34. K. K. Rebecca Lai, William J. Broad, and David E. Sanger, "North Korea Is Firing Up a Reactor. That Could Upset Trump's Talks with Kim," *New York Times* (March 27, 2018), https://www.nytimes.com/interactive/2018/03/27/world/asia/north-korea-nuclear.html.

35. The uranium-enrichment sites included the Yongbyon complex. Simon Denyer, "North Korea's Yongbyon Nuclear Complex at the Heart of Trump-Kim Summit," *Washington Post* (February 22, 2019), https://www.washingtonpost.com/world/asia_pacific/north-koreas-yongbyon-nuclear-complex-at-the-heart-of-trump-kim-summit/2019/02/22/ee99269a-352d-11e9-8375-e3dcf6b68558_story.html.
36. Will Ripley, "How Trump and Kim's Summit Dream Fell Apart," CNN Politics (March 2, 2019), https://www.cnn.com/2019/03/02/politics/trump-kim-summit-dream-ripley-intl/index.html.
37. The cooperation could occur when parties, by necessity, must combine threats with "assurances" (Kydd and McManus 2017).
38. Zachary Laub, "The Impact of the Iran Nuclear Agreement. Council on Foreign Relations," web page, May 8, 2018, https://www.cfr.org/backgrounder/impact-iran-nuclear-agreement.
39. Although North Korea would most certainly evaluate any agreement by evaluating its provisions—their symmetry, cost, intrusiveness, and reversibility—even some experts wonder whether leaving the Iran agreement per se would help or hurt chances of obtaining an agreement with North Korea. As one former US negotiator with North Korea expressed its likely reaction to a US withdrawal from the Iran deal, "I don't think they'd see it as negative or positive, because they think they're different from everybody else, anyway." Victor Cha, quoted in Karen DeYoung, Anne Gearan, and David Nakamura, "Trump Expected to Impede Iran Deal," *Washington Post* (May 8, 2018), A1.
40. Rick Noack, "Trump Just Contradicted Bolton on North Korea. What's the 'Libya Model' They Disagree On?," *Washington Post* (May 17, 2018), https://www.washingtonpost.com/news/world/wp/2018/05/16/whats-this-libya-model-north-korea-is-so-angry-about/.
41. Rick Noack, "How Kim-Trump Tensions Escalated: The More the U.S. Said 'Libya,' the Angrier North Korea Got," *Washington Post* (May 24, 2018), https://www.washingtonpost.com/news/world/wp/2018/05/24/the-more-pence-and-trump-say-libya-the-angrier-north-korea-gets/.
42. Thus, for example, "different types of leaders will perceive the same signal in multiple ways" (Yarhi-Milo, Kertzer, and Renshon 2018: 2171).
43. As Schelling (1960: 83–188) acknowledged, international conflicts are seldom of the "zero-sum" variety, where losses for one party amount to gains for another.
44. Jeffrey Kimball, "Did Thomas C. Schelling Invent the Madman Theory?," History News Network, The George Washington University, https://historynewsnetwork.org/article/17183.
45. The machine proved a flawed deterrent, however, with the destruction of the earth when parodied in the movie *Dr. Strangelove*. The Soviets built the machine without signaling they had "climbed into the backseat." To quote a famous line from the movie, "Why didn't you tell the world, eh?"
46. Seitz and Talmadge (2020) correctly conclude that the main problems with the madman strategy reduce to "signaling," "credibility," and "assurance."
47. David L. Stern, Karen DeYoung, and Karoun Demirjian, "Putin Wields ICBM Test as a Warning," *Washington Post* (April 21, 2022), A1.
48. Even excellent football teams can acquire dispositional reputations for weakness or softness after flopping consistently in big games.
49. If a person cannot dismiss evidence altogether, they must somehow address it. Cognitive theory establishes they address it in a manner that preserves the key belief.
50. For an excellent review of the state of research on reputations in international politics, see Jervis et al. (2021).
51. For that reason, positing the reputation debate as one between believers and "reputation skeptics," per Lupton (2020), does a disservice to the gradations of opinion concerning when and how much reputations matter.

52. As Shannon and Dennis (2007: 289) conclude, "Acts of firmness are discounted, reinterpreted, or situationally attributed to preserve the paper tiger image" among militant Islamists of their powerful nemesis.
53. See Harvey and Mitton (2017) for an opposing take on reputational effects in some contemporary cases. For a good review of the reputation literature, see Dafoe et al. (2014).

Chapter 7

1. Thus, I offer my research, in part, as a response to the long-standing argument (Cohen 1986) that the CMC is over-studied: that the crisis is effectively sui generis.
2. In this, I disagree with those who conclude that crisis participants divided between "hawks" and "doves" (Dobbs 2008b: 5). My thinking is closer to that of Blight et al., who divide the crisis participants among "hawks," "doves," and "owls"—the latter seemingly located between the hawkish and dovish ends of the spectrum. Indeed, I maintain that the "hybrid" in thinking left critical issues unaddressed. The owls (President Kennedy included) were reluctant to initiate violence but failed to recognize the potential liabilities of their seemingly cautious approach.
3. These members constituted the NSC's Executive Committee (ExComm) in subsequent (post–October 22) meetings.
4. With attending controversy over attribution and exact phrasing, the tapes were re-transcribed for subsequent works. The 2002 version of the book is used in the analysis here, though the level of detail required for the coding renders it insensitive to issues that have arisen over exact word choices.
5. The October 20 session was not recorded; the coding was based on the detailed notes taken for that session. The NSC met fifteen times, on ten different days. The final session occurred on October 28.
6. Among other things, that allows the attribution of alleged benefits and costs to specific options. Inasmuch as a "narrow option focus" is the basis of my argument, that approach allows "diversity" in thinking to register to create a fair (actually, tough) test of the thesis.
7. In the latter instance, the comment receives the same coding as the comment referenced.
8. Sometimes a speaker might *seem* to favor an option but only for purposes of argumentation.
9. It was thus coded here under Item 10j, "Building/retaining political support."
10. That reference to the domestic political environment sometimes provoked knowing chuckles nonetheless says much about the specter of hawkish resistance hovering over the proceedings.
11. On the influence of domestic politics on crisis deliberations, see Lebow (1990).
12. Indeed, the coding of an invasion as a "first resort" exaggerates the option's standing in temporal priority. When the participants spoke of options in general, without referring to their exact timing, those instances are coded here as "first" options. Given the late prevalence of invasion as a "second-wave" option in later deliberations, we can suppose, then, that even the "first-wave" line includes at least some references to a follow-on invasion.
13. Costs are not counted when they are discounted (Item 6a).
14. The shading between these sets of lines thus represents the vertical "distance" (difference) between the costs and net costs of the various options.
15. Although, to aid readability, the figure does not include the invasion option, the cost and net cost lines—like those for a blockade—place the findings close to the *x*-axis.
16. Then, the participants focused on air strikes and rejected negotiations as an option.
17. As we shall see, the pattern likely speaks to "polarization" in the discussion of various options: some participants saw costs where others saw benefits.
18. I use the term "escalatory" as shorthand for concerns about provocation, escalation, the conflict's spread, a nuclear launch, or general war.
19. The lines for the costs and benefits of a *specific option* reflect cumulative totals for comments coded, in each case, for that option and a given cost and/or benefit.

20. Even concerns expressed by NSC members about the costs of air strikes were often directed specifically at the full-blown air campaign against Soviet and Cuban military assets (including SAM sites and military aircraft) that the military chiefs were promoting.
21. On the historical roots of the Jupiters trade, see Bernstein (1980).
22. McNamara initially *seemed* to support a quick air strike against the Soviet missiles. He expressed concerns about (a) attacking the missiles once they became operational and (b) giving the Soviets prior warning of a US strike by indicating that the United States had detected the missiles. Yet McNamara also expressed doubts, very early on, about the strategic significance of the Soviet deployments; and he quickly became vocal about the escalatory potential of US military options.
23. Kennedy met with the JCS on October 19, at which point there was likely much prior discussion among the Chiefs, allowing them to promote a well-formed, consensus view.
24. Neutralizing time in this way provides a better sense of the different style of argumentation likely in a meeting with military "advocates."
25. Indeed, the JCS members were quite irreverent, even profane, when Kennedy left the room, with the tape recorder still running. The JCS members groused about what was said, and left unsaid, in the prior session.
26. Apart from the small size of the force, they offered the Soviets no real strategic advantage given a "timing" problem. If launching these weapons first, the Soviets would provide the warning of a pending ICBM attack that would trigger the launch of the US nuclear force against their Soviet targets.
27. The Soviets would eventually deploy submarines equipped to fire SLBMs proximate to the US shoreline. Those weapons would present a far graver threat to the United States given their relative invulnerability.
28. On these bombers, see Garthoff (1980).
29. For an early but prescient discussion of Soviet goals in the crisis, see Horelick (1964).
30. Regardless, if precarious control over these missiles were the issue, militarization of the conflict, before regularized rules and procedures were instilled to control of these missiles, was hardly advisable.
31. They well understood that a blockade, absent a favorable Soviet response, would not solve their problem (Gibson 2011: 373).

Chapter 8

1. For a summary of the proposals and counterproposals, see Arms Control Association, "History of Official Proposals on the Iranian Nuclear Issue," January 2014, http://www.armscontrol.org/factsheets/Iran_Nuclear_Proposals.
2. For a well-researched analysis of the Iranian government's goals and negotiating positions, see Tabatabai (2017).
3. Under the July 2015 agreement that followed, the issue was left to the IAEA to resolve. In December 2015, the IAEA issued its report, concluding that Iran had previously engaged in nuclear-weapons research. Iran's interlocutors expressed a desire to move forward nonetheless rather than dwell on Iran's prior activities; critics charged that failing to hold Iran accountable would create monitoring blind spots and sacrifice reference points for judging Iran's full compliance with the agreement.
4. Carol Morello and Karen DeYoung, "In Iran Nuclear Talks, Near-Collapse before Breakthrough," *Washington Post* (April 8, 2015), A6.
5. Michael Gordon, "Outline of Iran Nuclear Deal Sounds Different from Each Side," *New York Times* (April 4, 2015).
6. Unfortunately, for outside observers, the implicating and the exculpatory evidence were fundamentally inconclusive: leaders sometimes feign intransigence to appease domestic supporters, deflect challenges from the opposition, or reap benefits in the negotiations by alluding to the

constraints that the internal squabbling imposes on their negotiating flexibility. See Putnam (1988). In struggling to decipher mixed messages, Western diplomats had to decide whether to tread softly so as not to taint, embarrass, or provoke the accommodative faction.

7. The academic literature tends to focus instead on prerequisites for nuclear self-sufficiency. For that purpose, it relies on proxies for essential program requirements, opting for measures that are easily monitored using available data. See Jo and Gartzke (2007) and Meyer (1984). For a good critique, see Sagan (2010).
8. For insights on this position, see Hymans (2010).
9. Admittedly, *none* of the responses above to the testing standard are easily dismissed, for each could carry weight under the right circumstances. After all, Israel *did* acquire nuclear weapons without testing; and even crude, cumbersome nuclear devices are potentially deliverable and useful for *some* purposes.
10. Indeed, using a bomb's worth of 20-percent enriched uranium as a standard presumes that Iran would rush for a *single* bomb and would do so without knowing how much of that uranium it would actually need for a device—lacking relevant experience in uranium processing. On this, see Dan Williams, "Analysis: Obama Won't Trip over Netanyahu's Iran 'Red Line,'" *Reuters* (March 15, 2013), http://www.reuters.com/article/2013/03/15/ususaisraelobamairanidUSBRE92E06Q20130315.
11. Full participation requires adherence to the Comprehensive Safeguards Agreements for verifying the accuracy of country reporting and the Additional Protocols for verifying the completeness of reporting—the latter, by granting the IAEA access to undeclared sites to conduct inspections and sampling and by providing the agency with requested documentation. On the Protocols, see Mark Hibbs, "The Unspectacular Future of the IAEA Additional Protocol. Proliferation Analysis," Carnegie Endowment for International Peace (website), April 26, 2012, http://carnegieendowment.org/2012/04/26/unspectacularfutureofiaeaadditionalprotocol.
12. Proximity and capabilities alone can explain Israel's 1981 attack on Iraq's Osirak reactor and 2007 attack on a suspected Syrian nuclear reactor.
13. In this, the administration was arguably responding to Israel's manipulation of US domestic politics. Israel's incentive was to use all available political leverage, including public pressure upon the administration (before the close 2012 US presidential election) and a threat of immediate, unilateral military action, to get the United States to attack Iran—and, falling short, to get the United States to commit to attack after Israel's unilateral window of military opportunity had closed. Essentially, Israel tried to solve a "commitment problem" by getting the Obama administration to announce publicly—and, therefore, presumably irrevocably—its intention to strike Iranian nuclear facilities when the United States sought instead to keep its options open. Having staked a public position, the administration presumably could not easily back down with subsequent evidence that Iran's nuclear program had progressed. On "audience costs," see Fearon (1994).
14. See Müller and Schmidt (2010: 132). The extent of these activities in a country can remain unknown even to its leaders so that their words and behavior are potentially misleading. Saddam Hussein was constantly misinformed about the actual state of Iraqi capabilities by subordinates who feared for their personal safety. See Woods, Lacey, and Murray (2006).
15. In consequence, Iran's nuclear progress would remain open to dispute. When Iran announced to UN officials in January 2013 that it would install advanced (IR-2M) centrifuges at the Natanz installation, increasing the speed with which the country could acquire material for a bomb, outside analysts could still find reason for optimism: Iran would have difficulty under sanctions acquiring materials to construct these centrifuges, had slowed the growth of its uranium stockpile, and had not announced plans to introduce the centrifuges at the less-vulnerable Fordow facility, near the city of Qom. Likewise, with the disclosure of Iran's efforts in late 2011 to procure from China 100,000 ring-shaped magnets, which could support a quintupling of (IR-1) centrifuge numbers, observers still had reason to believe that Iran was

not any closer to acquiring a weapon. Iran converted some uranium into metallic form and kept its holdings of medium-enriched uranium below levels that were necessary to construct a bomb. Joby Warrick, "Iran to Enrich More Uranium," *Washington Post* (February 1, 2013), A8; and Warrick, "Iran Sought Banned Magnets," *Washington Post* (February 14, 2013), A1.

16. On this point, it must be said that many dozens of states have refused to accept the NPT Additional Protocols to avoid burdensome reporting and administrative requirements and protest the failure of nuclear-weapons states to disarm.
17. On decisional maladies, see Jervis (1982/83) and Jervis, Lebow, and Stein (1989). Put differently, the problem for a coercing party is that its compellence demands increase the target's deterrence concerns. See Lebow and Stein (1990: 353–356).
18. The key judgments of the estimate are available at National Security Archive: National Intelligence Council, *Iraq's Continuing Programs for Weapons of Mass Destruction* (Washington, DC: National Foreign Intelligence Board, 2002), http://www.gwu.edu/~nsarchiv/NSAEBB/NSAEBB80/wmd15.pdf.
19. It reached this conclusion despite the conclusion by the International Atomic Energy Commission, based on its inspections through 1999, that there was "no indication that Iraq possesses nuclear weapons or any meaningful amounts of weapon-usable nuclear material, or that Iraq has retained any practical capability (facilities or hardware) for the production of such material." The skepticism was backed by the US Department of State's Bureau of Intelligence and Research, which indicated that the evidence was "inadequate" to conclude that Iraq is pursuing an "integrated and comprehensive approach to acquire nuclear weapons." See Richelson (2007: 485).
20. Once the assumption that Iraq had these weapons took hold, the bias among intelligence organizations was toward reinforcement over change. The obvious question was, "Why wouldn't Iraq come clean, now, if it had nothing to hide?" That Iraq acted furtively—and had previously deceived the outside world—colored all assessments of Iraqi behavior. As a result, exculpatory evidence was mistaken for subterfuge, the lack of evidence was treated as confirmatory evidence, implicating evidence was viewed as more compelling than a lack of information where it should have been found, and alternative explanations for the evidence were ignored or discounted when the evidence also confirmed preferred explanations. See Commission on the Intelligence Capabilities of the United States Regarding Weapons of Mass Destruction (2005) and U.S. Senate Select Committee on Intelligence (2004). For an excellent summary and analysis, see Jervis (2010).
21. Mark Landler and Helene Cooper, "Obama Rebuffs Netanyahu on Setting Limits on Iran's Nuclear Program," *New York Times* (September 14, 2012), A7.
22. That question will remain should one or more countries attack Iran's nuclear facilities in the near or long term. After an attack, Iran might still retain parts of its nuclear infrastructure, close its facilities entirely to inspection, and commit fully to building a bomb to guard against future attacks.
23. For the pros and cons on attacking Iran, see, respectively, Kroenig (2012) and Kahl (2012).
24. For other metrics that serve as potential red-line standards, see Altman and Miller (2018).
25. For a more thorough examination of the agreement and its generalizability to other potential proliferators, see Kerr (2017).
26. For instance, see David Albright et al. (2014) and Institute for Science and International Security (2014).
27. David Ignatius, "Closer, but Still No Deal on Iran," *Washington Post* (April 23, 2014), A15.
28. How successfully Iran could hide evidence given satellite surveillance and residual radioactivity remained in dispute. For an excellent discussion of these issues, see Samore (2015: 38–39).
29. On this, see Austin Long, "If You Really Want to Bomb Iran, Take the Deal," *Monkey Cage* (blog), *Washington Times* (webpage), April 3, 2015, https://www.washingtonpost.com/blogs/monkey-cage/wp/2015/04/03/if-you-really-want-to-bomb-iran-take-the-deal.

30. Quoted in Jennifer Rubin, "Voting 'No' Is the Way to Prevent War," *Right Turn* (blog), *Washington Post* (webpage), July 15, 2015, https://www.washingtonpost.com/blogs/right-turn/wp/2015/07/15/voting-no-is-the-way-to-prevent-war/.
31. See, e.g., Fareed Zakaria, "Sen. Schumer's Illogical Case against the Iran Deal," *Washington Post* (August 13, 2015), https://www.washingtonpost.com/opinions/dear-sen-schumer-dont-vote-against-the-iran-nuclear-deal/2015/08/13/7b806630-41f4-11e5-846d-02792f854297_story.html.
32. To be sure, some critics have focused their skepticism on the far end of the agreement in conceding that "the deal would block the uranium enrichment, plutonium separation and covert paths to a nuclear bomb for the next 15 years." Dennis Ross and David H. Petraeus, "How to Put Some Teeth into the Nuclear Deal with Iran," *Washington Post* (August 25, 2015), https://www.washingtonpost.com/opinions/how-to-put-some-teeth-into-the-nuclear-deal-with-iran/2015/08/25/6f3db43c-4b35-11e5-bfb9-9736d04fc8e4_story.html.
33. Carol Morello, "Retired Generals and Admirals Urge Congress to Reject Iran Nuclear Deal," *Washington Post* (August 26, 2015), https://www.washingtonpost.com/world/national-security/retired-generals-and-admirals-urge-congress-to-reject-iran-deal/2015/08/26/8912d9c6-4bf5-11e5-84df-923b3ef1a64b_story.html.
34. For political reasons, supporters reluctantly acknowledge their reliance on such optimism. Gardiner Harris, "Deeper Mideast Aspirations Seen in Nuclear Deal with Iran," *New York Times* (July 31, 2015), http://www.nytimes.com/2015/08/01/world/middleeast/deeper-mideast-aspirations-seen-in-nuclear-deal-with-iran.html.
35. See Kelsey Davenport, "Iran Dismantling Centrifuges, IAEA Says," Arms Control Association (webpage), December 2015, http://www.armscontrol.org/taxonomy/term/153.
36. Alexandra Jaffe, "Colin Powell: Iran Deal Is a 'Pretty Good Deal,'" *NBC News* (September 6, 2015), http://www.nbcnews.com/storyline/iran-nuclear-talks/colin-powell-iran-deal-pretty-good-deal-n422551.
37. *Haaretz*, "36 Retired U.S. Generals and Admirals Announce Support of Iran Deal," *Haaretz* (Israel) (August 12, 2015), http://www.haaretz.com/world-news/1.670756.
38. Mark Landler, "Trump Abandons Nuclear Deal He Long Scorned," *New York Times* (May 8, 2018), https://www.nytimes.com/2018/05/08/world/middleeast/trump-iran-nuclear-deal.html.
39. See Jarrett Blanc and James M. Acton, "The Trump Administration and the Iran Nuclear Deal: Analysis of Noncompliance Claims," Carnegie Endowment for International Peace, October 12, 2017, https://carnegieendowment.org/2017/10/12/trump-administration-and-iran-nuclear-deal-analysis-of-noncompliance-claims-pub-73214; Salvador Rizzo and Meg Kelly, "Fact-Checking President Trump's Reasons for Leaving the Iran Nuclear Deal," *Washington Post* (May 9, 2018), https://www.washingtonpost.com/news/fact-checker/wp/2018/05/09/fact-checking-president-trumps-reasons-for-leaving-the-iran-nuclear-deal/.
40. Mark Landler, "Trump Abandons Nuclear Deal He Long Scorned," *New York Times* (May 8, 2018), https://www.nytimes.com/2018/05/08/world/middleeast/trump-iran-nuclear-deal.html.
41. That is, better than the Obama administration's deal—one of a number that Trump deemed "the worst deal ever."

REFERENCES

Acton, James M. 2020. "Why Warhead Ambiguity Can Cause Mischaracterization Pre-Launch." Washington, DC: Carnegie Endowment for International Peace. https://carnegieendowment.org/2020/04/09/why-warhead-ambiguity-can-cause-mischaracterization-pre-launch-pub-81450.

Acton, James M. 2018. "Escalation through Entanglement: How the Vulnerability of Command-and-Control Systems Raises the Risks of an Inadvertent Nuclear War." *International Security* 43(1): 56–99.

Albright, David, Olli Heinonen, and Andrea Stricker. 2014. "Five Compromises to Avoid in a Comprehensive Agreement with Iran." Report (June 3). Institute for Science and International Security. https://isis-online.org/uploads/isis-reports/documents/Five_Bad_Compromises_3June2014-final.pdf.

Albright, David, Olli Heinonen, and Andrea Stricker. 2018. "Breaking Up and Reorienting Iran's Nuclear Weapons Program." Report (October 29). Institute for Science and International Security. https://isis-online.org/uploads/isis-reports/documents/Reorienting_AMAD_Program_29Oct2018_Final.pdf.

Allyn, Bruce J., James G. Blight, and David A. Welch. 1989/90. "Essence of Revision: Moscow, Havana, and the Cuban Missile Crisis." *International Security* 14(3): 136–172.

Altman, Dan. 2021. "Red Lines: Enforcement, Declaration, and Ambiguity in the Cuban Missile Crisis." *Journal of Strategic Studies*. http://doi.org/10.1080/01402390.2021.2004397.

Altman, Dan, and Nicholas L. Miller. 2017. "Redlines in Nuclear Nonproliferation." *Nonproliferation Review* 24(3–4): 315–342.

Anderson, Justin V., and Amy J. Nelson. 2019. "The INF Treaty: A Spectacular, Inflexible, Time-Bound Success." *Strategic Studies Quarterly* 13(2): 90–122.

Argyris, Chris, and Donald A. Schon. 1978. *Organizational Learning: A Theory of Action Perspective*. Reading, MA: Addison-Wesley.

Arms Control Association (ACA). 2018. "Trump's Threat to Violate the Iran Nuclear Deal and How to Respond." *Issue Briefs* 10(5). https://www.armscontrol.org/issue-briefs/2018-04/trumps-threat-violate-iran-nuclear-deal-respond.

Asal, Victor, and Kyle Beardsley. 2007. "Proliferation and International Crisis Behavior." *Journal of Peace Research* 44(2): 139–155.

Avey, Paul C. 2019. *Tempting Fate: Why Nonnuclear States Confront Nuclear Opponents*. Ithaca, NY: Cornell University Press.

Baker, William D., and John R. Oneal. 2001. "Patriotism or Opinion Leadership? The Nature and Origins of the 'Rally 'Round the Flag' Effect." *Journal of Conflict Resolution* 45(5): 661–687.

Ball, Desmond. 1980. *Politics and Force Levels: The Strategic Missile Program of the Kennedy Administration*. Berkeley, CA: University of California Press.

Ball, Desmond, and Richard C. Toth. 1990. "Revisiting the SIOP: Taking War-Fighting to Dangerous Extremes." *International Security* 14(4): 65–92.

Baum, Matthew A. 2002. "The Constituent Foundations of the Rally-Round-the-Flag Phenomenon." *International Studies Quarterly* 46(2): 263–298.

Beardsley, Kyle, and Victor Asal. 2009. "Winning with the Bomb." *Journal of Conflict Resolution* 53(2): 278–301.

Bell, Mark S. 2019. "Nuclear Opportunism: A Theory of How States Use Nuclear Weapons in International Politics." *Journal of Strategic Studies* 42(1): 3–28.

Bell, Mark S., and Julia Macdonald. 2019. "How to Think about Nuclear Crises." *Texas National Security Review* 2(2): 41–64. https://tnsr.org/2019/03/how-to-think-about-nuclear-crises/.

Bennett, Bruce W. 2010. *Uncertainties in the North Korean Nuclear Threat*. Santa Monica, CA: RAND, National Defense Research Institute. https://www.rand.org/pubs/documented_briefings/DB589.html.

Bernstein, Barton J. 1980. "The Cuban Missile Crisis: Trading the Jupiters in Turkey." *Political Science Quarterly* 95(1): 97–125.

Bernstein, Barton J. 2000. "Understanding Decisionmaking, U.S. Foreign Policy, and the Cuban Missile Crisis." *International Security* 25(1): 134–164.

Betts, Richard K. 1987. *Nuclear Blackmail and Nuclear Balance*. Washington, DC: The Brookings Institution.

Biden, Joseph R., Jr. 2020. "Why America Must Lead Again: Rescuing U.S. Foreign Policy after Trump." *Foreign Affairs* 99(2): 64–76. https://www.foreignaffairs.com/articles/united-states/2020-01-23/why-america-must-lead-again.

Bin, Li. 2007. "Tracking Chinese Strategic Mobile Missiles." *Science and Global Security* 15(1): 1–30.

Blair, Bruce G. 1987. "Alerting in Crisis and Conventional War," pp. 75–120 in *Managing Nuclear Operations*, ed. Ashton B. Carter, John D. Steinbruner, and Charles A. Zraket. Washington, DC: The Brookings Institution.

Blair, Bruce G. 1993. *The Logic of Accidental Nuclear War*. Washington, DC: The Brookings Institution.

Blair, Bruce G. 2018. "Strengthening Checks on Presidential Nuclear Launch Authority." *Arms Control Today* 48(1): 6–13.

Blair, Bruce G., and Chen Yali. 2006. "The Fallacy of Nuclear Primacy." *China Security* (Autumn): 51–77.

Blight, James G., and Janet M. Lang. 1995. "Burden of Nuclear Responsibility: Reflections on the Critical Oral History of the Cuban Missile Crisis." *Peace and Conflict Journal of Peace Psychology* 1(3): 225–264.

Blight, James G., Joseph S. Nye, Jr., and David A. Welch. 1987. "The Cuban Missile Crisis Revisited." *Foreign Affairs* 66(1): 170–188.

Boyd, Dallas. 2019. "Avoiding Self-Inflicted Wounds to the Credibility of the US Nuclear Deterrent." *Nonproliferation Review* 26(1–2): 105–126.

Bracken, Paul. 1983. *The Command and Control of Nuclear Forces*. New Haven, CT: Yale University Press.

Brodie, Bernard. 1966. *Escalation and the Nuclear Option*. Princeton, NJ: Princeton University Press.

Brutger, Ryan, and Joshua D. Kertzer. 2018. "A Dispositional Theory of Reputation Costs." *International Organization* 72(3): 693–724.

Burr, William. 2005. "The Nixon Administration, the 'Horror Strategy,' and the Search for Limited Nuclear Options, 1969–1972." *Journal of Cold War Studies* 7(3): 34–78.

Carter, Ashton B. 1987. "Assessing Command System Vulnerability," pp. 555–610 in *Managing Nuclear Operations*, ed. Ashton B. Carter, John D. Steinbruner, and Charles A. Zraket. Washington, DC: The Brookings Institution.

Chase, Michael S. 2013. "China's Transition to a More Credible Nuclear Deterrent: Implications and Challenges for the United States." *Asia Policy* 16(July): 69–101.

Chayes, Abram, and Antonia Handler Chayes. 1995. *The New Sovereignty: Compliance with International Regulatory Agreements.* Cambridge, MA: Harvard University Press.

Cho, Seong-Ryoul. 2009. "North Korea's Security Dilemma and Strategic Options." *Journal of East Asian Affairs* 23(2): 69–102.

Christensen, Thomas. 2012. "The Meaning of the Nuclear Evolution: China's Strategic Modernization and US-China Security Relations." *Journal of Strategic Studies* 35(4): 447–487.

Cohen, Eliot A. 1986. "Why We Should Stop Studying the Cuban Missile Crisis." *The National Interest* 1985/86: 3–13.

Commission on the Intelligence Capabilities of the United States Regarding Weapons of Mass Destruction. 2005. "Case Study: Iraq," in *Report to the President of the United States, March 31, 2005,* 43–249. Washington, DC: Commission on the Intelligence Capabilities of the United States Regarding Weapons of Mass Destruction. https://fas.org/irp/offdocs/wmd_report.pdf.

Crescenzi, Mark J. C. 2007. "Reputation and Interstate Conflict." *American Journal of Political Science* 51(2): 382–396.

Crescenzi, Mark J. C. 2018. *Of Friends and Foes: Reputation and Learning in International Politics.* New York: Oxford University Press.

Cunningham, Fiona S., and M. Taylor Fravel. 2019. "Dangerous Confidence? Chinese Views on Nuclear Escalation." *International Security* 44(2): 61–109.

Dafoe, Allan, and Devin Caughey. 2016. "Honor and War: Southern U.S. Presidents and the Effects of Concern for Reputation." *World Politics* 68(2): 341–381.

Dafoe, Allan, Jonathan Renshon, and Paul Huth. 2014. "Reputation and Status as Motives for War." *Annual Review of Political Science* 17(1): 371–393.

Davenport, Kelsey. 2015. Iran, P5+1 Make Progress on Nuclear Text. *Arms Control Today* 45(5): 25–26. https://www.armscontrol.org/act/2015-06/news/iran-p51-make-progress-nuclear-text.

Davis, Lynn E. 1975. Limited Nuclear Options: Deterrence and the New American Doctrine. Adelphi Paper No. 121. London: International Institute for Strategic Studies.

Denmark, Abraham, and Caitlin Talmadge. 2021. Why China Wants More and Better Nukes: How Beijing's Nuclear Buildup Threatens Stability. *Foreign Affairs* (November 19). https://www.foreignaffairs.com/articles/china/2021-11-19/why-china-wants-more-and-better-nukes.

Dobbs, Michael. 2008a. *Why We Should Still Study the Cuban Missile Crisis.* Special Report 205. United States Institute of Peace. https://www.usip.org/publications/2008/06/why-we-should-still-study-cuban-missile-crisis.

Dobbs, Michael. 2008b. *One Minute to Midnight: Kennedy, Khrushchev, and Castro on the Brink of Nuclear War.* New York: Vintage Books.

Early, Bryan R., and Victor Asal. 2018. Nuclear Weapons, Existential Threats, and the Stability–Instability Paradox. *Nonproliferation Review* 25(3–4): 223–247.

Edelstein, David M. 2002. Managing Uncertainty: Beliefs about Intentions and the Rise of Great Powers. *Security Studies* 12(1): 1–40.

Eden, Lynn. 2004. *Whole World on Fire: Organizations, Knowledge, and Nuclear Weapons Devastation.* Ithaca, NY: Cornell University Press.

Escribà-Folch, Abel, and Daniel Krcmaric. 2017. Dictators in Exile: Explaining the Destinations of Ex-Rulers. *Journal of Politics* 79(2): 560–575.

Fearon, James D. 1994. "Domestic Political Audiences and the Escalation of International Disputes." *American Political Science Review* 88(3): 577–592.

Fearon, James D. 1997. "Signaling Foreign Policy Interests: Tying Hands Versus Sinking Costs." *Journal of Conflict Resolution* 41(1): 68–90.

Feaver, Peter D. 1992/93. "Command and Control in Emerging Nuclear Nations." *International Security* 17(3): 160–187.

Feiveson, Harold A. 1981. "The Dilemma of Theater Nuclear Weapons." *World Politics* 33(2): 282–298.

Friedberg, Aaron L. 1980. "A History of the U.S. Strategic 'Doctrine'—1945 to 1980." *Journal of Strategic Studies* 3(3): 37–71.

Freedman, Lawrence. 1983. *The Evolution of Nuclear Strategy*. New York: St. Martin's Press.

Friedman, Jeffrey A., Jennifer S. Lerner, and Richard Zeckhauser. 2017. "Behavioral Consequences of Probabilistic Precision: Experimental Evidence from National Security Professionals." *International Organization* 71(4): 803–826.

Friedman, Jeffrey A., and Richard Zeckhauser. 2012. "Assessing Uncertainty in Intelligence." *Intelligence and National Security* 27(6): 824–847.

Friedman, Jeffrey A., and Richard Zeckhauser. 2018. "Analytical Confidence and Political Decision-Making: Theoretical Principles and Experimental Evidence from National Security Professionals." *Political Psychology* 39(5): 1069–1087.

Garthoff, Raymond L. 1980. "American Reaction to Soviet Aircraft in Cuba, 1962 and 1978." *Political Science Quarterly* 95(3): 427–439.

Gavin, Francis J. 2009/10. "Same as It Ever Was: Nuclear Alarmism, Proliferation, and the Cold War." *International Security* 34(3): 7–37.

Gavin, Francis J. 2012. *Nuclear Statecraft: History and Strategy in America's Atomic Age*. Ithaca, NY: Cornell University Press.

Gavin, Francis J. 2020. *Nuclear Weapons and American Grand Strategy*. Washington, DC: The Brookings Institution.

George, Alexander L., 1971. "Comparison and Lessons," pp. 211–254 in *The Limits of Coercive Diplomacy: Laos-Cuba-Vietnam.*, ed. Alexander L. George, David K. Hall, and William E. Simons, Boston, MA: Little, Brown and Company.

Gibbons, Rebecca Davis, and Keir Lieber. 2019. How Durable Is the Nuclear Weapons Taboo? *Journal of Strategic Studies* 42(1): 29–54.

Gibson, David. R. 2011. Avoiding Catastrophe: The Interactional Production of Possibility during the Cuban Missile Crisis. *American Journal of Sociology* 117(2): 361–419.

Gottfried, Kurt, and Bruce G. Blair, eds. 1988. *Crisis Stability and Nuclear War*. New York: Oxford University Press.

Glaser, Charles L. 2014. *Analyzing Strategic Nuclear Policy*. Princeton, NJ: Princeton University Press.

Glaser, Charles L., and Steve Fetter. 2005. "Counterforce Revisited: Assessing the Nuclear Posture Review's New Missions." *International Security* 30(2): 84–126.

Glaser, Charles L., and Steve Fetter. 2017. "The Limits of Damage Limitation." *International Security* 42(1): 201–207.

Gordon, Michael R., and Bernard E. Trainor. 2006. *Cobra II: The Inside Story of the Invasion and Occupation of Iraq*. New York: Pantheon.

Gray, Colin S. 1979a. "Nuclear Strategy: A Case for a Theory of Victory." *International Security* 4(1): 54–87.

Gray, Colin S. 1979b. "The SALT II Debate in Context." *Survival* 21(5): 202–205.

Gray, Colin S., and Keith Payne. 1980. "Victory Is Possible." *Foreign Policy* 39 (Summer): 14–27.

Green, Brendan Rittenhouse. 2019. "Somewhere between 'Never' and 'Always.'" *Texas National Security Review* (July 2): 52–60. https://tnsr.org/roundtable/policy-roundtable-nuclear-first-use-and-presidential-authority/.

Green, Brendan Rittenhouse, and Austin Long. 2017a. The MAD Who Wasn't There: Soviet Reactions to the Late Cold War Nuclear Balance. *Security Studies* 26(4): 606–641.

Green, Brendan Rittenhouse, and Austin Long. 2017b. "The Limits of Damage Limitation." *International Security* 42(1): 193–199.

Grieco, Joseph M. 2001. "Repetitive Military Challenges and Recurrent International Conflicts, 1918–1994." *International Studies Quarterly* 45(2): 295–316.

Halperin, Morton H. 1966. *The 1958 Taiwan Straits Crisis: A Documented History*. Memorandum RM-4900-ISA. Santa Monica, CA: The RAND Corporation.

Harvey, Frank P., and John Mitton. 2017. *Fighting for Credibility: US Reputation in International Politics*. Toronto: University of Toronto.

Haslam, Jonathan. 1990. *The Soviet Union and the Politics of Nuclear Weapons in Europe, 1969–87*. Ithaca, NY: Cornell University Press.

Heginbotham, Eric, Michael S. Chase, Jacob L. Heim, Bonny Lin, Mark R. Cozad, Lyle J. Morris, Christopher P. Twomey, Forrest E. Morgan, Michael Nixon, Cristina L. Garafola, and Samuel K. Berkowitz. 2017. *China's Evolving Nuclear Deterrent: Major Drivers and Issues for the United States*. Santa Monica, CA: The RAND Corporation.

Herek, Gregory M., Irving L. Janis, and Paul Huth. 1987. "Decision Making during International Crises: Is Quality of Process Related to Outcome?" *Journal of Conflict Resolution* 31(2): 203–226.

Herrmann, Richard K. 1985. *Perceptions and Behavior in Soviet Foreign Policy*. Pittsburgh, PA: University of Pittsburgh.

Hoffman, Aaron M. 2006. *Building Trust: Overcoming Suspicion in International Conflict* Albany: State University of New York Press.

Hoffman, David E. 2009. *The Dead Hand: The Untold Story of the Cold War Arms Race and Its Dangerous Legacy*. New York: Doubleday.

Holsti, Ole R. 1962. "The Belief System and National Images: A Case Study." *Journal of Conflict Resolution* 6(3): 244–252.

Hopf, Ted. 1995. *Peripheral Visions: Deterrence Theory and American Foreign Policy in the Third World, 1965–1990*. Ann Arbor: University of Michigan Press.

Horelick, Arnold L. 1964. "The Cuban Missile Crisis: An Analysis of Soviet Calculations and Behavior." *World Politics* 16(3): 363–389.

Huth, Paul K. 1988. "Extended Deterrence and the Outbreak of War." *American Political Science Review* 82(2): 423–443.

Huth, Paul K., and Bruce Russett. 1988. "Deterrence Failure and Crisis Escalation." *International Studies Quarterly* 32(1): 29–45.

Hymans, Jacques E. C. 2006. *The Psychology of Nuclear Proliferation: Identity, Emotions, and Foreign Policy*. New York: Cambridge University Press.

Hymans, Jacques E. C. 2010. "When Does a State Become a 'Nuclear Weapons State'? An Exercise in Measurement Validation," pp. 102–123 in *Forecasting Nuclear Proliferation in the 21st Century*. Vol. 1, *The Role of Theory*, ed. William C. Potter and Gaukhar Mukhatzhanova. Stanford, CA: Stanford University Press.

Hymans, Jacques E. C. 2013. "The Threat of Nuclear Proliferation: Perception and Reality." *Ethics and International Affairs* 27(3): 281–298.

Hymans, Jacques E. C., and Matthew Gratias. 2013. "Iran and the Nuclear Threshold: Where Is the Line?" *Nonproliferation Review* 20(1): 13–38.

Institute for Science and International Security. 2014. "Defining Iranian Nuclear Programs in a Comprehensive Solution under the Joint Plan of Action." Report (January 15). *Institute for Science and International Security*. http://www.acamedia.info/politics/nonproliferation/references/isis/defining_iranian_nuclear_program_January_2014.pdf.

International Crisis Group. 2014. "Iran and the P5+1: Solving the Nuclear Rubik's Cube." Report 152 (May 9). *Middle East & North Africa*. https://www.crisisgroup.org/middle-east-north-africa/gulf-and-arabian-peninsula/iran/iran-and-p5-1-solving-nuclear-rubik-s-cube.

Janis, Irving L. 1972. *Victims of Groupthink: A Psychological Study of Foreign-Policy Decisions and Fiascoes*. Oxford, England: Houghton Mifflin.

Jervis, Robert. 1976. *Perception and Misperception in International Politics*. Princeton, NJ: Princeton University Press.

Jervis, Robert. 1979. Deterrence Theory Revisited. *World Politics* 31(2): 289–324.

Jervis, Robert. 1982/83. Deterrence and Perception. *International Security* 7(3): 3–27.

Jervis, Robert. 1984. *The Illogic of American Nuclear Strategy*. Ithaca, NY: Cornell University Press.

Jervis, Robert. 2010. *Why Intelligence Fails: Lessons from the Iranian Revolution and the Iraq War*. Ithaca, NY: Cornell University Press.

Jervis, Robert. 2013. "Getting to Yes with Iran: The Challenges of Coercive Diplomacy." *Foreign Affairs* 92(1): 105–115.

Jervis, Robert, Richard Ned Lebow, and Janice Gross Stein, eds. 1989. *Psychology and Deterrence*. Baltimore, MD: Johns Hopkins University Press.

Jervis, Robert, Keren Yarhi-Milo, and Don Casler. 2021. "Redefining the Debate over Reputation and Credibility in International Security: Promises and Limits of New Scholarship." *World Politics* 73(1): 167–203.

Jo, Dong-Joon, and Erik Gartzke. 2007. "Determinants of Nuclear Weapons Proliferation." *Journal of Conflict Resolution* 51(1): 167–194.

Kahl, Colin H. 2012. "Not Time to Attack Iran: Why War Should Be a Last Resort." *Foreign Affairs* 91(2): 166–173.

Kahneman, Daniel. 2011. *Thinking, Fast and Slow*. New York: Farrar, Straus and Giroux.

Kahneman, Daniel, and Amos Tversky. 1979. Prospect Theory: An Analysis of Decision under Risk. *Econometrica* 47(2): 263–291.

Kang, Choi, and Kim Gibum. 2017. A Thought on North Korea's Nuclear Doctrine. *Korean Journal of Defense Analysis* 29(4): 495–511.

Kaplan, Fred. 1982. Strategic Thinkers. *Bulletin of the Atomic Scientists* 38(10): 51–56.

Kaplan, Fred. 1983. *The Wizards of Armageddon*. New York: Simon & Schuster.

Katz, Amrom H. 1979. *Verification and SALT: The State of the Art and the Art of the State*. Washington, DC: Heritage Foundation.

Kelley, Judith G. 2017. *Scorecard Diplomacy: Grading States to Influence Their Reputation and Behavior*. Cambridge, UK: Cambridge University Press.

Kerr, Paul. 2017. "The JCPOA and Safeguards: Model or Outlier?" *Nonproliferation Review* 24(3–4): 261–273.

Kertzer, Joshua D. 2016. *Resolve in International Politics*. Princeton, NJ: Princeton University Press.

Kertzer, Joshua D. 2017. "Resolve, Time, and Risk." *International Organization* 71(Supplement S1): S109–S136.

Kier, Elizabeth, and Jonathan Mercer. 1996. "Setting Precedents in Anarchy: Military Intervention and Weapons of Mass Destruction." *International Security* 20(4): 77–106.

Knopf, Jeffrey W. 2010. "The Fourth Wave in Deterrence Research." *Contemporary Security Policy* 31(1): 1–33. https://www.tandfonline.com/doi/full/10.1080/13523261003640819.

Koblentz, Greg D. 2014. *Strategic Stability in the Second Nuclear Age*. Council Special Report No. 71. https://cdn.cfr.org/sites/default/files/pdf/2014/11/Second%20Nuclear%20Age_CSR71.pdf.

Kristensen, Hans M. 1997. "Targets of Opportunity." *The Bulletin of the Atomic Scientists* 53(5): 22–28.

Kristensen, Hans M. 2010. "Obama and the Nuclear War Plan." *Federation of American Scientists Issue Brief* (February). https://fas.org/programs/ssp/nukes/publications1/WarPlanIssueBrief2010.pdf.

Kristensen, Hans M., Robert S. Norris, and Ivan Oelrich. 2009. *From Counterforce to Minimal Deterrence: A New Nuclear Policy on the Path toward Eliminating Nuclear Weapons*. Occasional Paper No. 7 (April). Washington, DC: Federation of American Scientists/Natural Resources Defense Council. https://fas.org/pubs/_docs/occasionalpaper7.pdf.

Kroenig, Matthew. 2012. "Time to Attack Iran: Why a Strike Is the Least Bad Option." *Foreign Affairs* 91(1): 76–86.

Kroenig, Matthew. 2013. "Nuclear Superiority and the Balance of Resolve: Explaining Nuclear Crisis Outcomes." *International Organization* 67(1): 141–171.

Kroenig, Matthew. 2014. *A Time to Attack: The Looming Iranian Nuclear Threat*. NY: St. Martin's Press.

Kroenig, Matthew. 2017. "The Limits of Damage Limitation." *International Security* 42(1): 199–201.

Kroenig, Matthew. 2018. *The Logic of American Nuclear Strategy: Why Strategic Superiority Matters*. NY: Oxford University Press.

Kydd, Andrew H. 2007. *Trust and Mistrust in International Relations*. Princeton, NJ: Princeton University Press.

Kydd, Andrew H., and Roseanne W. McManus. 2017. "Threats and Assurances in Crisis Bargaining." *Journal of Conflict Resolution* 61(2): 325–348.

Lebovic, James H. 2005. "How Organizations Learn: U.S. Government Estimates of Foreign Military Spending." *American Journal of Political Science* 39(4): 835–863.

Lebovic, James H. 2007. *Deterring International Terrorism and Rogue States: U.S. National Security Policy after 9/11*. London: Routledge.

Lebovic, James H. 2013. *Flawed Logics: Strategic Nuclear Arms Control from Truman to Obama*. Baltimore, MD: Johns Hopkins University.

Lebovic, James H. 2017. "Unipolarity: The Shaky Foundation of a Fashionable Concept," pp. 476–494 in *The Oxford Encyclopedia of Empirical International Relations Theories*. Vol. 4, ed. William Thompson. New York: Oxford University Press.

Lebovic, James H. 2019. *Planning to Fail: The US Wars in Vietnam, Iraq, and Afghanistan*. New York: Oxford University Press.

Lebovic, James H. 2022. "The Selling of a Precedent: The Past as Constraint on Congressional War Powers?" *Journal of National Security Law and Policy* 12(3): 445–470.

Lebovic, James H., and Erik Voeten. 2009. "The Cost of Shame: International Organizations and Foreign Aid in the Punishing of Human Rights Violators." *Journal of Peace Research* 46(1): 79–97.

Lebow, Richard Ned. 1983. "The Cuban Missile Crisis: Reading the Lessons Correctly." *Political Science Quarterly* 98(3): 431–458.

Lebow, Richard Ned. 1987. *Nuclear Crisis Management: A Dangerous Illusion*. Ithaca, NY: Cornell University Press.

Lebow, Richard Ned. 1990. "Domestic Politics and the Cuban Missile Crisis: The Traditional and Revisionist Interpretations Reevaluated." *Diplomatic History* 14(4): 471–492.

Lebow, Richard Ned, and Janice G. Stein. 1990. "Deterrence: The Elusive Dependent Variable." *World Politics* 42(3): 336–369.

Levi, Barbara G., Frank von Hippel, and William H. Daugherty. 1987/88. "Civilian Casualties from 'Limited' Nuclear Attacks on the Soviet Union." *International Security* 12(3): 168–189.

Levy, Jack S. 1997. "Prospect Theory, Rational Choice, and International Relations." *International Studies Quarterly* 41(1): 87–112.

Lewis, Jeffrey G., and Scott D. Sagan. 2016. "The Nuclear Necessity Principle: Making U.S. Targeting Policy Conform with Ethics & the Laws of War." *Daedalus* 145(4): 62–74.

Lieber, Keir A., and Daryl G. Press. 2006a. "The Rise of U.S. Nuclear Primacy." *Foreign Affairs* (March/April). http://www.acamedia.info/politics/escalation/nass/The_Rise_of_U.S._Nuclear_Primacy_%7C_Foreign_Affairs.pdf.

Lieber, Keir A., and Daryl G. Press. 2006b. "The End of MAD? The Nuclear Dimension of U.S. Primacy." *International Security* 30(4): 7–44.

Lieber, Keir A., and Daryl G. Press. 2017. "The New Era of Counterforce: Technological Change and the Future of Nuclear Deterrence." *International Security* 41(4): 9–49.

Lieber, Keir A., and Daryl G. Press. 2020. *The Myth of the Nuclear Revolution: Power Politics in the Atomic Age*. Ithaca, NY: Cornell University Press.

Lindsay, James M. 2012. "TWE Remembers: Black Saturday—Near Calamities Abound as JFK Offers Khrushchev a Deal (Cuban Missile Crisis, Day Twelve)." Blog Post, October 27. https://www.cfr.org/blog/twe-remembers-black-saturday-near-calamities-abound-jfk-offers-khrushchev-deal-cuban-missile.

Lodal, Jan, Keir A. Lieber, Daryl G. Press, James M. Acton, Hans M. Kristensen, Matthew McKinzie, and Ivan Oelrich. 2010. "Second Strike: Is the U.S. Nuclear Arsenal Outmoded?" *Foreign Affairs* 89(2): 145–152.

Long, Austin, and Brendan Rittenhouse Green. 2015. "Stalking the Secure Second Strike: Intelligence, Counterforce, and Nuclear Strategy." *Journal of Strategic Studies* 38(1–2): 38–73.

Lortie, Bret. 2001. "A Do-It-Yourself SIOP." *Bulletin of the Atomic Scientists* 57(4): 22–29.

Lupovici, Amir. 2010. "The Emerging Fourth Wave of Deterrence Theory—Toward a New Research Agenda." *International Studies Quarterly* 3(1): 705–732.

Lupton, Danielle L. 2018. "Reexamining Reputation for Resolve: Leaders, States, and the Onset of International Crises." *Journal of Global Security Studies* 3(2): 198–216.

Lupton, Danielle L. 2020. *Reputation for Resolve: How Leaders Signal Determination in International Politics*. Ithaca, NY: Cornell University Press.

Luttwak, Edward N. 1987. *Strategy: The Logic of War and Peace*. Cambridge, MA: Harvard University Press.

Mahnken, Thomas G., and Gillian Evans. 2019. "Ambiguity, Risk, and Limited Great Power Conflict." *Strategic Studies Quarterly* 13(4): 57–77.

Mansfield, Edward D. 1993. Concentration, Polarity, and the Distribution of Power. *International Studies Quarterly* 37(1): 105–128.

March, James G., and Herbert A. Simon. 1993. *Organizations*. Cambridge, MA: Blackwell.

Matray, James I. 2002. "Dean Acheson's Press Club Speech Reexamined." *Journal of Conflict Studies* 22(1): 28–55.

May, Ernest R., and Philip D. Zelikow, eds. 2002. *The Kennedy Tapes: Inside the White House during the Cuban Missile Crisis*. New York: W. W. Norton & Company.

McDermott, Rose. 2001. *Risk-Taking in International Politics: Prospect Theory in American Foreign Policy*. Ann Arbor: University of Michigan Press.

McKinzie, Matthew G., Thomas B. Cochran, Robert S. Norris, and William M. Arkin. 2001. *The U.S. Nuclear War Plan: A Time for Change*. National Resources Defense Council.

McManus, Roseanne W. 2017. *Statements of Resolve: Achieving Coercive Credibility in International Conflict*. Cambridge, UK: Cambridge University Press.

Mercer, Jonathan. 1996. *Reputation and International Politics*. Ithaca, NY: Cornell University Press.

Meyer, Stephen M. 1984. *The Dynamics of Nuclear Proliferation*. Chicago, IL: University of Chicago Press.

Meyer, Stephen M. 1987. "Soviet Nuclear Operations," pp. 470–531 in *Managing Nuclear Operations*, ed. Ashton B. Carter, John D. Steinbruner, and Charles A. Zraket. Washington, DC: The Brookings Institution.

Monteiro, Nuno P. 2014. *Theory of Unipolar Politics*. New York: Cambridge University Press.

Morgan, Patrick M. 1983. *Deterrence*. Beverly Hills, CA: Sage.

Morgan, Patrick M. 1985. "Saving Face for the Sake of Deterrence," pp. 125–152 in *Psychology and Deterrence*, ed. Robert Jervis, Richard Ned Lebow, and Janice Gross Stein. Baltimore, MD: Johns Hopkins University Press.

Morgan, Patrick M. 2003. *Deterrence Now*. Cambridge: Cambridge University Press.

Müller, Harald, and Andreas Schmidt. 2010. "The Little-Known Story of Deproliferation: Why States Give Up Nuclear Weapons Activities," pp. 124–158 in *Forecasting Nuclear Proliferation in the 21st Century*. Vol. 1, *The Role of Theory*, ed. William C. Potter and Gaukhar Mukhatzhanova. Stanford, CA: Stanford University Press.

Narang, Vipin. 2015. "Nuclear Strategies of Emerging Nuclear Powers: North Korea and Iran." *Washington Quarterly* 38(1): 73–91.

Narang, Vipin, and Ankit Panda. 2020. "North Korea: Risks of Escalation." *SURVIVAL* 62(1): 47–54.

Nolan, Janne E. 1989. *Guardians of the Arsenal: The Politics of Nuclear Strategy*. New York: Basic Books.

Nolan, Janne E. 1991. "The INF Treaty," pp. 355–397 in *The Politics of Arms Control Treaty Ratification*,ed. Michael Krepon and Dan Caldwell. New York: St. Martin's.

Office of the Secretary of Defense. 2021. *Military and Security Developments Involving the People's Republic of China*. Annual Report to Congress. https://media.defense.gov/2021/Nov/03/2002885874/-1/-1/0/2021-CMPR-FINAL.PDF.

Pape, Robert A. 2005. "Soft Balancing against the United States." *International Security* 30(1): 7–45.

Paul, T. V. 2009. *The Tradition of Non-Use of Nuclear Weapons*. Stanford, CA: Stanford University Press.

Pauly, Reid B. C. 2015. "Bedeviled by a Paradox: Nitze, Bundy, and an Incipient Nuclear Norm." *Nonproliferation Review* 22(3–4): 441–455.

Payne, Keith B. 2011. "Maintaining Flexible and Resilient Capabilities for Nuclear Deterrence." *Strategic Studies Quarterly* 5(2): 13–29.

Payne, Keith B. 2016. "Why US Nuclear Force Numbers Matter." *Strategic Studies Quarterly* 10(2): 14–24.

Payne, Keith B. 2020. "The Great Divide in US Deterrence Thought." *Strategic Studies Quarterly* 14(2): 16–48.

Pillar, Paul R. 1983. *Negotiating Peace: War Termination as a Bargaining Process*. Princeton, NJ: Princeton University Press.

Pillar, Paul R. 2014. *Intelligence and U.S. Foreign Policy: Iraq, 9/11, and Misguided Reform* New York: Columbia University Press.

Pious, Richard M. 2001. "The Cuban Missile Crisis and the Limits of Crisis Management." *Political Science Quarterly* 116(1): 81–105.

Pipes, Richard. 1977. "Why the Soviet Union Thinks It Could Fight and Win a Nuclear War." *Commentary* 64(1): 21–34.

Pipes, Richard. 1980. "Militarism and the Soviet State." *Daedalus* 109(4): 1–12.

Press, Daryl G. 2005. *Calculating Credibility: How Leaders Assess Military Threats*. Ithaca, NY: Cornell University Press.

Press, Daryl G., Scott D. Sagan, and Benjamin A. Valentino. 2013. "Atomic Aversion: Experimental Evidence on Taboos, Traditions, and the Non-Use of Nuclear Weapons." *American Political Science Review* 107(1): 188–206.

Putnam, Robert D. 1988. "Diplomacy and Domestic Politics: The Logic of Two-Level Games." *International Organization* 42(3): 427–460.

Quester, George. 2006. *Nuclear First Strike: Consequences of a Broken Taboo*. Baltimore, MD: Johns Hopkins University Press.

Rapport, Aaron. 2015. *Waging War, Planning Peace: U.S. Noncombat Operations and Major Wars*. Ithaca, NY: Cornell University Press.

Reiter, Dan, and Paul Poast. 2021. "The Truth about Tripwires: Why Small Force Deployments Do Not Deter Aggression." *Texas National Security Review* 4(3): 34–53. https://tnsr.org/2021/06/the-truth-about-tripwires-why-small-force-deployments-do-not-deter-aggression/.

Reiter, Dan, and Allan C. Stam. 2002. *Democracies at War*. Princeton, NJ: Princeton University Press.

Richelson, Jeffrey T. 1986. "The Dilemmas of Counterpower Targeting," pp. 159–170 in *Strategic Nuclear Targeting*, ed. Desmond Ball and Jeffrey T. Richelson. Ithaca, NY: Cornell University Press.

Richelson, Jeffrey T. 2007. *Spying on the Bomb: American Nuclear Intelligence from Nazi Germany to Iran and North Korea*. New York: W. W. Norton & Company.

Riqiang, Wu. 2013. "Certainty of Uncertainty: Nuclear Strategy with Chinese Characteristics." *Journal of Strategic Studies* 36(4): 579–614.

Robb, Charles S., and Charles F. Wald. 2014. *Evaluating a Nuclear Deal with Iran*. Washington, DC: Bipartisan Policy Center. https://bipartisanpolicy.org/download/?file=/wp-content/uploads/2019/03/BPC-Evaluating-an-Iran-Nuclear-Deal.pdf.

Rosato, Sebastian. 2014/15. "The Inscrutable Intentions of Great Powers." *International Security* 39(3): 48–88.

Rosenberg, David Alan. 1983. "The Origins of Overkill: Nuclear Weapons and American Strategy, 1945–1960." *International Security* 7(4): 3–71.

Rovner, Joshua. 2011. *Fixing the Facts: National Security and the Politics of Intelligence*. Ithaca, NY: Cornell University Press.

Rovner, Joshua. 2017. "Two Kinds of Catastrophe: Nuclear Escalation and Protracted War in Asia." *Journal of Strategic Studies* 40(5): 696–730.

Sagan, Scott D. 1985. "Nuclear Alerts and Crisis Management." *International Security* 9(4): 99–139.

Sagan, Scott D. 1987. "SIOP-62: The Nuclear War Plan Briefing to President Kennedy." *International Security* 12(1): 22–51.

Sagan, Scott D. 1989. *Moving Targets: Nuclear Strategy and Nuclear Security*. Princeton, NJ: Princeton University Press.

Sagan, Scott D. 1993. *The Limits to Safety: Organizations, Accidents, and Nuclear Weapons*. Princeton, NJ: Princeton University Press.

Sagan, Scott D. 2000. "The Commitment Trap: Why the United States Should Not Use Nuclear Threats to Deter Biological and Chemical Weapons Attacks." *International Security* 24(4): 85–115.

Sagan, Scott D. 2003. "More Will Be Worse," pp. 46–87 in *The Spread of Nuclear Weapons: A Debate Renewed*, ed. Scott D. Sagan and Kenneth N. Waltz. New York: W. W. Norton & Company.

Sagan, Scott D. 2010. "Nuclear Latency and Nuclear Proliferation," pp. 80–101 in *Forecasting Nuclear Proliferation in the 21st Century*. Vol. 1, *The Role of Theory*, ed. William C. Potter and Gaukhar Mukhatzhanova. Stanford, CA: Stanford University Press.

Sagan, Scott D. 2021. "Nuclear Revelations about the Nuclear Revolution." *Texas National Security Review* 4(3): 135–138.

Sagan, Scott D., and Benjamin A. Valentino. 2017. "Revisiting Hiroshima in Iran: What Americans Really Think about Using Nuclear Weapons and Killing Noncombatants." *International Security* 42(1): 41–79.

Sagan, Scott D., and Allen S. Weiner. 2021. "The Rule of Law and the Role of Strategy in U.S. Nuclear Doctrine." *International Security* 45(4): 126–166.

Samore, Gary, ed. 2015. *The Iran Nuclear Deal: A Definitive Guide*. Cambridge, MA: Belfer Center for Science and International Affairs (BCSIA). https://www.belfercenter.org/publication/iran-nuclear-deal-definitive-guide.

Saunders, Elizabeth N. 2017. "No Substitute for Experience: Presidents, Advisers, and Information in Group Decision Making." *International Organization* 71(1): 219–247.

Schear, James A. 1985. "Arms Control Treaty Compliance: Buildup to a Breakdown." *International Security* 10(2): 141–182.

Schelling, Thomas C. 1960. *The Strategy of Conflict*. London: Oxford University Press.

Schelling, Thomas C. 1966. *Arms and Influence*. New Haven: Yale University Press.

Schelling, Thomas C., and Morton H. Halperin. 1961. *Strategy and Arms Control*. New York: The Twentieth Century Fund.

Schilling, Warner R. 1981. "U.S. Strategic Nuclear Concepts in the 1970s: The Search for Sufficiently Equivalent Countervailing Parity." *International Security* 6(2): 48–79.

Schlosser, Eric. 2013. *Command and Control: Nuclear Weapons, the Damascus Accident, and the Illusion of Security*. New York: Penguin Books.

Schulte, Paul. 2012. "Tactical Nuclear Weapons in NATO and Beyond: A Historical and Thematic Examination," pp. 13–74 in *Tactical Nuclear Weapons and NATO*, ed. Tom Nichols, Douglas Stuart, and Jeffrey D. McCausland. Carlisle Barracks, PA: Strategic Studies Institute, Army War College.

Scott, Len, and R. Gerald Hughes, eds. 2015. *The Cuban Missile Crisis: A Critical Reappraisal*. New York: Routledge.

Sechser, Todd S., and Matthew Fuhrmann. 2017. *Nuclear Weapons and Coercive Diplomacy*. Cambridge: Cambridge University Press.

Seitz, Samuel, and Caitlin Talmadge. 2020. "The Predictable Hazards of Unpredictability: Why Madman Behavior Doesn't Work." *The Washington Quarterly* 43(3): 31–46.

Shannon, Vaughn P., and Michael Dennis. 2007. "Militant Islam and the Futile Fight for Reputation." *Security Studies* 16(2): 287–317.

Shlaim, Avi. 1976. "Failures in National Intelligence Estimates: The Case of the Yom Kippur War." *World Politics* 28(3): 348–380.

Slocombe, Walter. 1981. "The Countervailing Strategy." *International Security* 5(4): 18–27.

Smith, Gerard. 1981. *Doubletalk: The Story of the First Strategic Arms Limitation Talks*. Garden City, NY: Doubleday.

Snyder, Glenn H. 1961. *Deterrence and Defense*. Princeton, NJ: Princeton University Press.

Snyder, Glenn H., and Paul Diesing. 1977. *Conflict among Nations: Bargaining, Decision Making, and System Structure in International Crises*. Princeton, NJ: Princeton University Press.

Solingen, Etel. 2009. *Nuclear Logics: Contrasting Paths in East Asia and the Middle East*. Princeton, NJ: Princeton University Press.

Solingen, Etel, ed. 2012. *Sanctions, Statecraft, and Nuclear Proliferation*. New York: Cambridge University Press.

Stein, Janice Gross. 1992. "Deterrence and Compellence in the Gulf, 1990-91: A Failed or Impossible Task?" *International Security* 17(2): 147–179.

Steinbruner, John D. 1974. *The Cybernetic Theory of Decision: New Dimensions of Political Analysis*. Princeton, NJ: Princeton University Press.

Steinbruner, John D. 1987. "Choices and Tradeoffs," pp. 535–554 in *Managing Nuclear Operations*, ed. Ashton B. Carter, John D. Steinbruner, and Charles A. Zraket. Washington, DC: The Brookings Institution.

Tabatabai, Ariane. 2017. "Negotiating the 'Iran Talks' in Tehran: The Iranian Drivers that Shaped the Joint Comprehensive Plan of Action." *Nonproliferation Review* 24(3–4): 225–242.

Talbott, Strobe. 1985. *Deadly Gambits: The Reagan Administration and the Stalemate in Nuclear Arms Control*. New York: Vintage Books.

Talmadge, Caitlin. 2015. *The Dictator's Army: Battlefield Effectiveness in Authoritarian Regimes*. Ithaca, NY: Cornell University Press.

Talmadge, Caitlin. 2017. "Would China Go Nuclear? Assessing the Risk of Chinese Nuclear Escalation in a Conventional War with the United States." *International Security* 41(4): 50–92.

Tang, Shiping. 2005. "Reputation, Cult of Reputation, and International Conflict." *Security Studies* 41(1): 34–62.

Tannenwald, Nina. 2007. *The Nuclear Taboo: The United States and the Non-Use of Nuclear Weapons since 1945*. Cambridge: Cambridge University Press.

Terriff, Terry. 1995. *The Nixon Administration and the Making of U.S. Nuclear Strategy*. Ithaca, NY: Cornell University Press.

Tertrais, Bruno. 2010. "The Future of Extended Deterrence: A Brainstorming Paper," pp. 7–14 in *Perspectives on Extended Deterrence*. Foundation for Strategic Research, No. 03/2010. https://inis.iaea.org/search/search.aspx?orig_q=RN:47061163.

Tertrais, Bruno. 2014. "Drawing Red Lines Right." *Washington Quarterly* 37(3): 7–24.

Thompson, William R. 2006. "Systemic Leadership, Evolutionary Processes, and International Relations Theory: The Unipolarity Question." *International Studies Review* 8(1): 1–22.

Tingley, Dustin H., and Barbara F. Walter. 2011. "The Effect of Repeated Play on Reputation Building: An Experimental Approach." *International Organization* 65(2): 343–365.

Trachtenberg, Marc. 1988/89. "A 'Wasting Asset': American Strategy and the Shifting Nuclear Balance, 1949–1954." *International Security* 13(3): 5–49.

Trager, Robert F., and Dessislava P. Zagorcheva. 2005. "Deterring Terrorism: It Can Be Done." *International Security* 30: 87–123.

US Department of Defense. 2018. *Nuclear Posture Review*. https://fas.org/wp-content/uploads/media/2018-Nuclear-Posture-Review.pdf.

US General Accounting Office (GAO). 1991. "Strategic Weapons: Nuclear Weapons Targeting Process." Fact Sheet for Congressional Requesters (September), GAO/NSIAD-91-319FS. https://www.gao.gov/assets/90/89136.pdf.

US Senate. 1979a. *The SALT II Treaty, Report of the Committee on Foreign Relations*. Committee on Foreign Relations. Washington, DC: U.S. Government Printing Office.

US Senate. 1979b. *Military Implications of the Treaty on the Limitation of Strategic Offensive Arms and Protocol Thereto (SALT II Treaty)*. Parts 1–4. Committee on Armed Services. Washington, DC: U.S. Government Printing Office.

US Senate Select Committee on Intelligence. 2004. *Report on the U.S. Intelligence Community's Prewar Intelligence Assessments on Iraq*. Washington, DC: 108th Congress. http://fas.org/irp/congress/2004_rpt/ssci_iraq.pdf.

Vail, Christopher. 2016. "The Legality of Nuclear Weapons for Use and Deterrence." *Georgetown Journal of International Law* (48): 839–872.

Van Cleave, William R. 1986. *Fortress USSR: The Soviet Strategic Defense Initiative and the U.S. Strategic Defense Response*. Palo Alto, CA: The Hoover Institute Press, Stanford University.

Van Cleave, William R. 1989. "Honored in the Breach." *Bulletin of the Atomic Scientists* 45(4): 33.
Van Evera, Stephen. 1999. *Causes of War: Power and the Roots of Conflict*. Ithaca, NY: Cornell University Press.
Wagner, R. Harrison. 1993. "What Was Bipolarity?" *International Organization* 47(1): 77–106.
Wagner, R. Harrison. 2000. "Bargaining and War." *American Journal of Political Science* 44(3): 469–484.
Waltz, Kenneth N. 1979. *Theory of International Politics*. Reading, MA: Addison-Wesley.
Weisiger, Alex, and Keren Yarhi-Milo. 2015. "Revisiting Reputation: How Past Actions Matter in International Politics." *International Organization* 69(2): 473–495.
Welch, David A. 1989. "Crisis Decision Making Reconsidered." *Journal of Conflict Resolution* 33(3): 430–445.
Wells, Samuel F., Jr. 1981. "The Origins of Massive Retaliation." *Political Science Quarterly* 96(1): 31–52.
Whitlark, Rachel Elizabeth. 2019. "Should Presidential Command over Nuclear Launch Have Limitations? In a Word, No." *Texas National Security Review* 2(3): 139–144.
https://tnsr.org/wp-content/uploads/2019/08/TNSR-Journal-Vol-2-Issue-3-Whitlark.pdf.
Winter, David G. 2003. "Asymmetrical Perceptions of Power in Crises: A Comparison of 1914 and the Cuban Missile Crisis." *Journal of Peace Research* 40(3): 251–270.
Wirtz, James J. 2012. *Deterring the Weak: Problems and Prospects*. Proliferation Papers. The Institut Français des Relations Internationales (Ifri) Security Studies Center.
Wohlforth, William C. 1999. "The Stability of a Unipolar World." *International Security* 24(1): 5–41.
Wohlstetter, Albert. 1959. "The Delicate Balance of Terror." *Foreign Affairs* 37(2): 211–234.
Wohlstetter, Albert. 1974a. "Is There a Strategic Arms Race?" *Foreign Policy* 15 (Summer): 3–20.
Wohlstetter, Albert. 1974b. "Is There a Strategic Arms Race? (II) Rivals but No 'Race.'" *Foreign Policy* 16 (Autumn): 48–92.
Wohlstetter, Albert, and Richard Brody. 1987. "Continuing Control as a Requirement for Deterring," pp. 142–196 in *Managing Nuclear Operations*, ed. Ashton B. Carter, John D. Steinbruner, and Charles A. Zraket. Washington, DC: The Brookings Institution.
Woods, Kevin, James Lacey, and Williamson Murray. 2006. "Saddam's Delusions: The View from the Inside." *Foreign Affairs* 85(3): 2–26.
Woolf, Amy F. 2021. "Defense Primer: Command and Control of Nuclear Forces." *In Focus*. US Congressional Research Service (September 30). https://sgp.fas.org/crs/natsec/IF10521.pdf.
Yarhi-Milo, Keren. 2013. "In the Eye of the Beholder: How Leaders and Intelligence Communities Assess the Intentions of Adversaries." *International Security* 38(1): 7–51.
Yarhi-Milo, Keren. 2018. *Who Fights for Reputation: The Psychology of Leaders in International Conflict*. Princeton, NJ: Princeton University Press.
Yarhi-Milo, Keren, Joshua D. Kertzer, and Jonathan Renshon. 2018. "Tying Hands, Sinking Costs, and Leader Attributes." *Journal of Conflict Resolution* 62(10): 2150–2179.
Yetiv, Steve A. 2013. *National Security Policy through a Cockeyed Lens: How Cognitive Bias Impacts U.S. Foreign Policy*. Baltimore, MD: Johns Hopkins University Press.
Zartman, I. William, and Guy Olivier Faure. 2005. "The Dynamics of Escalation and Negotiation," in *Escalation and Negotiation in International Conflicts*, ed. I. William Zartman and Guy Olivier Faure. Cambridge: Cambridge University Press.
Zeisberg, Mariah. 2015. *War Powers: The Politics of Constitutional Authority*. Princeton, NJ: Princeton University Press.
Zhang, Baohui. 2007. "The Modernization of Chinese Nuclear Forces and Its Impact on Sino-U.S. Relations." *Asian Affairs: An American Review* 34(2): 87–100.
Zhang, Baohui. 2008. "The Taiwan Strait and the Future of China's No-First-Use Nuclear Policy." *Comparative Strategy* 27(2): 164–182.

INDEX

For the benefit of digital users, indexed terms that span two pages (e.g., 52–53) may, on occasion, appear on only one of those pages.

Tables and figures are indicated by *t* and *f* following the page number